MERCHANTS OF KNOWLEDGE

STANFORD **OTTOMAN WORLD** SERIES

MERCHANTS *of* KNOWLEDGE

INTELLECTUAL EXCHANGE IN THE OTTOMAN EMPIRE AND RENAISSANCE EUROPE

Robert G. Morrison

STANFORD UNIVERSITY PRESS
Stanford, California

Stanford University Press
Stanford, California

Library of Congress Cataloging-in-Publication Data
Names: Morrison, Robert G., 1969- author.
Title: Merchants of knowledge : intellectual exchange in the Ottoman Empire and Renaissance Europe / Robert G. Morrison.
Other titles: Stanford Ottoman world series.
Description: Stanford, California : Stanford University Press, [2025] | Series: Stanford Ottoman world series | Includes bibliographical references and index.
Identifiers: LCCN 2024047778 (print) | LCCN 2024047779 (ebook) | ISBN 9781503636323 (cloth) | ISBN 9781503642683 (paperback) | ISBN 9781503642690 (epub)
Subjects: LCSH: Jewish scholars—Greece—Ērakleion—History. | Islamic learning and scholarship—Turkey—History. | Astronomy, Renaissance. | Philosophy, Renaissance. | Turkey—Relations—Europe. | Europe—Relations—Turkey. | Europe—Intellectual life—Turkish influences.
Classification: LCC DR479.E85 M67 2025 (print) | LCC DR479.E85 (ebook) | DDC 949.6/031—dc23/eng/20250121

LC record available at https://lccn.loc.gov/2024047778
LC ebook record available at https://lccn.loc.gov/2024047779

Cover design: Lindy Kasler
Cover art: Wikimedia Commons

CONTENTS

Note on Translation and Transliteration vii

Preface ix

Map of the Eastern Mediterranean ca. 1500 CE xiv

Introduction 1

One **The Network of Merchant-Scholar Families of Candia** 15

Two **Astrology, a Shared Hermeneutic** 30

Three **Transactions of Astronomical Tables and Instruments** 60

Four **Aristotelianism across Borders** 87

Five **Aristotelianism and *Qabbalah*** 119

Six **Theoretical Astronomy between Renaissance Italy and the Ottoman Empire** 150

Seven **Tricks of the Trade in Mechanics and Medicine** 184

Conclusion 212

Acknowledgments 217

Notes 221

Bibliography 275

Index 309

NOTE ON TRANSLATION AND TRANSLITERATION

Unless otherwise noted, all translations of foreign language sources are my own.

Names were spelled inconsistently in the Venetian archives. I have standardized the spelling of names that appeared more than once. The names of Jews in other sources are spelled according to their standard English and Hebrew equivalents.

Arabic, Ottoman Turkish, and Modern Turkish words have been transliterated according to the *International Journal of Middle East Studies* style guide. With Hebrew words, I have differentiated the consonants, except for sin and samek̲, but have not distinguished the lengths of vowels.

Following the convention in Jewish studies, titles of works of Hebrew secondary literature are given in English with the notation (Heb.).

All English translations of complete verses from the Bible come from the 1917 Jewish Publication Society translation.

PREFACE

Early in my career, a senior colleague kindly invited me to join a research group called Before Copernicus that would explore the broader, multidisciplinary background of the astronomy of Nicholas Copernicus (d. 1543). Copernicus is famous for his proposal of a sun-centered arrangement of the planets, but there is no consensus on why he proposed such an arrangement. Why would a scholar of Islam and Judaism like me be interested in Copernicus? For over a century, historians of science have been aware of similarities between science in Islamic societies and Copernicus's work. The possibility that transregional exchange with Islamic societies contributed to Copernicus's theories was first considered by historians in the 1950s. At the beginning of this millennium, the eminent general historian of astronomy John North concluded that "it is hardly possible to doubt that Copernicus was aware of some text or another in which they [Islamic theories] were to be found."[1] Still, Copernicus did not cite a single source that was composed within two hundred years of his own lifetime, and he never traveled to an Islamic society nor read an Islamic language. No new direct evidence of contacts between Renaissance Europe and the Islamic world that are relevant to the context of Copernicus's astronomy had been uncovered in the previous few decades.

At the time of the invitation, I was finishing a book on Islam and science in fourteenth-century Iran. In a question that mocked academic hyperspecialization, I was asked whether I would be comfortable moving from studying the fourteenth century to studying the fifteenth century. When the Before Copernicus group first met, it became clear to me that several members of the group knew much about Islam and science in the fifteenth century, but no one was working on science in the Jewish communities of the time. With the intellectual hubris and, perhaps, the community spirit recognizable to those who work in small academic departments, I volunteered. Within a year, it was clear to me that fifteenth-century Jewish communities were an understudied conduit of Islamic astronomy from the thirteenth and fourteenth centuries to scholars in Renaissance Italy. I came across a cluster of transregional Jewish scholars connected to Candia (present-day Heraklion) on Crete and, by 2011, was convinced that I needed to write a book about them.

Though I had already known about some of the scholars as individuals, serendipity led me to the connections between them. Once I realized that a scholar's last name could be spelled in a few different ways, references by one scholar to another in multiple disciplines emerged. Then, while browsing the stacks of Stanford University's library, I found a book with a few photographs of documents that showed that these same Jewish scholars interacted with Venetian officials. Such information about the nonacademic lives of scholars of this period is rarely found and is a key foundation of my argument. Finally, a reviewer of my second book pointed out an important similarity, which I had overlooked, between the contents of a Jewish text and Copernicus's work. Hence, the contents of Jewish texts, in addition to the roles played by Jewish intermediaries, mattered. The unexpected twists and turns of my own research process have increased my appreciation of the many contingencies of transregional exchange.

In this book, I plumb the narrative of intellectual exchange in the Eastern Mediterranean in the critical period between 1450 and 1550 by linking well-documented instances of exchange with the less documented and, thus, explaining and reframing episodes of exchange that have been contested. Intellectual exchange was facilitated by this transregional cluster of Jewish scholars who came from families engaged in commerce that bridged the Ottoman Empire, Crete, and the Veneto. In many cases, outstanding scholars from those families participated in commerce. The commercial activities of these families depended on socially and geographically wide-ranging contacts, and these contacts, in turn, promoted intellectual exchange. Since commerce was the scaffolding for intellectual exchange, and because the sale of books and the ability to translate redounded to scholars' commercial advantages, the most important scholars in these families were, in a direct sense, merchants of knowledge. Though anyone who travels is a vector of exchange, these individuals' mobility and transregional connections stood out. This book's title is also a metaphor in that by labeling the most important scholars from these families "merchants of knowledge," I couch the intellectual exchange in terms of the scholars' own prerogatives, whether intellectual growth or the accumulation of social capital or both, not in terms of any agenda of Renaissance humanism or any telos of the history of science.

Identifying the scholarly intermediaries has helped me identify relevant exchange that occurred in multiple fields. Not all these disciplines are relevant to dominant narratives of scientific progress. Rather, I find that the threads of transregional conversations in a variety of disciplines intertwined. For instance, matters of astrology impinged upon philosophy, *Qabbalah*, and theoretical astronomy. As the scholars I study were engrossed in several disciplines, exchange flowed in multiple directions. I elucidate how Jewish scholars from the Eastern Mediterranean came to be interested in texts written in Latin, as knowledge from Latin texts contributed to Jewish intellectual life. Why else would erudite Jewish scholars from the Eastern Mediterranean seek out Christian scholars from the Veneto? Latin texts and their contents were in demand at the Ottoman court. Moving in the other direction, the merchants of knowledge were sought after in the Italian Peninsula as teachers of *Qabbalah* and Aristotelian philosophy, areas pertinent to Judaism and Christianity. Jewish scholars were competent in combinations of Arabic, Hebrew, Turkish, and Latin, and these linguistic skills enabled them to market their knowledge in ways that advanced their own aims; in that respect too, these scholars were merchants of knowledge.

By considering the full breadth of knowledge that interested these scholars, I am able to demonstrate some instances of exchange, which the historical actors did not record, that may have influenced Copernicus's theoretical astronomy. Not only do I name an intermediary for this exchange, but I also answer a more important question that is insufficiently investigated in earlier work on the relationship between the Renaissance and Islamic societies: namely, the European context for exchange of theoretical astronomy. It is easier to study how European Christian scholars engaged with and subsequently accepted or dismissed the contributions of scholars from Islamic societies in subjects such as philosophy, astrology, and medicine because there is a great deal of evidence showing that scholars from the Eastern Mediterranean and Europe pursued these fields in ways that were mutually intelligible. In theoretical astronomy, however, it has been harder to detect a transregional conversation acknowledged by the historical actors. The possible absence of such interaction in theoretical astronomy might imply that since pre-Copernican Renaissance astronomy was not as sophisticated as Islamic astronomy, any European Renaissance astronomers who encountered the astronomy of Islamic societies would have unhesitatingly appropriated it. But without a transregional conversation about theoretical astronomy, how would a merchant of knowledge have even *located* counterparts potentially interested in the advanced theoretical astronomy of late medieval Islamic societies? Scholars could not pass on knowledge that the recipient did not understand or, more important, in which the potential recipient was uninterested. I argue in this book that transregional conversations in other fields, such as astrology and Aristotelian philosophy, upon which theoretical astronomy impinged, became the substrate for the exchange of theoretical astronomy.

By exposing this wide-ranging intellectual and commercial exchange, I rebut the presumption of a clean division between the intellectual life of Renaissance Europe and Islamic societies, which is a corollary of the idée fixe that the Renaissance marked the autochthonous birth of modern Europe.[2] The merchants of knowledge were more than messengers who traveled between regions that were otherwise separate. Rather, the merchants of knowledge created their own distinct intellectual life in which the contents of Arabic, Latin, and occasionally Turkish and Greek texts were appropriated and integrated into Hebrew compositions. European scholars hunted down these Hebrew texts directly and the Ottoman sultans acquired a few texts composed by merchants of knowledge and their contacts. The merchants of knowledge did not bridge a cultural divide; they blended cultures in a distinctive way that elites from Istanbul and the Veneto valued. This remarkable interest from non-Jews in Jewish intellectual life bolstered Jewish scholars' status as intermediaries.

All this exchange was possible because the Renaissance Veneto and the Ottoman Empire were connected diplomatically and commercially. The military power of the Ottoman Empire reached its zenith in the sixteenth century and was a formidable military threat to a nascent Europe. Europe, conversely, defined its identity against the Ottoman Empire. But political and military competition between the Ottoman Empire and Europe did not mean a lack of contact. Even the anti-Ottoman rhetoric that some Renaissance humanists put forth turns out, as Nancy Bisaha has found, to have been an attempt to compensate for borrowing.[3] The exchange of objects such as observational instruments, manuscripts, and mechanical devices between merchants of knowledge and their contacts brought socioeconomic benefits that far exceeded these objects' monetary value. Commercial exchange fostered intellectual exchange.

Merchants of Knowledge is unprecedented because in it one finds evidence that Mediterranean intellectual life, between 1450 and 1550, was interdependent and interdisciplinary. Depicting the Renaissance as a solely European Christian phenomenon in which the rediscovery of classical Greek texts lit the path to modernity contradicts the historical record. Contemporary developments in Islamic societies and Jewish cultures were vital. Yet intermediaries brought Jewish and Islamic knowledge to Europe only because Ottoman elites took an interest in Latin scholarship on its own terms, not due to military and political pressure as would be the case a few centuries later. Hence, the Renaissance portrayed in *Merchants of Knowledge* is not, from the perspective of the Ottoman Muslim contacts of the Jewish merchants of knowledge, hegemonic. It is closer to a Renaissance that suits a West that is waking up to the cultural and political implications of its own diversity.

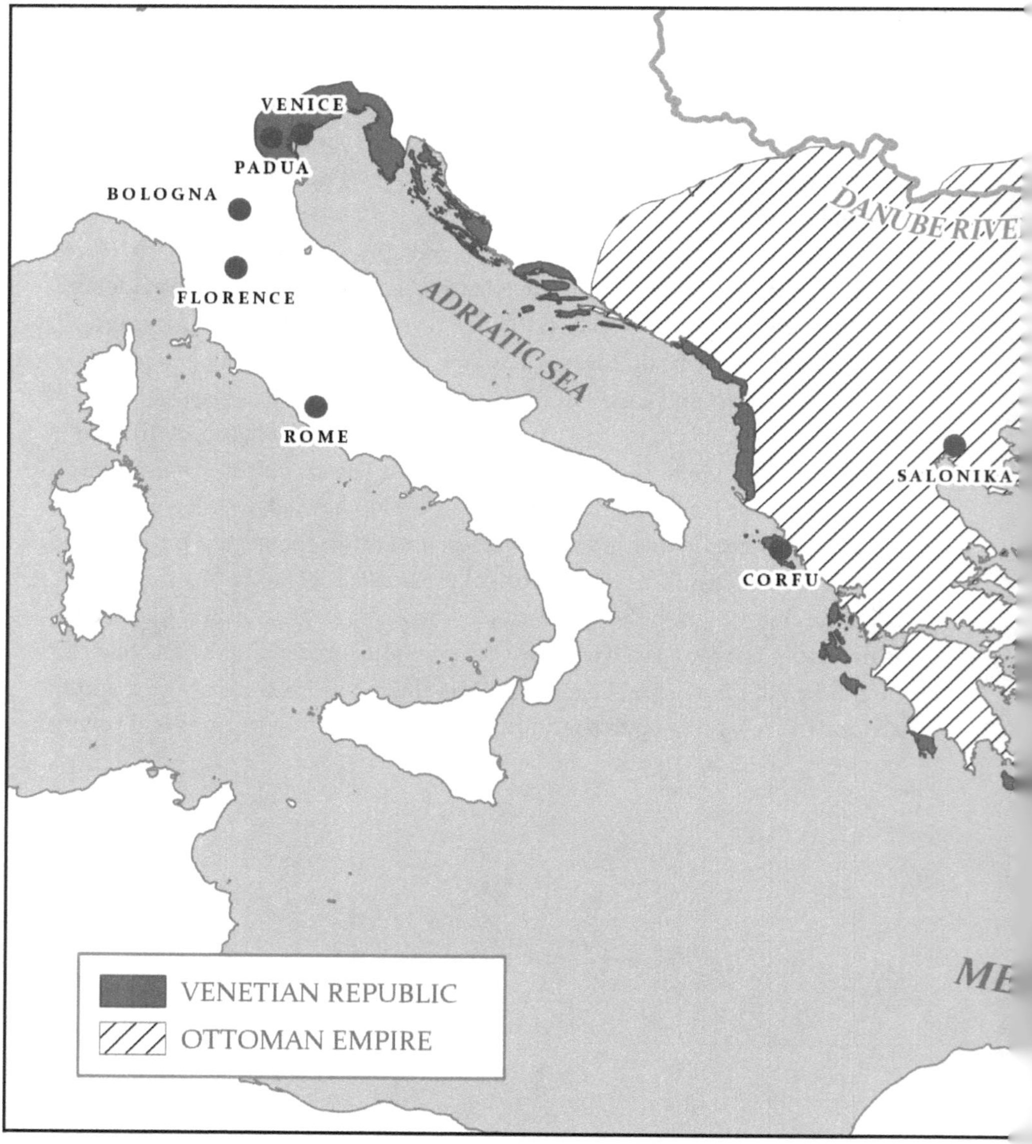
VENICE
PADUA
BOLOGNA
FLORENCE
ROME
ADRIATIC SEA
DANUBE RIVE
SALONIKA
CORFU
ME
VENETIAN REPUBLIC
OTTOMAN EMPIRE

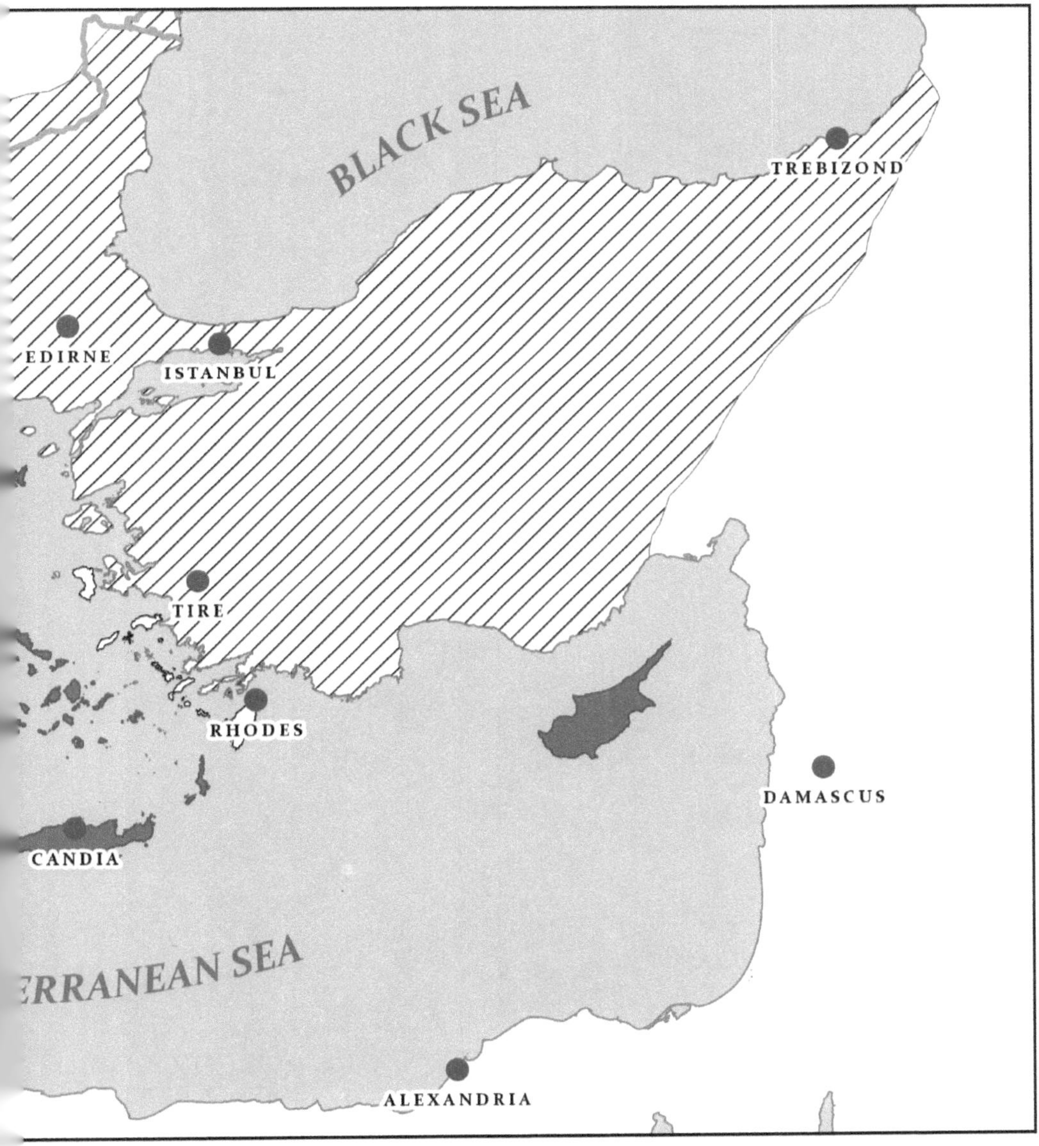

The eastern Mediterranean ca. 1500 CE. Map by Aaron Gilbreath.

MERCHANTS OF KNOWLEDGE

INTRODUCTION

The years between 1450 and 1550 witnessed the Ottoman conquest of Constantinople and an increasing awareness of the West as a distinct entity. Because of the formidable cultural and military strengths of the Ottoman Empire and because the Renaissance is seen as the beginning of the rise of the West, the extent and impact of transregional intellectual exchange during these years may come as a surprise. Yet intellectual exchange in Eurasia occurred at other unexpected times. The amazing efflorescence of science in Islamic societies began at the end of the first Islamic century (730 CE), just as the Islamic Empire was rapidly expanding and just when one might think that Muslims would have felt that they had little to learn from the Byzantine and Sassanian societies that they conquered. But the passage of knowledge into Arabic through profuse translations from Greek, Persian, and Sanskrit is indisputable. Though the reasons why victors appropriated the cultures of the vanquished remain contested, there is a consensus that the reasons for the translations must be sought in the culture of the nascent Islamic Empire because knowledge has never been fated to travel. Hence, a principal goal of this book is to search in Mediterranean societies for the causes of the transregional intellectual exchange occurring between 1450 and 1550. The specific motivations of the scholars I study are different from the motivations of the scholars involved with the rise of science in Islamic societies, but, in both cases, identifying their exact motivations is the key to understanding why exchange occurred.

My concern with the motives of the historical actors explains why I prefer to write of intellectual exchange than of the transmission of ideas. The term "transmission" connotes a teleology and unidirectionality that does not suit the facts of this book.[1] I call the most prominent scholars merchants of knowledge because they recognized the concrete social and economic benefits of all sorts of exchange, intellectual and commercial. Exchange happened because exchange benefited the merchants of knowledge and their contacts, not because of the exigencies of the present.

Appreciation of the dynamics of exchange depends on a full awareness of what was exchanged in this economy of knowledge. In the century I study, many different commodities, such as wine, animal skins, and wool, traveled throughout the Eastern Mediterranean between Candia (present-day Heraklion) on the island of Crete, the Veneto (Venice and its environs, which include Padua), and the Ottoman Empire. Indeed, the expansion of the Renaissance economy meant political expansion into the Eastern Mediterranean.[2] Payments for goods were sometimes transferred by chirographs, documents that could be authenticated by matching both halves on the wavy line along which they were torn. Payments were recorded with officials in multiple languages in multiple locales. Letters of exchange, a way to draw on a distant bank, were another means of payment. All this commercial exchange depended also on shipping, and shipments needed to be insured against loss and piracy. Insuring boats was a way for Candiote Jews to forge relationships with Christians. Since some of the Jewish scholarly intermediaries were also businessmen and most others were related to *commerçants*, trade was the background for the intellectual exchange that is the focus of the book. Transregional contacts were the rule, not the exception. These connections were facilitated by the increased rationalization of banking and trade in Europe.[3] Recordkeeping and financial instruments became remarkably sophisticated.[4] Translation facilitated payments and records just as translation was a mode of intellectual exchange. Both scholarly and commercial translation—e.g., of chirographs or of letters of exchange—brought economic benefits.

Most important, intellectual exchange was transactional. A scholar gave in one field and took in another. Trading also provides a conceptual model for my study of the intellectual exchanges. For the Venetians and

Ottomans, trade was almost as much of a priority as political dominance.[5] Just as one cannot understand an economy without considering all of the goods and services transacted, one cannot understand some intellectual developments, such as the critical advances in theoretical astronomy, without studying all of the intellectual life, including *Qabbalah* and astrology, of the time.

Migration and Markets

The far-flung exploits of the merchants of knowledge were an outgrowth of the general patterns of the movement of Jewish populations in the wake of successive expulsions from Iberia. According to Elijah Capsali (d. 1555), Sultan Bayezit II (r. 1481–1512) famously welcomed Jews fleeing the expulsions from the Iberian Peninsula.[6] A conservative assessment of the number of Jews who made it to the Ottoman Empire from the Iberian Peninsula is thirty thousand.[7] These migrants were greeted by new professional opportunities. In the Ottoman Empire, the linguistic skills of the newly arrived Jews proved advantageous, as they could access knowledge other scholars could not. Some of these Jews became physicians at the Ottoman court, a position that afforded them valuable access to the sultan.[8] Well-placed Jewish scholars were also economic intermediaries. Through Yacup Pasha, the Jewish physician of Bayezit's father, Sultan Mehmed II (d. 1481), a Venetian gained the right to produce alum, a chemical sometimes used in dying, in old and new Phocaea (modern Foça) on the Aegean coast.[9] While the Jews were moving east, the Ottomans were looking west. Mehmed II was so interested in European goods that there was an atelier in Istanbul for European artisans.[10]

There were relevant population shifts within Asia Minor as well. Mehmed II initiated his *sürgün* (resettlement) policy in 1456 in order to repopulate Istanbul after the devastation of his 1453 conquest of the city.[11] The *sürgün* had a massive impact on Romaniot Jews, the Jews of the former Byzantine lands. Talented Romaniot Jews came to Istanbul from Anatolia. Romaniot Jews have not received the scholarly attention bestowed upon Sephardic Jews, but Romaniots play the more significant role in this book. And Jews were not the only ones uprooted. Due to the Ottoman conquest of Istanbul, Christian scholars from Asia Minor departed Constantinople, sometimes for Crete.[12]

Crete was another destination for Jewish immigration from the continent. There had been a Jewish community on Crete since the first century BCE. The historian Socrates of Constantinople (d. ca. 450 CE) took note when a messianic figure prompted the conversion of some Cretan Jews to Christianity after he failed to lead them "through a dry sea to the land of promise," i.e., to Palestine.[13] Following the temporary, partial conquests of Crete by Umayyad armies, Muslims conquered the island completely under the reign of the Abbasid caliph al-Maʾmūn (d. 833).[14] The city of Candia, the name of which derives from the Arabic *khandaq* (trench), was founded by Muslims in the ninth century. Armies of the Byzantine Empire reconquered Crete from the Muslims in 961. Crete began to be ruled by Venice from a base at Candia in 1204. Under Venetian rule, Cretan Jews enjoyed the commercial privileges of Venetian citizens without being full citizens of the Venetian Republic.[15] By the thirteenth century, Jews from Europe were coming to Crete, sometimes on journeys further east.[16]

Jewish migration to Candia from the west increased due to the Black Death in the mid-fourteenth century. The arrival of Jews from the Iberian Peninsula intensified after 1391 and transformed the Candiote Jewish community.[17] In notarial records, some Jews on Candia were identified with the terms Spagnolo or Sefaradi, indicating their ongoing identification with Iberia. By the end of the fourteenth century, Jewish physicians who were on Venetian Crete had diverse backgrounds, hailing from the West, Arab lands, and the Byzantine Empire.[18] In 1536, there were four synagogues on Candia, one of which was Ashkenazic.[19] Jews came to Candia from as far afield as the land of the Tatars,[20] as well as from Italy and Arabic-speaking lands.[21] Greek was the lingua franca of the Candiote Jews, despite their diverse origins, though they usually composed texts in Hebrew.[22]

Venice has a reputation as an important city for transregional exchange, but it was not a flourishing center of Jewish life between 1450 and 1550. Nearby Padua, however, had a venerable community of Jewish merchants and scholars. The flight of Jews from persecution related to the Black Death led to the earliest Jewish settlement in Padua no later than 1369. These first Jewish residents were mandated to establish pawnshops.[23] The earliest Candiote Jew known to have resided in Padua was Musetto b. Judah Malbiegonato da Candia in 1402.[24] Within a few decades, Jews from Candia arrived in Padua for business, the university, and the yeshivah, a school of Jewish learning.[25] The faculty of medicine at the University of Padua graduated its first Jewish physician in 1409.[26] Sometimes an individual working in Padua was a businessman and a scholar. For example, between 1426 and 1433, Salamon of Candia possessed shares in lending institutions of credit (*banchi feneratizi*) while he resided in Padua as a student of arts and medicine.[27] Members of distinguished Candiote families studied at the Padua yeshivah,[28] and famous Ashkenazic rabbis from the yeshivah in Padua traveled to Candia.[29] Connections between Padua and Candia were foundational for the intellectual and commercial activities I examine in this book.

Scholarship on Transregional Exchange

Anyone researching transregional exchange among Jews in the Mediterranean cannot escape the penumbra of Goitein's *A Mediterranean Society.* Since then, nonetheless, some remarkably innovative monographs on premodern Jewish commercial networks have appeared.[30] There have also been studies of transregional Jewish scholars. For example, Havah Tirosh-Rothschild studied David b. Judah Messer Leon (d. ca. 1526), who moved from Italy to the Ottoman Empire to Salonika.[31] David's work was informed by Islamic and Renaissance thought and bridged regional Jewish cultures. Mercedes García-Arenal examined how Samuel Pallache (d. 1616) journeyed from Iberia to Morocco and then back to Europe as a commercial agent, posing at times as a Christian.[32]

Others have researched transregional intellectual exchange and networks in the premodern Islamic world. In *The World of Murtaḍā al-Zabīdī (1732-91): Life, Networks and Writings,* Stefan Reichmuth described intricately the social and intellectual networks of an eighteenth-century South Asian scholar who adopted a Yemeni identity and moved to the Middle East. The notes of interactions that Zabīdī kept are one of Reichmuth's main sources. Zabīdī, like many figures I study in this book, excelled in multiple disciplines and took approaches that paralleled those current in Europe. Samer Akkach examined a similar time period in *ʿAbd al-Ghani al-Nabulusi: Islam and the Enlightenment.* Like the merchants of knowledge, Nābulusī (d. 1731) lived during a transitional period, during the Scientific Revolution and before the *nahḍa.*

Concentrating on the seventeenth century, Khaled El-Rouayheb demonstrated beyond a shadow of a doubt in *Islamic Intellectual History in the Seventeenth Century: Scholarly Currents in the Ottoman Empire and the Maghreb* that not only were scholars in the Maghrib and the central Ottoman lands connected but that these scholarly exchanges invigorated Sufism, *kalām* (Islamic philosophical theology), and the study of logic. Sometimes scholars in different regions disagreed with each other, and regional schools of thought endured. In *Merchants of Knowledge*, we will see that the specifics of astrology, philosophy, and *Qabbalah* varied regionally. With regard to the sixteenth century, Helen Pfeifer, in *Empire of Salons: Conquest and Community in Early Modern Ottoman Lands*, found that men's literary salons were sites of intellectual exchange between the Arab lands recently conquered from the Mamlūks and the Persianate realms of the Ottoman Empire.[33] In *Intellectual Networks in Timurid Iran*, Evrim Binbaş has uncovered an entire network of fifteenth-century scholars who pursued occult sciences. He found cases in which one member of the network shaped the career of another member.[34] As was true for the figures studied in this book, political upheavals enhance intellectual connections.[35] Sites of intellectual exchange were where social status was flaunted and negotiated.[36]

From Avner Ben-Zaken's *Cross-Cultural Scientific Exchanges in the Eastern Mediterranean, 1560–1660*, I was inspired to look hard for ways in which Muslim scholars sought out the contents of Latin texts in multiple disciplines and to illustrate how exchange was at least bidirectional, if not multidirectional. I, nevertheless, accept the critiques of that book.[37] As I focus in my book on the dynamics of intellectual exchange, my conceptualization of scholarly intermediaries as merchants of knowledge is influenced by E. Natalie Rothman's *Brokering Empire: Trans-Imperial Subjects between Venice and Istanbul*. In one chapter, Rothman explored how translation was a critical activity of transregional subjects; I explain how some merchants of knowledge were involved with both scholarly translation and written translation for commercial purposes. *Merchants of Knowledge* is distinct because I focus on a period that is a century earlier and on the details of the scholarly disciplines in which exchange developed.

As the chronological scope of this book is, compared to most of these books, earlier, the sources are different. I have chosen to structure my book around disciplinary conversations, relying on texts composed by the merchants of knowledge and their contacts, rather than around stories about transregional scholars. In that respect, I have been inspired by how, in *Islam Translated: Literature, Conversion, and the Arabic Cosmopolis of South and Southeast Asia*, Ronit Ricci explicated how the translation of a single Arabic book, *The Book of One Thousand Questions*, into multiple Southeast Asian languages linked Muslims and disciplines of Islamic knowledge. I investigate how intellectual connections between Jews, Christians, and Muslims in the Eastern Mediterranean facilitated translations. Ricci depicted an "Arabic cosmopolis"; I portray a shared world comprising fields as seemingly far apart as *Qabbalah* and theoretical astronomy.

Merchants of Knowledge also differs from the aforementioned books because I seek to demonstrate how the Renaissance, often seen as the birth of the West, involved numerous exchanges with Islamic societies. The perception of the Renaissance as a time of rebirth and recovery of the greatness of classical antiquity in the mythology of the modern West obscures the significant role of scholars from the Eastern Mediterranean in Renaissance intellectual life. This is not the first book to redress the impression that the presumed (re)birth of the West did not involve Jews and Muslims from the Eastern Mediterranean. Dag Nikolaus Hasse, in his 2016 *Success and Suppression: Arabic Sciences and Philosophy in the Renaissance*, demonstrated that the "win on points" of Renaissance humanists, whose intellectual agenda championed the contributions of classical antiquity, over intellectuals who valued the work of Islamic societies contributes mightily to the aforementioned perception.[38] Hasse unpacked the cultural dynamics of when and why Renaissance intellectuals appreciated Arabic knowledge encountered via Latin translations and when and why they preferred what they found in the Greek. The debates were surprisingly nuanced, and compelling arguments for the value of Arabic learning existed. Hasse documented how the win on points came to be perceived as a lopsided tally. In all cases, though, Hasse restricted himself to exchanges that the historical actors acknowledged, most of which depended on texts produced before 1200.

George Saliba, in *Islamic Science and the Making of the European Renaissance*, advanced in unraveling cases of scholarly exchange involving texts produced after 1200 between Renaissance Europe and Islamic societies that none of the historical actors acknowledge.[39] Hasse described Saliba's research as concerned with the "putative subcutaneous impact of Arabic sciences on the West."[40] The argument advanced in *Islamic Science*, however, was that modern claims of Europeans' independent discovery, not the argument for exchange of Islamic science from after 1200, were what is putative. Once the scope of exchange is widened in this book to include more disciplines and commerce, Saliba's conclusions become even more difficult to deny. But from the perspective of the historical actors, his claims are decentered. The exchange of theories in astronomy important for the European Scientific Revolution was not the dominant concern of the scholarly intermediaries, the merchants of knowledge.

The Available Sources

A distinctive inventory of primary sources exists. A plenitude of texts, in a number of disciplines, composed by the merchants of knowledge and their contacts, are available as manuscripts and incunabula. These texts yield a comprehensive picture of the deep, sophisticated conversations in a variety of disciplines. In some of these texts, we find references to transregional intellectual exchange and to knowledge gleaned from texts in other languages. Sometimes a single scholar composed in more than one language; at other times, the same scholars translated. Their multilingualism cannot be but evidence for their intent and ability to reach multiple communities.

Moving from the body of the texts to the flyleaves, some manuscripts contain statements of sale that establish the connection between figures. By deciphering statements of sale, Umberto Cassuto uncovered the transaction of 155 Jewish manuscripts around 1540 by Elijah Capsali,[41] who, at the time, was the head of the Candiote Jewish community, to three agents: Azalino, Battista, and Sacellani.[42] The identity of the purchaser remains unknown.[43] These manuscripts ended up in the library of the Fugger family of merchant bankers.[44] The Fugger mercantile network, by the early 1500s, established branches in both Venice and Padua,[45] and under Jakob Fugger (d. 1525), the Fuggers' trading operations reached Istanbul.[46] The manuscripts purchased by the agents became part of the personal collection of Ulrich Fugger (d. 1584), which was transferred to the Palatine Library in Heidelberg after his death. The Palatine Library suffered from looting during the Thirty Years War (1618–48), so, in 1623, the remaining manuscripts joined the collection of the Vatican Library. This purchase, sale, and transfer of manuscripts exemplifies the intersection of commercial and scholarly exchange.

The intellectual exchanges documented in manuscripts transpired in the context of the extensive commercial activities of the families of the scholarly intermediaries. The principal primary source for these activities is the records of the court of the Venetian Duke of Candia, as well as the notarial records, both available at the Archivio di Stato di Venezia (ASV).

These records tell us a great deal about the social conditions in which intellectual exchange occurred. The court functioned as a shared space because Jews and Muslims and Christians from outside the Venetian Republic were willing to use it. The two key archival collections for this book are the *Notai di Candia* (NDC), the records of the Venetian notary, and *Duca di Candia* (DDC), the records of the court of the Duke of Candia. When Candiote Jews resorted to the court to resolve communal and familial matters, the issues were no longer purely intramural. Most important, copious records of business transactions demonstrate that meaningful contact between Jews and non-Jews was unexceptional and happened more frequently than evidence from the discrete episodes of intellectual exchange found in manuscripts alone suggests. The records remind us that books and manuscripts, the stuff of intellectual exchange, were objects of purchase and sale in the milieux of the merchants of knowledge.

The archival documents offer up fascinating details about how translation mattered for commercial exchange.[47] Chirographs were not necessarily in languages known to court officials. For example, in May 1526, Moses Mavrogonato gave an oath before the ducal court in Candia, in Latin, to corroborate a Hebrew chirograph in the hand of Samuel b. Jonathan Romano.[48] The resolution of questions about estates also involved documents in multiple languages. In 1540, Levi Mavrogonato appeared before the ducal court in order to resolve a dispute about the will of his brother David.[49] He petitioned a certain Mechir Sacerdote, a Jew who was previously an agent for the deceased. According to Levi, Mechir was obligated for claims on his late brother's estate. Levi, to support his own case, brought a will written in Greek, by the imperial notary George Lima, on May 28, 1538. In response, Mechir furnished a later will, from February 1539, which was in the hand of Bernolai Bonbari, the consul of Alexandria, who was a resident of Constantinople at the time. The will presented by Mechir was validated (*roborato*) by Jacob Canalis, the baylo of Constantinople in June 1540. Thus, the competing claims of the creditors could be evaluated. Either documents were translated, or, if the absence of a reference to translation is significant, then the Mavrogonatos were proficient in multiple languages. Linguistic proficiency mattered in other areas. In 1469, David Mavrogonato (d. 1470), a translator for Greek priests,[50] became an envoy from the Venetian Republic to Istanbul.[51] Commercial and political relations would have ceased without translation.[52]

Trust between the families, their business contacts, and the court was crucial for this commercial exchange beyond confessional lines.[53] Jews were not of equal status with Christians in the Venetian Republic. Jews complained that they were taxed like foreigners, though they felt that they deserved to be treated like Venetian subjects.[54] Even someone like David Mavrogonato, whose taxes were minimized due to his exceptional service to the Venetian Republic, gained only the trappings of citizenship on Crete, but not the title.[55] Jews were resented by the Christian population on Crete, which was frequently in their debt, and, in 1449, Jews were accused of crucifying Christians' sheep.[56] Nonetheless, the recourse of Candiote Jews to the secular Venetian court allowed them "to function as a traditional Jewish community without the anxieties and reservations of other medieval Jewish communities."[57]

Jews on Candia turned to the court even for matters of Jewish family law, an arena in which they were not obligated to access the court. Conflicted marriages were not kept under wraps. For instance, we read that, in 1515, Hestera, the wife of Lazarus Salonichico (who was said to be of North African origin)[58]

> needs to go to a house with the aforesaid Lazarus, her husband, and to obey the commands of the husband, according to what Jewish laws say about this . . . according to the customs of husbands with wives, and through this it can be known if she becomes pregnant.[59]

In another case, in 1514, agents of the late wife of Elia (Liacho, i.e. Elijah) Nomico, probably the son of the famous book collector Levi b. Elijah Nomico (fl. fifteenth century), went before the ducal court to secure rights for her daughter, Sarulla, who had fallen out with her stepmother, Elia's current wife.[60] Sarulla, having reached the age of maturity, wished to take possession of one of Elia's houses as provided for in the will (*testamentum*) of her late mother. Elia contested the claim.

Sometimes Candiote Jews assisted the court in factfinding. In 1525, Immanuel Ḥen and Menasheh Delmedigo collaborated as the deputies (*legati*) of the Duke of Candia in civic matters. Both men, who were identified as physicians, were directed to intervene in the case of a woman, Plecti, married to a Jew, Sabathi Balaza.[61] Ḥen and Delmedigo reported the wife's complaints about his awful activities and clothes (*pessimae operationae et costumi*), which made it impossible for the married couple to cohabit. The duke enjoined a payment from Sabathi to Plecti of three ducats per month for three months and recommended that Sabathi send her a petition for divorce according to Jewish law.

Though the focus of this book is Jewish scholarly intermediaries who were linked to Candia, they inhabited an island with a diverse population. In addition to Jews from North Africa, such as the aforementioned Lazarus Salonichico, we read frequently in the ducal court documents about Jews from Ottoman lands. For example, among the ranks of Jewish taxpayers in a 1505 census found in the ducal court's records, there was a Lazari Turco and his son Acharos, as well as a Jocuda filio Michael Turco.[62] In these cases, the term *turco* denoted connections to the Ottoman Empire[63] and comprised non-Muslim Ottoman subjects and converts to Islam from Christianity.[64] For example, a document dated February 13th, 1453, tells of calfskins that were to be sold to benefit some orphans under the supervision of a noble.[65] One of the signatories of the document was a certain Michael Vrachuli, who was identified as a Turk (*turco*). Some Turks became official residents of Candia; Manoli, a Venetian who converted to Islam, was one.[66]

Muslims interacted with the Venetian government in a variety of ways. In a straightforward sense, Muslims trusted the ducal court to record business agreements. For example, in 1504 or 1505, a Turk named Mustapha Celichi, who became an adopted citizen of Candia (*adoptus Candidae*), concluded an agreement with a boat owner, Stamtio Grina of Corfu (though residing in Candia), who was shipping honey water and other commodities.[67] Venetians likewise trusted Turks; In May 1488, Marinus Signolo appeared at the court and approved of (*probo*) Mustafa Cana, a boat owner identified as a Turk, about to pilot a ship to Chios.[68] In 1554, the Duke of Candia paid a Turk named Zafer, who was the captain of a *schirazzo* (a square-rigged cargo vessel) traveling to Rethymno.[69] The court attempted to accommodate linguistic differences with Muslims as it did with Jews. On May 20, 1540, a Turkish *schirazzo* captain named Mehmed appeared in the chancellery and requested an interpreter, Jemalio Schrenza.[70] And sometimes, social relations developed between Turks and Christians on Candia. A certain Turk, probably a non-Muslim Ottoman, named Zammis entrusted his twelve-year-old son Giorgio to a Christian resident of Candia, Michaele Plaidemo, in 1501–2. Plaidemo was obligated to provide footwear, clothing, food, and shelter. If Giorgio absconded, Zammis would be obligated to pay a penalty.[71] The court of Candia helped make the city into an entrepôt where Jewish and Christian merchants and Muslim sailors benefited from relationships with each other. The Fuggers' agents would have no trouble locating prominent families with manuscripts.

Outline of Chapters

Different dimensions of exchange emerge in each chapter of *Merchants of Knowledge*. I begin by describing, in chapter 1, the network of merchant-scholar families from which the merchants of knowledge hail. The merchant-scholar families are those whose members owned or copied the trove of manuscripts that were sold to agents between 1539 and 1541 and ended up in Ulrich Fugger's (d. 1584) personal collection.[72] Many of the anecdotes I have provided in this chapter involved members of the merchant-scholar families. This complex of sales, however, is not the only example of commerce in books and manuscripts discussed in the book. In multiple cases, members of the merchant-scholar families gained socioeconomically from their talents in translation for both commerce and scholarship. Commerce also fostered trust among diverse constituencies. The chapter concludes by introducing the merchants of knowledge whose activities are the subject of the rest of the book. Individual merchants of knowledge traversed the boundary between the Venetian Republic and the Ottoman Empire via the transregional connections established by the merchant-scholar families.

In each of the other six chapters of the book, I describe exchanges between the merchants of knowledge and their contacts in a particular field. In chapter 2, I investigate how scholars from across the Mediterranean shared both technical information about judicial astrology and a discourse about astrology as a cultural hermeneutic. Astrology was widely practiced even though some of the merchants of knowledge and their contacts took positions in other disciplines that militated against acknowledging the practicability of astrological forecasting. Then, in chapter 3, on practical astronomy, I cover the production and exchange of tables, which were necessary for astrological forecasting. The merchants of knowledge and other Jewish scholars conveyed information from late medieval Islamic tables to Christian scholars. Conversely, a Muslim scholar relied on a merchant of knowledge to slake his thirst for tables published in European languages. Also in this chapter, I link the less visible exchange of information in tables with the documented, transregional passages of astronomical instruments.

In chapter 4, devoted to philosophy, I commence by covering exchange between the merchants of knowledge and Jews and Christians in the Veneto, where Jews were in demand as instructors. Philosophy was central to Romaniot Jews' own intellectual life and was a field in which they relied on both Latin and Arabic sources. Philosophy became the context for the exchange of discrete texts in other fields. In chapter 5, then, I examine exchange in the field of *Qabbalah*. By broadening the examination of the conversation to comprise *Qabbalah*, I clarify how *Qabbalah* both impinged on philosophy and astrology and was a field in which Jews and Christians came together. In these exchanges between the merchants of knowledge and their Christian contacts, Hebrew and even Arabic sources were taken seriously. These were disciplines in which the historical actors acknowledged exchange. In all four of these chapters, I present extensive evidence that both Christians and Muslims were interested in Romaniot Jewish intellectual life and that the contributions of the merchants of knowledge depended heavily on Arabic, Turkish, and Latin knowledge. Given all these motivations for multidirectional intellectual exchange, the documented exchange of texts in the aforementioned areas is not surprising.

Some of the information studied in chapter 3 came from the same astronomers of Islamic societies who produced theories whose exchange has been harder to document. In chapter 6, where I focus on the passage of advanced theoretical astronomy from Islamic societies to the Veneto, the explanation of exchange unacknowledged by the historical actors culminates. We learn that a merchant of knowledge who knew of these sophisticated theories of Islamic astronomy voyaged from Istanbul to Venice during the years when Copernicus studied medicine at the University of Padua. These innovative Islamic theories were relevant to another debate in astronomy, informed by the exchange of philosophy covered in chapter 4, in which European scholars connected to the merchants of knowledge were clearly involved and for which there was more evidence of transfer. Thus, exchanges in some fields set the stage for exchanges in other fields. Although Copernicus was not a named contact of the merchants of knowledge, building blocks of his theories clearly arrived in Europe from the Eastern Mediterranean because of the merchants of knowledge.

In chapter 7, I analyze exchange that mostly followed the passage of theoretical astronomy to the Veneto. For example, sometime after 1507, the same merchant of knowledge who traveled to Venice authored, in Ottoman Turkish, a text on pharmacology. In that text, there are descriptions of pharmacological computations from Latin medical texts previously unknown to readers of Islamic languages. The merchant of knowledge must have learned about Latin pharmacological computus in the Veneto. The same scholar recalled, in a different text, a handheld device that aided transcription. This device was first encountered at the Ottoman sultan's court, but only while visiting Venice in 1500 did the merchant of knowledge encounter someone who could explain its operation. Devices were transregional, and the mastery of devices combined intellectual and material exchange with the accrual of social capital from demonstrating the device before the sultan's court. Multidirectional exchange was contingent upon the prerogatives of the merchants of knowledge and their contacts, not the mythology of Renaissance humanists or a whiggish history of science. A complete understanding of intellectual exchange requires recognizing how translation and the demystification of technology were transactional.

ONE

THE NETWORK *of* MERCHANT-SCHOLAR FAMILIES *of* CANDIA

By the middle of the fifteenth century, Candia was an important intellectual center of Romaniot Jewry, a part of the Venetian Republic, and a hub for trade with Ottoman lands. There was a confluence of languages and cultures that suited the talented. Intellectual exchange flourished in the context of the transregional commercial and social exchanges recorded in the Venetian archives. In this chapter, we meet the merchant-scholar families to which the merchants of knowledge belonged.

The merchant-scholar families were those Cretan families whose members owned, authored, or copied the trove of manuscripts that were sold to agents[1] between 1538 and 1541 and ended up in Ulrich Fugger's (d. 1584) personal collection.[2] Accumulating goods necessitated wealth and expertise in languages.[3] The Fuggers' acquisition of books, which was followed a few decades later by the construction of Hans Jakob Fugger's *wunderkammer* in Augsburg, required agents with knowledge of the subject matter of the books.[4] Besides Venice and Padua, Rome and Florence were sites of other important offices for the Fuggers' book acquisitions in Italy.[5] The book collector Ulrich Fugger was a scholar and one-time chamberlain of the pope. Other Fuggers shared a scholarly bent. For example, Jakob Fugger (d. 1525; "the Rich") was interested in financing a college for monks at Augsburg that would offer instruction in Latin, Greek, and Hebrew.[6] For those manuscripts that could not be purchased, Ulrich Fugger's brother,[7] Johann Jakob Fugger (1516–75; a.k.a. Hans Jakob) employed Jewish copyists in Venice.[8] Elijah Levita (d. 1549), a Hebrew instructor for Jewish families in Padua[9] and proofreader for the Venetian printer Daniel Bomberg,[10] served the Fuggers as well.[11] The Fuggers were connoisseurs.

As Ulrich Fugger knew what he bought, the Fuggers' decision to acquire manuscripts from Crete was no coincidence. Crete was, for decades, a destination for Renaissance scholars. Arsenios Apostolis (d. 1535), a philologist who was born on Crete, returned to Crete from Venice in 1497–49 to collect manuscripts for Aldus Manutius and the Aldine Press.[12] The range of titles acquired by the Fuggers is evidence that they were well apprised of the deep Jewish intellectual life, particularly in the areas of philosophy and *Qabbalah*, found on Candia.[13] Their acquisitions included manuscripts of texts that had already been printed. For example, the eventual Vatican MS Ebr. 257 is a manuscript of Joseph Albo's *Seiper ha-ʿiqqarim* (The Book of Roots), a text of religious philosophy. The manuscript was sold in 1489 to Elqanah Capsali, the father of the merchant of knowledge Elijah Capsali, the mediator of the sale of manuscripts to the agents.[14] Though the book was printed in 1485 by Gershom Soncino in Venice,[15] the manuscript was nonetheless read carefully throughout.[16]

Likewise, the acquisition of the future Vatican MS Ebr. 373 by Ulrich Fugger suggests that his agents were aware of the differences between the manuscript versions, in addition to printed ones, of Ibn Sīnā's *Canon on Medicine* (Ar. *al-Qānūn fī al-ṭibb*). That manuscript contained the partial 1408 translation of the *Canon* by Joseph Lorqi. In 1491, a Hebrew translation of the *Canon*, which reflected the 1279 translation of Nathan Ha-Meʾati, but revised by Joseph Lorqi, was printed in Naples.[17] A 1492 printed edition, from Naples, reflected just the partial Lorqi translation. Because there are far more surviving printed versions of the Meʾati/ Lorqi translation, the purpose of the procurement of the future Vatican MS Ebr. 373 may have been to compare and contrast the printed Meʾati/ Lorqi translation with the printed edition of the less available partial translation by Lorqi.

The large purchase at the end of the 1530s was preceded by the purchase of at least one manuscript from the merchant-scholar families. That is, a library established in Venice following the death of Cardinal Domenico Grimani in 1523, which became an early source of the Fuggers' Hebrew library, contained a codex owned by David Capsali.[18] Grimani's acquisition of the manuscript was facilitated by how Grimani studied with the merchant of knowledge Elijah Delmedigo (d. 1497).[19]

The merchant-scholar families who dealt manuscripts to the Fuggers' agents through Elijah Capsali were intricately linked to each other, for over a century in some cases. The first three merchant-scholar families I list (the Delmedigos, Capsalis, and Balbos) were the most wealthy and prominent. The Capsalis and Delmedigos were connected through marriage. Participation in civic and economic life brought the Balbos and Cohen Ashkenazis, the next merchant-scholar families I list, into contact with the Delmedigos and Capsalis. The two merchant-scholar families who follow, the Algazis and the Galeanos, were not as wealthy or as prominent in civic life yet were highly significant scholarly intermediaries. A member of the Algazi family was a pupil of a Capsali and another Algazi was a pupil of a Galeano. Members of these interlinked merchant-scholar families sustained the intellectual exchange that I describe in the rest of the book. When appropriate, I identify the merchants of knowledge in each merchant-scholar family so we can see how the transregional careers of the merchants of knowledge resembled the paths of others in their families. The merchant-scholar families were tied to each other, and to Christians and Muslims, through commerce. Though intellectual intermediaries sought to impress, while commerce was, well, transactional,[20] transactions of manuscripts marked the intersection of commercial and intellectual exchange. There were four merchant-scholar families—the Ḥen, Astruc, Nomico, and Casani families—that lacked a merchant of knowledge but interacted with members of the merchant-scholar families that did. The activities of these four families exemplified the intersection of commercial and scholarly activity, contact with non-Jews, and transregional activity. Additional biographical information about the merchants of knowledge and their contacts follows in later chapters. In this chapter, I aim to show how the links between the merchant-scholar families, whether or not those links were built by scholars, were the scaffolding for exchanges among the merchants of knowledge and between them and their contacts.

The Delmedigos

The earliest people with the family name Delmedigo to come to Candia arrived in the mid-fourteenth century from Negroponte.[21] By the mid-fifteenth century, many prominent and wealthy people on Candia shared the family name Delmedigo and enjoyed transregional contacts. It is not certain that they were all related. Two or three individuals with the name Judah Delmedigo arrived on Crete by about 1360.[22] In 1438, Moses b. Abba ha-Zakein Delmedigo, the *condestabulo*[23] (head) of the Jewish community in Candia, recorded a debt of twenty ducats owed to him with the Paduan notary.[24] A Venetian ducat contained approximately 3.5 grams of gold, which is worth $264.25 at the time of writing.[25] A different Moses Delmedigo had a son, Elijah (d. 1497), who was a merchant of knowledge.[26] Though other Delmedigos traveled to Padua solely for commerce, Elijah made his way from Candia to Padua in 1480 in order to study medicine and supported himself by teaching philosophy. Elijah Delmedigo attracted famous Christian students, including Pico della Mirandola (d. 1494).[27] By the first half of 1493,[28] Elijah Delmedigo returned to Crete and became the *condestabulo.*[29]

Echoing how Elijah Delmedigo earned money from scholarly translation, Elijah's son Moses[30] benefited from his or someone else's ability to translate in at least two cases. First, in 1502, the Venetian notary Nicholas Corersi noted the receipt of fifty gold ducats from Moses Delmedigo and two other Jews of Candia.[31] Nicholas added that he received two payments of twenty-five ducats in 1500 and returned to the Jews ten ducats. From a marginal notation we learn that a letter of exchange (a way to draw on deposits in a distant bank) for the fifty ducats was sent from Constantinople and was remitted in Candia. Either Moses wrote the letter in a European language, or a Hebrew letter was translated. Second, in 1541, Moses Delmedigo corresponded in Hebrew, from Damietta (in Egypt) on the way to Candia, to say that he sent fifty ox skins but had not yet received the expected shipment of wine in return from Elijah Casani.[32] The letter was recorded in the archives in Hebrew, meaning that someone must have explained to the Venetian officials what was written.

The descendants of the other Moses Delmedigo, i.e., Moses b. Abba ha-Zakein Delmedigo, were also active in transregional commerce and scholarship.[33] A Menaḥem (Manuel) b. Samuel Delmedigo, who from 1502 bore the title of "doctor of medicine," also was a teacher (*maestro*) of a young Elijah

Capsali, the future *condestabulo*, chief rabbi of Candia, chronicler and contact of the Fuggers' agents. Meyuḥas Delmedigo[34] owned two manuscripts, sold to the Fuggers' agents in 1539–42.[35] Eliezer Delmedigo studied in Cairo[36] with Joseph Caro (d. 1575), author of the magisterial legal compendium *Shulḥan ʿaruḵ*.[37] Along with scholarship, money mattered. Manuel on several occasions tried to recoup debts from private individuals in Padua and from banks around the Veneto.[38] Delmedigos were wealthy. In a tax register from March 6, 1542, Rabbi Julio Delmedigo paid the most taxes, at 142 iperpera (from the Greek ὑπέρπυρον),[39] and Rabbi Salamo Delmedigo the second most, at ninety-five iperpera.[40] Though the relations between all the Delmedigos are not fully understood, wealth, civic prestige, learning, and connections with non-Jews were found in two branches of the Delmedigo family or in two different families with the same name. While Elijah Delmedigo distinguished himself as a transregional scholar, other Delmedigos were transregional merchants.

The Capsali Family

The Capsalis came to Crete by the thirteenth century.[41] Commercial opportunity, as was the case with the Delmedigos, attracted a member of the Capsali family, Elia Capsali, to Padua by 1427.[42] After the Ottoman conquest of Istanbul, the Capsalis bridged the Ottoman Empire with Candia. The merchant of knowledge Moses Capsali (d. ca. 1495) was born on Crete but ascended to the status of the leading rabbi of Constantinople by 1450. He enjoyed a place at the *dīwān* of Mehmed the Conqueror.[43] Moses Capsali's nephew, Elqanah Capsali, married the daughter of Isaac b. Ephraim Delmedigo, Pothula.[44] Their son was the merchant of knowledge Elijah Capsali (d. 1555), and Manuel became one of Elijah Capsali's teachers.

While Elijah Capsali was growing up, his family was wealthy. The two figures who paid the highest tax of ninety iperpera in the tax census of May 27, 1505, were Elqanah Capsali and Menasheh Delmedigo.[45] Samuel Capsali, Elijah Capsali's uncle, was next at seventy iperpera.[46] Relations between Delmedigos and Capsalis were strengthened in the commercial realm. For example, in June of 1505, one Zachary Spanopoulo appeared in the chancellery and pledged on behalf of the boat owner (*padrono*) Georgio Rizo for the trip to Chios.[47] On December 12, 1505, Samuel Capsali claimed repayment of that pledge to Spanopoulo to insure Georgio Rizo's trip to Chios. Although the boat ended up being captured by pirates, it returned safely to its home port, and Capsali petitioned as early as October 17 through Menasheh Delmedigo for six gold ducats back.[48] The risk of these commercial activities was reflected in how Samuel Capsali guaranteed the money for the risk (*arisicum*) of the journey.[49] Stationary members of merchant-scholar families financed the mobility of others.

Elijah Capsali studied in Padua with Elijah Delmedigo's son, Judah.[50] Capsali was also a student at the yeshivah in Padua, under Judah Minz, between 1508 and 1510. By 1515, Elijah Capsali served as *condestabulo*, an office the duties of which included negotiating onerous tax burdens.[51] The geographic range of his epistolary activity was broad. Elijah Capsali corresponded with rabbis in Istanbul, such as Elijah ha-Levi, and one of his teachers from Padua, Israel Ashkenazi, wrote to Elijah Capsali from Jerusalem.[52] In 1523, he composed *Seider Eliyahu zuṭa*, a history of the Jews of Asia Minor. Commerce did not escape Elijah Capsali's attention. He traded wine with a certain Rabbi al-Ashqar.[53] Being from a family with a wealth of non-Jewish contacts as well as bonds with Candiote merchant-scholar families, Elijah Capsali was well situated to mediate the sale of manuscripts that defined the group of merchant-scholar families.

The Balbo Family

The Balbos are the other most prominent merchant-scholar family in the mid-fifteenth century mentioned in this book. The Balbo family must have arrived on Crete over a century before because, in 1334, Anastasia, the widow of Judah Balbo, left a will with a provision for funds to be used for the dowries of poor Jewish women.[54] By the late fourteenth century, members of the family had become prolific money lenders.[55] The merchant of knowledge Michael b. Shabbetai Balbo became a chamberlain (*camerarius*) for the Candiote Jewish community as well as *condestabulo*. On October 23, 1454, Michael b. Shabbetai Balbo, Ephraim Delmedigo, Isaac Todesco, and the *condestabulo* David Capsali signed a document establishing the schedule for the Jewish community's tax payments.[56] These connections to the Delmedigos and Capsalis paid dividends for Michael's intellectual pursuits. Michael, in his polemic against metempsychosis, wielded his social prominence against his *arriviste* adversary. That epistolary debate was a significant development in mid-fifteenth-century Jewish intellectual life in the

Eastern Mediterranean and will be studied later in this book.[57] He sent a record of the debate to Rabbi Moses Capsali in Istanbul. Both manuscript records of the debate fell into the Fuggers' possession; Michael's record became Vatican MS Ebr. 105.

Michael Balbo's son Shabbetai (Sabatheo) was more prominent[58] than his brothers Solomon and Isaac.[59] In 1488, Menasheh Delmedigo, then the *condestabulo* of the Jewish community on Crete, recorded the election of Sabatheo (i.e., Shabbetai) b. Michael Balbo as the next *condestabulo*.[60] The statement was undersigned by the *condestabulo*'s chamberlains, including Samuel b. Fraym (Ephraim?) Delmedico (i.e., Delmedigo). Shabbetai continued to interact with Delmedigos. In October 1503, he accompanied his wife, Soltana, to the ducal court as her agent.[61] Mossaninus Delmedigo petitioned for the lifting of an order, imposed at Shabbetai Balbo's urging, to get him (Mossaninus) to satisfy his creditors by auctioning off some houses. In the end, Mossaninus Delmedigo was ordered to do just that. Despite that conflict, the families were joined by marriage. Another Balbo, Yocuda Balbo (fl. 1500–50),[62] married Ghana, the sister of Pothula Delmedigo, the mother of the merchant of knowledge Elijah Capsali.[63] Multiple threads stitched these three elite merchant-scholar families together. Shabbetai possessed the eventual Vatican MS Ebr. 257, which he sold to Elqanah Capsali and which was absorbed into the Fuggers' collection.[64] Scholarly acumen and civic stature coincided. The next four smaller merchant-scholar families also contain members who flourished in commerce, civic life, and scholarship, in various measures, and who interacted with the Balbos, Capsalis, and Delmedigos, as well as with Christian officials.

The Cohen Ashkenazi Family

The first member of the Cohen Ashkenazi family to arrive on Candia, in the 1460s, was the merchant of knowledge Moses b. Samuel Cohen Ashkenazi. He may have been of German origin.[65] He was Michael Balbo's opponent in the heated controversy over metempsychosis. In that same decade, Michael Cohen Ashkenazi authored *Urim wᵉ-ṯummim*, a tantalizing astrology text, and pursued the socioeconomic benefits of his deep knowledge of that science. On the flyleaves of the unique manuscript of *Urim wᵉ-ṯummim*, we find Moses Cohen Ashkenazi's forecasts for Antonio Vittori, the son of Benedetto Vittori, the Duke of Candia in the 1450s.[66] Another prediction on the flyleaves was about a shipment of wine sent in 1464. Cohen Ashkenazi also produced a forecast for a boy born in Venice in 1454.[67] By treating matters of commerce in his forecasts for the Vittoris, Moses Cohen Ashkenazi attempted to gain influence with the same family with whom his rival merchant of knowledge Michael Balbo hoped to negotiate communal debt. The merchants of knowledge could be social climbers.

One of Moses's sons, the merchant of knowledge Saul Cohen Ashkenazi (d. ca. 1523), was a student of the merchant of knowledge Elijah Delmedigo.[68] He became prominent on Candia, signing four internal edicts of the Candiote Jewish community between 1513 and 1521.[69] Another of Moses's sons, Samuel Cohen Ashkenazi, served as *condestabulo* in either 1503 or 1508.[70] Given this matrix of connections, it is easy to imagine how the unique manuscript of *Urim we-ṯummim* reached the Fuggers' agents later in the sixteenth century.

The Algazi Family

No member of the Algazi family arrived on Candia before 1505.[71] But three members of the family, Isaac Algazi, a Joseph Algazi, *and* a Joseph Algazi *medicus* were mentioned in a 1521 tax census.[72] Despite the recent arrival of the Algazis, the skills of the second Joseph Algazi attracted the Venetians' attention. On August 21, 1522, we read that "Maestro Joseph Algazi, a Spanish Jewish physician (*medicus phisicus*), appeared at the chancellery and would be announced as soon as he treated (*medega*) the sick people (*amaladi*) written below."[73] One of the Josephs died in the plague of 1523.[74] The family accumulated a modest amount of wealth. The widow of Isaac Algazi, the son of one of the Josephs,[75] and Samuel Algazi (d. 1588) were listed in a tax census from 1541.[76] They paid less taxes than some Delmedigos and Capsalis. Samuel became a well-known scholar in Italy. He was a pupil of the merchant of knowledge Elijah Capsali, then of Judah Delmedigo, and owned Vatican MS 62 and Vatican MS 236, both of which ended up in the Fuggers' collection.[77] In the 1530s, the merchant of knowledge Abraham Algazi was a student of a merchant of knowledge from the Galeano family.[78] Abraham sold the future Vatican MS Ebr. 201, a text of qabbalistic scriptural commentary, to his teacher in 1539. All these connections matter because the Algazis, though less renowned, were the Candiote family most closely connected to the Galeanos.

The Galeano Family

Of all the merchant-scholar families, the Galeanos were the most recent arrivals on Crete, turning up between 1523 and 1525 from the Ottoman Empire. The merchant of knowledge Moses b. Judah Galeano (d. > 1542) referred to himself as al-Tīrawī, i.e., from Tire, a small city in the modern province of Izmir.[79] Thus, his family may have come to Istanbul as a result of the *sürgün*. He has been shown to be identical with the author Mūsā Jālīnūs of Arabic and Turkish works, so I will refer to him as Galeano/Jālīnūs.[80] In the last few decades of the fifteenth century, Galeano/Jālīnūs's uncle, Moses b. Elijah Galeano, produced several Hebrew works in Istanbul.[81]

Galeano/Jālīnūs's trip to the Veneto occurred sometime between 1497 and 1503, possibly during the second Ottoman-Venetian War, which spanned 1499–1503. Galeano/Jālīnūs's contacts with Ottoman elites intensified after his return from the Veneto. In 1505/6, he translated a Latin astronomy text into Arabic for the chief military judge (*al-qāḍī bi-l-ʿaskar*) ʿAbd al-Raḥmān Muʾayyadzādah (d. 1516).[82] Soon after, Galeano/Jālīnūs authored a text on pharmacology in Ottoman Turkish and dedicated it to Sultan Bayezit II's (r. 1481–1512) chief physician.[83] One can imagine that Galeano/Jālīnūs flaunted his linguistic range at the sultan's court to compete with the Sephardic Jews who had arrived a couple of decades earlier.

Corazzol proposed a 1525 *terminus post quem* for Galeano/Jālīnūs's arrival on Crete from Istanbul because Galeano/Jālīnūs did not appear in a 1523 list of Jews permitted to leave the Zudeca of Candia during an episode of the plague that ended in 1525.[84] The death of his teacher Elijah Mizraḥi in Istanbul in 1526 and the death of a patron, the sultan's chief physician, Ahī Çelebī, in 1524, offered additional impetuses for Galeano/Jālīnūs to leave Istanbul just after 1525. The unique linguistic competencies that Galeano/Jālīnūs honed in scholarly translation brought him to the attention of the colonial authorities soon after his arrival in Candia. In 1529, Galeano/Jālīnūs translated a letter addressed to the lordship (*signoria*) of Candia written in Turkish in Turkish (i.e., Arabic) script.[85] At least one manuscript that Galeano/Jālīnūs owned, Vat. MS Ebr. 201, was purchased by the Fuggers' agents.[86]

As a bearer of two names, perhaps it is not surprising that Galeano/Jālīnūs was not always within the purview of record keepers. By 1542, Galeano/Jālīnūs's son Jonah reached the age of maturity.[87] Since Galeano/

Jālīnūs and his wife Spagnola appeared in a 1542 tax census,[88] but not in any thereafter, they must have left Candia soon after 1542.[89] Galeano/Jālīnūs was the only merchant of knowledge to have an impact on the intellectual life of Candia, Istanbul, and the Veneto.

The Ḥen (Lat. Gratian) Family and Their Connection to the Casanis

There were merchant-scholar families who lacked a merchant of knowledge. Consider the Ḥen family, members of which reached Candia from the Iberian Peninsula by the fourteenth century.[90] A Sheʾaltiel Gratian, who was still in Iberia in the fourteenth century, was a scholarly correspondent of the renowned Catalonian scholar Profayṭ Duran (ca. 1350–1415).[91] A century later, Immanuel Ḥen witnessed the following intercommunal transaction involving Menasheh Delmedigo. On June 19, 1503, Demetrius Zilaiti appeared before the notary as the deputy of Petrus Trivisa.[92] According to a notarized letter, Levi, son of Michael Delemelech, was to be paid nine ducats and seven iperpera, which was about half of the nineteen ducats and six iperpera that Levi and another Jew named Gratia b. Ieste owed the agent of D. Viti Venerio. Levi was seized and incarcerated on account of that debt. Menasheh Delmedigo acknowledged that he had pledged to give the nine ducats and seven iperpera to Demetrius Zilaiti. Thus, the claim against Levi was resolved.

Involvement in financial affairs went along with scholarly engagement. Immanuel Ḥen's son, Sheʾaltiel, was a physician[93] and, like his father, an owner of Vatican MS 103, a codex on *Qabbalah* and scriptural commentary, which was sold to the Fuggers' agents in 1541.[94] Sheʾaltiel showed up frequently in the court documents in the 1540s in other capacities,[95] for example petitioning the Duke of Candia so that he might collect money from Liachum Casani.[96] The Casanis were another family who owned manuscripts transacted to the Fuggers.[97] The Casanis were in Candia by the thirteenth century[98] and occupied prominent positions in the Candiote Jewish community[99] through which they came into contact with other merchant-scholar families such as the Delmedigos.[100] In 1548, Sabati Casani was *condestabulo*.[101]

There were other members of the Ḥen family on Candia besides the father-son dyad of Immanuel and Sheʾaltiel; Maestro (i.e., physician) Leo Gratian appeared in a tax census from 1542.[102] Families lacking notable scholars nevertheless included members who possessed manuscripts, played important civic roles, and were noticed by the Fuggers' agents.

The Astruc Family and Their Connection to the Nomicos

The Astruc family, like the Ḥens, contained members with scholarly interests but no merchants of knowledge. The Astrucs came to Candia from the Iberian peninsula by the mid-fourteenth century.[103] Solomon Astruc, who served as *condestabulo*,[104] copied in 1451 a text of *Qabbalah* with diagrams for Jeremiah b. Moses Nomico.[105] The Nomicos were the owners of texts within a codex transacted to the Fuggers.[106] That codex, Vatican MS Ebr. 249, consisted of scriptural commentary and philosophy, parts of which were copied by Jeremiah Nomico in the 1450s.[107] Though Nomicos were on Candia in the thirteenth century,[108] their connections were insufficient to save Jeremiah Nomico from imprisonment in 1452 on the charge of crucifying a sheep.[109] The important book collector Levi Nomico may have been a relative of Michael Balbo.[110] While members of the Nomicos had scholarly interests, no member of that family functioned as a merchant of knowledge.

By 1503, another member of the Astruc family, Lye Astruc, occupied the position of *sansarius*,[111] a semi-official tax collector. In May 1503, Lye collected the *messitaria*, a commercial transaction tax, from Moses Russo after he, along with his sons and heirs, pledged to pay Isaac b. Solomon Denachu for nineteen *cantharsos*es (drinking jugs with handles) and five and a third rolls of unsalted brown skins.[112] Solomon Astruc was connected to Moses b. Elijah Delmedigo through civic life in the following way. On May 28, 1516, two Jews from Rethymno engaged Francis Phylletus to represent them in the appraisal (*taxandum telluris ordinarium*) of their land.[113] The Candiotes Samuel Theotonico, Solomon Astruc, and Moses Delmedigo, together with Jews from Chania (also on Crete), intervened with the Venetian authorities to be sure that the Jews of Rethymno were taxed at appropriate intervals and at appropriate rates. The fact that an Astruc was entrusted by other Jews with negotiating a significant financial obligation indicates the family's standing.

Astrucs, like members of other merchant-scholar families, treasured books. In 1517, a woman named Bona, along with an agent of Lazarus Astruc, turned up at the notarial court to return some property of a Jewish physician Emmanuel who was about to travel to Canea.[114] Emmanuel had been vouchsafed some books by Lazarus. The court determined that all the books that were deposited with Emmanuel should be located and left with Solomon Delmedigo.

The cluster of merchant-scholar families was defined by the sale of manuscripts to agents of the Fuggers. But the sale occurred *after* all the intellectual exchange that I describe in the following chapters. Since families not directly involved in that intellectual exchange passed on manuscripts to agents of the Fuggers, the cluster of merchant-scholar families was based in commercial and civic activity in addition to intellectual exchange. To the merchant-scholar families, the Fuggers would have appeared as another merchant-scholar family, though distinctly outside the cluster of merchant-scholar families spotlighted in this chapter.

An Open Network of Merchant-Scholar Families

What was the nature of the connections between the members of the Jewish merchant-scholar families? Some of the families, such as the Balbos and the Capsalis, had long-standing roots in the Eastern Mediterranean. Others, such as the Algazi and Cohen Ashkenazi families, had appeared there more recently. Recent arrivals nevertheless successfully cultivated connections with more established Candiote Jewish families as well as with Christians on Candia. At the same time, the strength of the ties of the merchant-scholar families to Candia had its limits. We read in the archives of the ongoing connection of some members of the merchant-scholar families to Iberia. Also, there is no evidence that the merchant-scholars, when away from home, exhibited a consciousness of being in any diaspora more specific than the Jewish Diaspora.[115] Claude Markovits noted the distinction between a network on one hand and exchange that went along with migration on the other hand. While the connections between the families were engendered by previous migrations, the families did not perceive themselves to be emigrating, migrating, or immigrating.[116] Padua and Istanbul were, instead, cities with long-standing Jewish communities that enjoyed long-standing ties, sometimes strengthened through marriage, to Crete. We should, however, consider the possibility that the merchant-scholar families formed a network through which goods, information, and people circulated. Candia would be the hub of the network, and Istanbul and Padua would be outposts of the network.[117]

Information, however, played a much different role for the merchant-scholar families than the role Markovits and other historians found it played in some other networks. Jessica Goldberg, in her research on merchants' correspondence from the eleventh century, uncovered significant exchange of insider knowledge of routes, orders to sell, and news of the community.[118] Any time that such information left the network, the commercial advantage of the merchants in the network was at risk. Similarly, Sebouh Aslanian, in his study of the Armenian New Julfa network, presented a "retinue of circulating clerics who helped maintain the identity of the communities dispersed along the network."[119] Those clerics were akin to the information that Markovits described because were the clerics to serve non-Armenians, then the colonies and the network would, almost by definition, lose their identity. Access to insider information (Markovits and Goldberg) or to the clerics (Aslanian) demarcated the boundary of the network. In contrast, given the right conditions of exchange, the merchant-scholars were delighted to share, impart, or sell information to anyone. Even knowledge about Judaism, which circulated among *but also beyond* the merchant-scholar families, was commodified. There was no evidence that the merchant-scholar families perceived themselves to have a privileged position when it came to the exchange of goods, books, or knowledge.[120] We do not know whether the merchant-scholar families guarded commercial information since we do not possess their correspondence.

Among the merchant-scholar families, translation fueled the circulation of information. Three cases involving the Astrucs, a merchant-scholar family, show how knowledge of multiple languages aided commerce. In the first, Elias b. Solomon Astruc represented his mother (Donortae) before the ducal court in 1470–72 with regard to his father's will. To resolve matters, a document (*cedula*) had to be translated from Hebrew (*Judaico*) into Latin.[121] In the second anecdote, from September 1489, a chirograph, a document with multiple copies of which each party possessed one, needed to be deciphered. The copies were separated with a wavy cut so the copies could be authenticated in the future by seeing if they fit together.[122] At the urging of Lazarus Astruc, the interpreter Mordechai Astruc arrived and swore that the Hebrew handwriting in the chirograph was in fact that of Lazarus

Astruc. In the third anecdote, Solomon Astruc's son Judah witnessed the transcription and translation of another chirograph, written in Egypt, in which Abraham Zafar's widow, Raina, stated that some land of his has been sold and was then in the possession of Moses Spano.[123] The Latin attestation did not reproduce the contents of the Hebrew text, meaning that oral assurances must have been received. The ability to use Hebrew and the connection to Candia traced the open border of the network of merchant-scholar families. Their non-Jewish contacts lacked at least the knowledge of Hebrew or the connection to Candia or perhaps both.

The recorded close and constant interactions between the merchant-scholar families indicated that the merchant-scholar families had a sense of connection. Such connections enabled the more sparsely documented but consequential intellectual exchanges of the merchants of knowledge. Thus, I have chosen to categorize the relationship between the merchant-scholar families as an open network, a term which highlights the connections and intersections of the merchant-scholar families with each other, their contacts in the Veneto and Istanbul, and the significance of the cluster of merchant-scholar families as a context for intellectual exchange. The merchant-scholar families did not police the boundaries of the network.

The role that translation played for the merchant-scholar families in both commerce and scholarship is another reason why I prefer to write of intellectual exchange than of transmission of ideas. The most prominent scholars in the merchant-scholar families were merchants of knowledge because they were aware of the concrete social and economic benefits of all sorts of exchange, intellectual and commercial. Exchange happened because it benefited the historical actors, not because of the exigencies of the present.

Conclusions

The intellectual world to be described in *Merchants of Knowledge* was both transregional, because it spanned the Veneto, Crete, and Istanbul, and integrated, because the merchants of knowledge, within the merchant-scholar families in the nexus, pursued agendas of research and translation that shared a broader context. The nonprovincial outlook of the merchants of knowledge was mirrored in the everyday lives of their business-minded family members. In some cases, the merchants of knowledge joined in commercial transactions. Because the merchants of knowledge participated in the wider world, the intellectual connections to be studied in the rest of the book were not isolated encounters. At the end of this chapter, we read how translation eased commercial transactions; in the rest of the book, we shall find that intellectual exchange was transactional.

Knowledge, however acquired, was always a commodity, and in the ensuing chapters, each devoted to a discrete discipline, I explain the market for the knowledge that circulated through the network of the merchants of knowledge. The transregional connections that I introduced in this chapter would be of no use for intellectual exchange if the contacts of the merchants of knowledge were uninterested in the knowledge they had. In chapter 2, the merchants of knowledge mediated exchange of a field, astrology, that pervaded economic and political decision-making as well as religious thought. Indeed, astrology was the most comprehensive intellectual field in the world of the merchants of knowledge. The Christian and Muslim contacts had higher social status, but the language skills of the merchants of knowledge afforded them access to a wider variety of texts and techniques. Their knowledge was prized and applicable in many contexts.

TWO

ASTROLOGY, *a* SHARED HERMENEUTIC

Horoscopes have never been published in the *New York Times* because, as an employee of that paper once explained, astrology "is not supported by science."[1] One hundred years ago, articles on astrology were grouped with ghost stories, and both were printed solely to entertain readers. In the 1980s, that goal of amusement merged with the duty to inform when *New York Times* reporters publicized Nancy Reagan's consultation of astrologers with an article entitled "Not a Slave to the Zodiac, Reagan Says."[2] *New York Times* reporters have also covered astrology in a more nuanced fashion, elucidating its connection to astronomy and high culture. In 2006, an article appeared about the rerelease of Gustav Holst's orchestral suite *The Planets.*[3] The 1916 version of the suite had seven movements for the seven known planets, each movement inspired by the planet's astrological significance. For example, the movement devoted to Mars was subtitled "The Bringer of War." The suite has been used in countless films. The rerelease included a short piece entitled "Pluto," as Pluto was recognized as a planet, and thus accorded a role by astrologers, after Holst wrote the suite. But the rerelease was finalized before Pluto's reclassification as a dwarf planet earlier in 2006 (also covered in the *Times*), which stripped Pluto of its astrological role. The fleeting emendation of Holst's work aside, the planets have retained their symbolism. We are ever seeking answers in the heavens.

Astrology has never been discussed in a unitary fashion; indeed, historians distinguish two types of astrology. The first, natural astrology, acknowledges that connections between the heavens and the earth effect terrestrial events such as meteorological phenomena. If these connections are rendered intelligible through symbols such as the zodiacal constellations, the second type—judicial astrology, i.e., forecasting—comes into play.[4] Unless specified, references to astrology in this book are to judicial astrology, because it was judicial astrology that preoccupied the merchants of knowledge and their contacts the most. It figured into multiple important decisions.

The forecasting techniques known to the merchants of knowledge and their contacts originated in Hellenistic Egypt.[5] The foundation of an astrological forecast was the ecliptic, the sun's circular path around the earth divided equally into the twelve zodiacal constellations. The ecliptic was the background against which scholars mapped the paths of the planets. The merchants of knowledge and their contacts used the ecliptic for four types of forecasts: genethliac forecasts, historical forecasts, elections, and interrogations. Genethliac (horoscopic) astrology encompassed predictions about the client's life based on the configuration of the heavens at birth. The ascendant, or the point on the ecliptic circle that was rising at the time of birth, was the most significant indicator in this forecasting method. Natal charts also required the determination of the locations of the twelve places (Gr. *topoi*),[6] twelve divisions of the ecliptic, which were counted from the ascendant moving counterclockwise. Each place represented a different dimension of an individual's life: e.g., life (#1), wealth (#2), family (#3), etc. Since each place, known to the merchants of knowledge and their contacts as a house, did not have to be of the same angular measure, their computation was a topic of discussion and debate among authors of astrology texts.[7] The places are mentioned in earlier works of astrology dating back to Late Antiquity (e.g., Ptolemy's *Tetrabiblos*) but often without specific instructions for how to compute them.

Astrologers answered more than just personal queries. Forecasting the results of a battle, for instance, by summing the natal charts of all involved would be impractical and imprecise. Predictions about the futures of nations and religious communities, as well as large-scale historical events, were the subject of historical astrology, which originated in the Sassanian Empire (224–651 CE). Historical astrology's novel premise was that at the time of a planetary conjunction—when an earthbound observer saw two planets superimposed in their paths—the rays of influence of each planet were conjoined. Thus, the overall effects of each planet merged. The effects differed depending on the aspects of the planets. If the angle between the sign in which one planet was found was 60 or 120 degrees from the sign in which another planet was found, the merged effect was auspicious. If the angle was 90 or 180 degrees, the merged effect was ominous. In historical astrology, the effect of a conjunction or an aspect varied depending on where it occurred against the background of the zodiacal constellations.

Two other types of predictions were practiced by some of the merchants of knowledge and their contacts. Elections helped the questioner determine the best time to do something, such as when to start a journey or found a city, whereas interrogations answered a question about the present, such as the location of a lost item. While all types of predictions depended on knowing the houses (also places), ascendant, and zodiacal signs, the coexistence of genethliac astrology, historical astrology, interrogations, and elections implied that astral causes were understood to be complex. The fact that even the most privileged actors sought postnatal forecasts meant that outcomes were not fully discernible at birth.

The modern perception of astrology as a pseudoscience should not obscure the ubiquity of its practice and the extent of its intellectual influence in the world of the merchants of knowledge and their contacts. In the Venetian Republic, the Ottoman Empire, and elsewhere in Europe, astrological forecasts were a tool of political discourse that helped make events more intelligible and rationalized processes, such as disease or political upheavals.[8] The Ottoman sultans of the late fifteenth and early sixteenth centuries legitimated their rule through historical astrology, while to the west, the clock tower in St. Mark's Square in Venice featuring the signs of the zodiac was commissioned by the Venetian Senate and constructed in 1499. Though the Istanbul Observatory was closed in the late sixteenth century, the practice of astrology did not abate in the Ottoman Empire[9] or in Northern Italy.[10]

Astrologers plied their science to take advantage of opportunities to minimize the anxieties of the powerful. As Robert Westman explained, "Astrological discourse was eminently suited to addressing the anxieties of rulers, and the unstable early modern political system created an ideal climate in which such anxieties flourished."[11] Astrology's political significance led the merchants of knowledge to rub shoulders with prominent figures. For example, we read in the previous chapter that the merchant of knowledge Moses Cohen Ashkenazi jotted down forecasts for Antonio Vittori, the son of the Duke of Candia, Benedetto Vittori. Even events that would be categorized as religious by moderns, such as the coming of the Apocalypse, could be predicted by astrologers.[12] Little escaped astrologers' attention, as astrology "concerned itself with explaining phenomena in all those areas that in the twentieth century fell under the disciplines of political science, economics, psychology, sociology, weather forecasting, and medicine."[13] No discipline was more important.

A Transregional Conversation about Astrology and Other Methods of Forecasting

Knowledge of astrology was at a premium during the time of the merchants of knowledge. Yet during the reign of Sultan Mehmed II from 1453 to 1481, the technical level of astrology and astronomy in the Ottoman Empire was not high, and it was necessary for the sultans to turn to experts from the Persianate Timurid Empire for expertise in astronomy and astrology.[14] Things began to change under his son Bayezit II (d. 1512), when Ottoman scholars began to travel for their studies in the sciences. Though the eventual court astrologer ʿAbd al-Raḥmān began learning astronomy and astrology in Ottoman lands, he journeyed to Iran in the 1490s to further his studies.[15] In his *Kitāb ḥifẓ al-ṣiḥḥa* (Book on the Preservation of Health), ʿAbd al-Raḥmān explained the theoretical connection between medicine and astrology. Astrologers studied how heat and moisture emanated (*fāʾiḍa*) from the sun and moon to the human body. Physicians were concerned with how human longevity depended on the balance of moisture and heat in the body.[16] Following this reasoning, the astrological exaltation (*saʿāda*) of the moon and sun during gestation affected this balance of moisture and heat. ʿAbd al-Raḥmān completed his treatise on medicine in February 1502.[17] Though ʿAbd al-Raḥmān was neither a merchant of knowledge nor one of their contacts, his career coincided with those of the merchants of knowledge and, as we will see, reflected similar impetuses to travel. The merchants of knowledge did not initiate the transregional conversation about forecasting in which they participated.

The conversation comprised debates over the best method of forecasting. The merchant of knowledge Galeano/Jālīnūs wrote at a time of great apocalyptic speculation in the Ottoman Empire and astrology was not the only method of forecasting *en vogue* at the Ottoman court. Another method, to which Galeano/Jālīnūs dedicated a Hebrew treatise titled *Seiper ha-goralot* (The Book of Lots), was geomancy.[18] In geomancy (Heb. *ḥokmat ha-nequdot*, Ar. *ʿilm al-raml*), one forecasted on the basis of configurations of marks made on a page or in sand, or of cast dots such as pebbles. Either the pebbles or the dots were interpreted to form one of sixteen figures.[19] Each figure had a variety of traits, and because astrology had been introduced into geomancy by the time Galeano/Jālīnūs wrote, the traits included those associated with the planets.[20] For example, three vertical rows of two capped by a single dot was interpreted to resemble a human and was related to Jupiter because it was considered an auspicious, ruling planet. The same configuration, but with the single dot at the bottom, was related to Saturn, a less auspicious planet. Galeano/Jālīnūs justified composing a text on geomancy because of how astrological forecasts were complicated by the required computations in mathematical astronomy.[21] Because the planets were thought to order humans' thoughts and deeds, an additional drawback of astrological interrogations was that the actor asking the question might wait to ask until the heavens indicated the answer they desired.[22] Geomancy was, in contrast, simpler. In addition, the casting of the dots could be free of astral influences if the geomancer could ascertain that the impulses that led to a client's question were not an effect of astral influences.[23]

In *Seiper ha-goralot* (The Book of Lots), Galeano/Jālīnūs indicated his familiarity with Islamic geomancy. Though his chief source was Rabbi Judah Ḥarizi's (d. 1225) work on geomancy,[24] he also evidently consulted the works of the Muslims Abū Saʿīd al-Ṭarābulsī,[25] Abū Saʿīd's student al-Zanātī,[26] and Zakariyyāʾ al-Madanī.[27] Galeano/Jālīnūs's allegation[28] that Muslim geomancers mistakenly believed their art to have originated with the prophet Daniel coincided with the claim of Ḥaydar-i Remmal, the geomancer of Sultan Süleiman the Magnificent (d. 1566), that his (geomantic) art was that of Daniel.[29] Galeano/Jālīnūs's claim about Ottoman Muslim geomancy was backed up by the sources.

Though Galeano/Jālīnūs presented geomancy as an alternative to astrology, he acknowledged that astral forces were real and that forecasts, however complicated, might be possible. In *Seiper ha-goralot*, he wrote that the human soul was hewn (*ḥuṣab*) from heavenly substances (*ʿaṣamim ʿelyonim*) and actualized in matter by the renewal, through the intermediation of the planets, of a predisposition in matter.[30] Though Galeano/Jālīnūs could be critical of astrology, he was not wholly opposed. In his longest, most interesting text, entitled *Puzzles of Wisdom* (Heb. *Taʿalumot ḥokmah*) and completed in 1537,[31] Galeano/Jālīnūs prioritized the hermeneutic value of astrological forecasting. Because astrologers discussed possible things in terms of their known ordered causes, he reasoned, they could demonstrate that that which initially appeared to be coincidental was actually the result of God's work. Galeano/Jālīnūs emphasized, quoting Psalms 37:23, that these known causes extended to individuals' actions and were ordered by God.[32] Since Galeano/Jālīnūs contended in *Puzzles* and *Seiper ha-goralot* that astrology would not help one avoid God's will, it makes sense that, after his arrival on Crete, he acquired a text, *Urim wᵉ-tummim*, about how astrology helps one make sense of God's will.[33] The title, *Urim wᵉ-tummim*, is a reference to components of the breastplate of the High Priest at the ancient Temple in Jerusalem. In *Urim wᵉ-tummim*, the merchant of knowledge Moses Cohen Ashkenazi unleashed an unprecedented, systematic integration of astrology and Judaism. He composed the text between 1465 and 1472 on Crete.

If Galeano/Jālīnūs's intellectual formation occurred outside Crete in the Ottoman Empire, why did Galeano/Jālīnūs become interested in *Urim wᵉ-tummim*? Part of the explanation is simple: Cohen Ashkenazi built on the oeuvre of Abraham Ibn Ezra (d. 1164 or 67), a poet, linguist, Bible commentator, and scholar of science and philosophy who was a pivotal figure for Romaniot Jewish scholars. But in addition, some of the contents of *Urim wᵉ-tummim* reflected Cohen Ashkenazi's acquaintance with Islamic and, more specifically, contemporary Ottoman sources and discourses on astrology. Thus, the intellectual milieu of Cohen Ashkenazi overlapped with the Ottoman context of Galeano/Jālīnūs's formation.

A consideration of the details of astral influences on cities in particular climatic zones reveals Cohen Ashkenazi's knowledge of Ottoman astrology. For example, Cohen Ashkenazi listed Jurjān and Ṭabaristān among cities in the second clime and Turkestan and East Khurāsān among those in the seventh clime.[34] Since Ibn Ezra did not mention these Iranian and Central Asian locales in *Seiper ha-ʿOlam* (The Book of the World),[35] Cohen Ashkenazi must have had access to Islamic sources, either written or oral, in order to know that those cities were in the second and seventh climes. When Cohen Ashkenazi listed which zodiacal signs ruled each city, he made four references to a Turkish scholar (*ḥakam tugar*); I do not know if he always meant the same one.[36] Cohen Ashkenazi cited the statement of a Turkish scholar,

נסיטוריסי (Nesitorisi?), that the sun ruled the second clime, contradicting Ibn Ezra who held that Jupiter ruled the second clime.[37] Cohen Ashkenazi's sources may have been oral.

In any case, the references to the Ottoman scholar are corroborated by the contents of Ottoman almanacs. Tunç Şen has demonstrated that the almanac (sing. *taqwīm*), produced yearly, was the most important textual genre for astrological forecasting in the Ottoman Empire during the fifteenth and sixteenth centuries.[38] Almanacs contained an ephemeris and predictions for a given year, and they were produced at the Ottoman court by the astrologers and presented to the sultan at the time of the vernal equinox.[39] European astrologers also produced annual prognostications for rulers,[40] and a fourteenth-century Byzantine astronomical/astrological text had the same format as later Ottoman almanacs.[41] Notably, the author of an Ottoman almanac for 854 AH/ ca. 1450 CE adopted the same position as the Turkish scholar Cohen Ashkenazi referenced: that the sun governed the second clime.[42] An unnamed Turkish scholar cited in *Urim wᵉ-ṯummim* and the author of the ca. 1450 almanac also concurred regarding which planets ruled the third, fourth, fifth, sixth, and seventh climes.[43] The fact that authors of almanacs contemporary with Cohen Ashkenazi corroborated information that Cohen Ashkenazi attributed to an Ottoman scholar or scholars substantiates the case for exchange between an Ottoman scholar or scholars and Cohen Ashkenazi. In one case, Cohen Ashkenazi used the word *subaşılar*,[44] an Ottoman Turkish term for security chiefs or military leaders, without citing a source.[45] Even the minutiae of Ottoman astrology traveled.

While Cohen Ashkenazi appropriated details from Ottoman astrology that had traveled to Crete, Galeano/Jālīnūs refined forecasting techniques in the wake of critiques from a scholar at the Ottoman court in Istanbul. Astrologers had grappled for centuries with the fact that twins, who presumably share a natal chart, often live different lives. In response, Galeano/Jālīnūs contended that astrologers erred when they placed too much weight on the moment of birth in a forecast.[46] He elaborated:

> We question the astrologers about the problem of twins and of a native in a locale the latitude of which is 66 degrees, at the moment when six zodiacal signs rise together.[47] How can the astrologers say that judgments about the newborn and his successes, characteristics, constitution, and teachings are dependent on the sign at the moment of his

> birth? Here are Esau and Jacob, who is holding on to the heel of Esau. One is a simple man residing in tents and the other is a hunting man, a man of the field.[48] Thus, were a man born in the place where the six zodiacal signs rose as one in the same moment, which of the six is his constellation? Thus, they questioned me [viz. Galeano/Jālīnūs] in the name of the excellent Ishmaelite scholar Kamāl Pāsha about how Enoch's rectification (*Mo'znei Ḥanok̲*) is true for the moments of conception and birth, as I [viz. the questioner of Galeano/Jālīnūs] tested it [many] times, and it lied.[49]

The questioner wondered how astrologers could possibly explain the different lives of Jacob and Esau, given that they had almost the same moment of birth. The procedure known as Enoch's rectification was used to determine the location of the ascendant at the time of conception based on the moon's location at the time of birth.[50] That procedure appeared useless in explaining the different courses of the lives of twins. Galeano/Jālīnūs responded that the birth of the first twin hastened the birth of the second:

> On the basis of all of this I [Galeano/Jālīnūs] respond that what is true in essence lies in accident because with Enoch's rectification, [viz. conception and birth] might come earlier for the reason of the cold of Mars or Mercury, as Ibn Ezra mentioned in *Nativities*, because they hasten the time of birth.[51] Thus the truth is found regarding those two effects [viz. conception and birth], about which the scholar who had invented Enoch's rectification lied, so he was pleased with the response. Thus, it is possible that the time periods of the birth, conception, and pregnancy of twins differ because one's pregnancy might be dependent upon the other's pregnancy. But the exit of the second twin is by accident and not at its time because when the womb opens for the exit of the first, the other must also exit.[52]

The procedure of Enoch's rectification, Galeano/Jālīnūs concluded, did not precisely yield the times of the birth and conception of twins. He argued that twins were born at nearly the same time because the birth of the first twin, not the stars, precipitated the birth of the second. Without the birth of the first twin, the second might not have been born for a few weeks. Hence, the frequent differences in twins' lives did not entail the dismissal of astral influences on the lives of twins. More importantly, however, the details of the vignette demonstrate that Jews' discourses about astrological forecasting overlapped with those of Muslims. While Ottoman critics of astrology, such as Kamāl Pāsha,[53] made valid points, Galeano/Jālīnūs could still defend astrology against the critiques.

If Kamāl Pāsha was, in fact, the philosopher Ibn Kamāl Pāsha, i.e., Kamāl Pāshazādah (d. 1534), his skepticism about astrology is best understood in the context of a larger debate about forecasting methods. Kamāl Pāshazādah practiced lettrism, a method of forecasting that involved interpreting the numerical values of words, the occult properties of letters, and the hidden meaning of certain names.[54] He provided political advice to Sultan Selim (r. 1512–20) that was based on lettrism.[55] In this context, the debate was actually about methods of forecasting, not about the licitness of forecasting. Galeano/Jālīnūs himself recognized that geomancy was an alternative approach to prognostication that was free of some of the drawbacks of astrology.

Forecasts were not confined to individuals' horoscopes and neither were critiques of them. In *Puzzles*, Galeano/Jālīnūs also addressed how historical astrology was based on premises that were mistaken or not fully understood.[56] The astral forces that were believed to affect nations, cities, and religions did not operate in a vacuum. For example, if one were to forecast that

> the land would be well-sown and there would be plenty, it is not possible to judge that if it were not like this, then there would have been famine. Famine is neither the contradiction nor the opposite of proper sowing. They are not arguing over truth and falsehood because it is possible that there be plenty in the land even if they sow poorly or hunger even if they sow well. This is like when the city is one of those lands in which the majority of sustenance comes from afar. Neither plenty nor famine are dependent on sowing the land because they [viz. plenty and famine] depend on that which is brought to it.[57]

Weighing a horoscope of good fortune for one city against a configuration portending ill for a second city, the food source for the first city, was complicated. Galeano/Jālīnūs also acknowledged the possibility of competing astral forces:

> Thus, it is possible that during a period of a good year for the city, evils come to it from a more universal (*kolelet*) configuration, like at the di-

> rection of the grand, middle, or small conjunctions that indicate the rise of a religion or kingdom among the people of that city. Or [evils may come] from the configuration of the heavens at the moment of its [the city's] foundation, as it arrived at a conjunction or aspect with a maleficent planet or with a solar or lunar eclipse that indicates it [evil], or plague or war or famine and the like. Hence, a difficulty of judgment is that knowledge of many causes, in addition to their combination, is necessary.[58]

Still, these complexities were sources of nuance rather than reasons to forswear astrology. These critiques of astrology were constructive critiques because for many of the merchants of knowledge and for many of their contacts, astrology was too big to fail.

At the Ottoman court, where Galeano/Jālīnūs was present, the discipline would have been nearly impossible to dismiss because historical astrology provided a powerful explanation for political transitions. Ottoman sultans were acclaimed using the lexicon of astrology as "The Lord of the [Auspicious] Conjunction," or *ṣāḥib-i qırān*, meaning that an auspicious conjunction of planets brought them to the throne.[59] Galeano/Jālīnūs's contemporary Mīrim Çelebī (d. 1525) praised Sultan Bayezit II (d. 1512) as *ṣāḥib-i qırān* in a commentary on Ulugh Beg's *zīj* (an astronomical handbook with tables).[60] Doubting historical astrology would have been akin to doubting the sultan's legitimacy, hence Galeano's/Jālīnūs's nuanced assessment.

Astrological forecasts continued to matter after a sultan's ascent to the throne. We read, in an almanac, of the planets' role in military events forecast for ca. 1450: "The ascendant of the year is in the sign of Scorpio and the lord [of the year], which is Mars, is at the same time at sextiles to Venus and is at the same time at quartiles to Jupiter. . . . May the triumph of the aforementioned Great King be strengthened in his white hand, may the warriors be supported and victorious."[61]

The aspects of Mars to Venus and to Jupiter accounted for the sultan's successes and tribulations in the upcoming year. Conjunctions were believed to affect the fortunes of others in the empire as well. The same almanac author explained that to know "the conditions of grand ministers and scholars and judges and imams of religion, may God improve their circumstances: Jupiter was at the center of the fifth [house] and in conjunction with Mercury and at quartiles with Mars and the lot of happiness. It indicates that this group's conditions shall be elevated and double."[62] Historical astrology was a robust source of counsel and a powerful mode of political explanation. Galeano/Jālīnūs, because he craved acceptance at the Ottoman court, aimed to present in *Puzzles* a more sophisticated view of the predictive value of conjunctions than his competitors. Existing practices at the Ottoman court could be refined. More generally, the imbrication of astrology and decision-making accounted for Galeano/Jālīnūs and Cohen Ashkenazi's interest in the subject and connections with Ottoman scholars. In that light, the notable overlaps between *Urim w^{e}-ṯummim*, Ottoman almanacs, and the astrological contents of *Puzzles* make sense.

Debates about Astrology

The ubiquity of astrology throughout the network of the merchants of knowledge and their contacts rendered controversies about its practicability all the more pressing. Most notably, the merchant of knowledge Elijah Delmedigo's student Pico della Mirandola (d. 1494), near the end of his life, departed from his earlier acceptance of judicial astrology to staunchly oppose forecasting. In his *Disputationes adversus astrologiam divinatricem*, which appeared posthumously, Pico specified that astrologers' error was allowing known connections between the heavens and the earth to lead to "an argument to persuade that we should resolve and believe that all human things, great and small, are bound by planets and stars as leaders and kings"—meaning that astral causes were dominant and ineluctable.[63] Pico alleged, in addition, that astrologers misrepresented the connections between the planets and the earth as "occult influences" (*occultis influxibus*), or, in other words, that each planet had a hidden, yet discernible and differentiable, effect on terrestrial events. Though Pico's language in the *Disputationes* was cutting, some modern scholars have interpreted the *Disputationes* as a call for a reform.[64] Because the merchants of knowledge and their contacts, including Pico, lived in a world in which divine power was rarely questioned, the conclusion that natural forces were instruments of God's will was inescapable.[65] Hence, Pico's opposition might be understood as loyal opposition. While they were divided on viable methods of forecasting, and even on the practicability of forecasting, the merchants of knowledge and their contacts coalesced around the need to explore the relationship between astrology and religious thought.

The shortcomings, ubiquity, and irreplaceability of astrology led to centuries of give-and-take among Muslims, Jews, and Christians. These never-ending conversations provide further evidence of the discipline's pervasiveness. Though antipathy to divination was expressed in the Qurʾān, early commentators on the Qurʾān conceded that divination yielded a few correct predictions of Muḥammad's advent.[66] Qurʾānic reiterations of the creation and control of the heavens as indications of God's greatness encouraged astrologers in Islamic societies. Al-Naẓẓām (d. 835–45), an early *mutakallim* (theologian), permitted astrological forecasting,[67] and the *faylasūf* (philosopher) al-Kindī (d. 870) advanced a compatibilist theory that accommodated astral determinism to the Qurʾān's message of human agency and responsibility.[68] The tenth-century Brethren of Purity (Ikhwān al-Ṣafāʾ), in their third epistle entitled *On "Astronomia,"* included several chapters in which they justified astrology.[69] Still, by the tenth century, astrology attracted consistent and widespread criticism from elite Muslim scholars.[70] Philosophers were concerned with astrologers' errors, with the impingement of forecasts on humans' free will, and with possible contradictions between humans' forecasts and God's unique knowledge of particulars. Theologians concluded that God's foreknowledge of humans' actions was beyond human ken, and many theologians denied that God depended on fixed, knowable chains of intermediate causes involving stars and planets. During the early centuries of Islam, as during the period of the merchants of knowledge, abstract arguments against astrology struggled and ultimately failed to eliminate a practice of immense import.[71]

Intellectual luminaries from the postclassical period (>1200 CE) of Islam were more open to astrology. In astrology, the planets functioned as intermediate causes, and though Ghazālī (d. 1111), in *Tahāfut al-falāsifa* (The Precipitance of the Philosophers), cautioned against an uncritical acceptance of intermediate causes,[72] Fakhr al-Dīn al-Rāzī (d. 1210) classified astrology with the mathematical sciences in *Jāmiʿ al-ʿulūm* (The Compendium of the Sciences).[73] Niẓām al-Dīn al-Nīsābūrī (d. ca. 1330) interpreted Q79:5 ("by those that direct an affair") to mean that the heavens controlled events on earth, though only with the permission (*idhn*) of God, the Creator.[74] Opposition to astrology never pervaded all Islamic texts or all segments of Islamic societies.[75] Just as early Abbasid caliphs wielded astrology as a tool of political legitimacy,[76] the Ottoman sultans employed astrologers (*munajjims*) whose forecasts lent religious authority to their reigns.[77]

Jewish discussions of astrology began centuries after the rise of Judaism. Despite astrology's significance in surrounding ancient Near Eastern cultures, the authors of the Jewish Bible had little to say on the subject.[78] Only by the time of the composition of the Babylonian Talmud in the first several centuries CE did astrology attract the interest of Jews. The authors of the Talmud equivocated on whether astrology was licit. The statement that Israel is not subject to the stars (Babylonian Talmud, *Shabbat*, 156a) implied that other nations were subject to astral forces, but there was also rabbinic opposition to forecasting. We read in the Babylonian Talmud, *Pesaḥim*, 113b: "We do not ask astrologers."[79]

In post-Talmudic Jewish discussions of astrology, the prevalence of astrological forecasting was balanced against provisions for the free will presumably necessary to choose religious observance. Abraham Ibn Ezra was the most influential Hebrew author on astrology in the ambit of the merchants of knowledge. Significantly, Ibn Ezra did not dispute the notion that astrological forces affected the Jews. For example, he believed that the conjunction of Saturn and Jupiter caused the Jews to remain in exile.[80] As individuals, however, Jews could free themselves from astrological forces through piety.[81] In a tract entitled *Seiper ha-Moladot* (The Book of Nativities), Ibn Ezra asserted that "there is no doubt that the righteous person is more protected than the scholar regarding the judgments coming from the stars."[82] In comments on Psalm 111:10 ("The fear of the Lord is the beginning of wisdom"), Ibn Ezra elucidated how controlling one's corporeal urges is the first step to becoming wise.[83] If the soul's control over the body strengthened, then the soul could repulse astrally determined diseases.[84] One could also be "completely saved by God's intervention from the harms preordained in his personal horoscope."[85] Astrology afforded the reader a way to understand God's involvement in the lives of non-Jews and Jews, observant or not.[86] God might also intervene in causal chains rather than through the stars. For Ibn Ezra, astrology was not just a practice to be judged acceptable but also an element of his religious thought.

Maimonides's (d. 1204) "Letter on Astrology," addressed to rabbis in Southern France, was the most devastating Jewish attack on astrology's principles and methodology known to the merchants of knowledge and their Jewish contacts.[87] The letter noted that natural astrology was valid but that the foundations of astrological forecasting were groundless and needlessly risked leading people to abandon their religious obligations.[88]

One should not forsake the law, Maimonides continued, on the basis of the "sayings of some individual sages in the Talmud and *Midrashot* whose words appear to maintain that at the moment of a man's birth, the stars will cause such and such to happen to him."[89] In other words, nondisparaging references to astrology in the rabbinic texts should not be understood as permissions. Maimonides challenged astrologers to rationalize the paradox of twins' shared birth times and places yet separate fortunes. God's judgment, Maimonides held, was the only plausible reason for the particular fortunes not only of twins but of all humans.[90] Maimonides's argument makes sense from a modern perspective but would have destabilized the intellectual milieu at the Ottoman court where Galeano/Jālīnūs was present.

The final medieval Jewish scholar of astrology known to the merchants of knowledge, as well as to Pico, was Levi b. Gerson (1288–1344).[91] Levi accepted the fact that humans' insufficient command of astronomy meant that astrologers often failed.[92] Though Levi produced a forecast for the Saturn-Jupiter conjunction of 1345, which became available in Latin,[93] he did not presume in *The Wars of the Lord* that free will could be fully overridden by astral forces nor that the future of the Jews could be fully determined by astral causes.[94] But astrology remained a valuable religious hermeneutic for Levi b. Gerson.[95] While Maimonides was unable to rationalize in the *Guide* why twelve loaves of bread were placed weekly in the ancient Temple in Jerusalem,[96] Levi reasoned that because the temple symbolized the unity of the cosmos, the existence of twelve zodiacal signs indicated that the table linked two components of the physical universe.[97] That insight of Levi did not depend on the practicability of any given forecast.

Christian scholars' positions resembled those of their Jewish counterparts. Some early Christian theologians feared that astrological forecasting delegitimized God's power. For example, the theologian Hippolytus of Rome (d. 235/36) was concerned with how astrology's explanation for humans' differences and the variation of climes restricted God's omnipotence.[98] Yet, more frequently, the issue for Christian thinkers was that astrology disrupted and hindered one's relationship with God. In order to lay bare the faults of ancient Roman culture, Augustine (d. 430) began chapter 5 of *The City of God* by refuting judicial astrology. He used the paradox of the different lives of twins to argue that the stars were not the independent causes of anything; the stars did not even signal one's fate to any meaningful extent. The heavens could affect physical things, he argued, but not the human soul, and God might punish people and nations for their freely willed actions.[99] In his *Confessions*, Augustine recounted how he transcended astrology to arrive at a fuller knowledge of God.[100] Humans, he insisted, must recognize the urgency of religious guidance. Astrology was a useless danger.

Through the Latin translation of Abū Maʿshar's (d. 886) *Great Introduction* (Ar. *al-Mudkhal al-kabīr*) by Johannes Hispalensis in 1133 and by Hermann of Carinthia in 1140, astrology introduced natural philosophy into the Latin West.[101] Subsequently, "no reputable physician of the later Middle Ages would have imagined that medicine could be successfully practiced without" astrology.[102] Hence, Christian theologians reoriented their perspective on astrology so that the moral choices of Christians were believed to remain immune from stellar influences.[103] In the Byzantine Empire, astrology underwent a resurgence in the Palaiologan period (1259–1453), during which the anonymous Byzantine author of a fourteenth-century apology for astronomy averred that astrology was a path to contemplating God. In 1338, the hesychast Gregory of Palamas sharply disagreed.[104] However, these and other, less vehement dismissals of astrology in the name of Christianity masked an acceptance of the field in the Byzantine Empire.[105] The centuries of discussions of astrology among Muslims, Jews, and Christians paved the way for the explorations of astrology as a mode of religious thought by the merchants of knowledge and their contacts.

A Comparison and Contrast of Rebuttals of Astrology

Conversations about forecasting among the merchants of knowledge and their contacts were not restricted to methods of forecasting but included critiques of astrology. Though religion informed critiques of astrology, the boundaries of religious traditions did not constrain the exchange of critiques of astrology. For instance, Pico argued that Christianity, not the cycles of the planets, was the key to understanding one's life. To highlight the problems of historical astrology,[106] Pico attacked a text allegedly written by Ibn Ezra entitled *De redemptione Israel*.[107] *De redemptione* was most likely Bar Ḥiyya's (d. 1135) *Scroll of the Revealer*, a text cited by Cohen Ashkenazi in *Urim wᵉ-ṯummim*.[108] In it, Bar Ḥiyya used the celestial configurations at Moses's birth as the basis for a prediction about the coming of the Messiah in 1464, when Saturn and Jupiter would be together in Pisces. Then, in the 1490s, Isaac Abravanel, a contact of the merchant of knowledge Saul Cohen Ashkenazi, forecast the imminent return of the Messiah on the basis of the *Scroll of the Revealer*.[109] In his critique, Pico remarked on the difficulty of determining exactly when Moses was born[110] and demonstrated that the calculations in the *Scroll of the Revealer* about the intervals between the beginning of the world, the birth of Moses, and the birth of Jesus were inaccurate.[111] Isaac's son Judah Abravanel learned of Pico's critique of historical astrology and, ostensibly as a result, avoided the subfield.[112] The exchange of criticisms of astrology was bidirectional: a Christian contact (Pico) of the merchants of knowledge dissuaded from astrology a Jewish scholar, Judah Abravanel, whose *Dialoghi d'amore* was read by Saul Cohen Ashkenazi, Moses Cohen Ashkenazi's son.[113]

Other similar critiques of astrology were common across confessional boundaries. For instance, Galeano/Jālīnūs found numerous logical flaws in astrology, such as in the practice of attributing certain qualities to certain houses.[114] Likewise, Pico exposed astrology's fallacious reasoning throughout the *Disputationes*. For example, section II.6 is entitled "How fallacious is this art, should it be either on account of the plurality or the divergence of astrologers' opinions."[115] Section III.2 begins, "Five reasons with which astrology can be seen to be strengthened."[116] Section III.8 contains "an objection against the aforementioned."[117] In section VI.6, "the reasons for various aspects are recited and disproved."[118] Sections X.2 and X.4 communicate how "the explanations of the astrologers are void and ridiculous."[119] Section X.10 addresses how "the explanations for the figures and the boundaries [of the signs] have already been rebutted."[120] Langermann suggested that Galeano/Jālīnūs's use of logic as an organizing principle for *Puzzles* is evidence of his contacts with Renaissance Christians.[121]

But the evocation of similar logical fallacies was an insufficient basis for Galeano/Jālīnūs to dismiss the entire discipline in the way that Pico did. These variant outcomes can be explained by Galeano/Jālīnūs's and Pico's divergent intellectual agendas. Pico's disagreements with astrology extended to mathematical astronomy, astrology's most precise and prestigious foundation. Pico noted that astrology suffered from the fact that there was no consensus among astronomers on the length of the time the sun took to pass from the beginning of one spring to the next.[122] By contrast, Galeano/Jālīnūs wrote on mathematical astronomy, projecting confidence in that discipline. More important, he framed astrology as a tool for explaining, not predicting. The case of rebuttals of astrology demonstrates that exchange did not always entail agreement.

The Integration of Astrology and Religious Thought

For the merchants of knowledge and their contacts, Abraham Ibn Ezra's integration of astrology into much of his oeuvre, including his massive scriptural commentary, was foundational. Given the proximity of earlier medieval Jewish and Christian theological positions on astrology, Christians' interest in Ibn Ezra's astrology—through Latin versions—is not surprising.[123] A Christian scholar on the edge of the merchants of knowledge's web of contacts, Sebastian Münster (d. 1552), knew of the astrological sections of Ibn Ezra's scriptural commentary. Münster, who published a Hebrew mathematics text by Galeano/Jālīnūs's teacher Elijah Mizraḥi (d. 1526) with a partial Latin translation, also published a Latin translation of most of Ibn Ezra's *Long Commentary* on Exodus 20 in 1527.[124] Awareness of long passages from Ibn Ezra's *Long Commentary* on Exodus 20 among Christians can be traced back to Peter of Limoges (fl. second half of the thirteenth century). In the portion that Münster translated and published, Ibn Ezra probed the relationship between the Ten Commandments and the planets, finding correspondences between certain commandments and certain planets. For example, the ninth commandment not to covet and the tenth commandment not to desire the wife of one's neighbor corresponded "to the Moon, which is in charge of wives and unstable things."[125] Jewish and Christian readers of Ibn Ezra's writings learned that the import of the Bible was cosmic.

The interest of the Christian contacts of the merchants of knowledge in Ibn Ezra's work was not limited to materials available in Latin; rather, Hebrew manuscripts of Ibn Ezra's scriptural commentary, previously owned by the merchant-scholar families on Candia, eventually joined the Fuggers' collection.[126] The Fuggers' acquisition of a manuscript formerly owned by Michael Balbo[127] of a commentary by Mordechai Kumaṭiano (d. 1482)[128] on Abraham Ibn Ezra's *Yᵉsod mora᾽* (The Foundation of Awe) treating the rationales for Jewish laws indicates that Christian scholars hoped to pursue the development of Ibn Ezra's thought in the work of subsequent Jewish thinkers. Access to the work of earlier Jewish thinkers might have been what guided Christians to the merchants of knowledge.

Reflecting the integration of astrology into all areas of life, the merchants of knowledge and their contacts spilled the most ink on the role of astrology in religious thought. In commentaries on *Yᵉsod mora᾽* and the Bible, Kumaṭiano analyzed the relationship between religious observance and astral causes. Because the human soul was not believed to mix with its substrate in the way that other things do, Kumaṭiano agreed with Ibn Ezra that the human soul could be liberated from astral causes by following God's commands.[129] Kumaṭiano argued that salvation from astral causes was the result of one's cleaving to God; when that cleaving ceased, such as when Cain murdered Abel, astral causes would take over.[130] Kumaṭiano acknowledged that, at times, even pious Jews could not redirect or eschew the influence of the stars. After all, generations of Israelites must remain in exile for the sins of an earlier generation.[131] In his comments on Exodus 34:10 in his scriptural commentary, Kumaṭiano argued that a knowledge of astrology enlightened one to the benefits of piety in warding off adverse outcomes.[132] Yet Kumaṭiano observed, in comments on Exodus 3:13, that Moses asked for a shortcut (*dereḵ qᵉṣara*) adapted to the people who were in his charge. Such a shortcut could have been the awareness of the astrally determined boundaries of one's life, knowledge which optimized adherence to Jewish law.[133] But astrology was not always in the service of Jews, Kumaṭiano wrote. Though God may have chosen an opportune time to command Abraham to depart from celestial worship, God, as Kumaṭiano observed *apud* Genesis 17:1 ("And when Abram was ninety years old and nine, the Lord appeared to Abram, and said unto him: 'I am God Almighty; walk before Me, and be thou wholehearted' "), also ordered Abraham not to consult the heavens before acting.[134]

Was there scriptural evidence that astrology ought to inform Jewish practice? Kumaṭiano's most comprehensive answer in the commentary on *Yᵉsod mora*ʾ came in an excursus on Exodus 28:30. The verse reads:

> And thou shalt put in the breastplate of judgment (*ḥoshen ha-mishpaṭ*) the Urim and the Tummim; and they shall be upon Aaron's heart, when he goeth in before the Lord; and Aaron shall bear the judgment of the children of Israel upon his heart before the Lord continually.

The biblical author prescribed that the High Priest of the Temple wear a breastplate (*ḥoshen*) with twelve stones, known as *urim* and *tummim*, over a garment called the *eipod*.[135] Each stone was engraved with the name of a tribe of ancient Israel.[136] Ibn Ezra proposed in his scriptural commentary that the high priest's breastplate containing the *urim* and *tummim* was an astrological instrument.[137]

In the excursus, Kumaṭiano delved into the use of the *urim* and *tummim* as a divinatory instrument. He began by providing the necessary background in geometry and astronomy.[138] After defining key terms such as the zodiac and the celestial equator—that is, the projection of the earth's equator onto the stars—Kumaṭiano analogized the stones to the zodiacal signs. The four rows of stones on the breastplate, he argued, corresponded with the four seasons, as there were three stones in each row and three signs in each season. The woven band (*ḥeisheḇ*) bisecting the *eip̱od* represented the celestial equator. The word *urim*, which means "fires" or "lights," denoted the sun and the moon. He argued as follows to conclude that the word *tummim* referred to the five planets. He explained that "the squares of the numbers from one to four preserve the squares of the numbers from nine to six, each according to its distance from five. Thus it [five] revolves about itself."[139] Kumaṭiano meant that as one moved along the number line in either direction from five, the squares shared the same digit in the units column. For example, the squares of four and six are sixteen and thirty-six respectively, and the squares of three and seven are nine and forty-nine respectively.[140] There were five planets because the number five was the midpoint. In all these respects, Kumaṭiano concluded, the breastplate of the high priest was a microcosm representing the zodiacal signs and planets. In that respect, it might be an instrument for divining the effects of cosmic forces on humans.

Yet Kumaṭiano conceded, in his scriptural commentary, that there were reasons to question the interpretation that the high priest's use of the *urim* and *tummim* was divination.[141] If the configuration of the heavens was decisive, what was the point of the high priest purifying his heart before approaching God's presence? If the configuration of the heavens was the decisive cause, the high priest's insights ought to be astrally determined. Kumaṭiano's response is thought provoking. He allowed that there was no question that the high priest's high station correlated with his wisdom. Through the purification of his heart, the high priest's soul was fully connected to God and separated from the distractions of the body, which was governed by the stars. Studying astronomy and astrology thus increased one's awareness of how God's providence (*hashgaḥah*) operated and, in turn, saved one's soul from astral influences. Hence, the study of astrology was good for the soul, and a pious soul would be free. Kumaṭiano, who in his commentary on *Yᵉsod mora'* wrote at the tail end of astrology's resurgence during the Palaiologan period of the Byzantine Empire, thus advanced Ibn

Ezra's project of integrating astrology into religious thought.[142] Kumaṭiano, like the merchants of knowledge and their other contacts, was understandably concerned with the theological ramifications of such a widespread practice. The continuing conversation on that topic explains the exchange, by the merchants of knowledge and their contacts, not only of manuscripts of Kumaṭiano's commentary on *Yesod mora*ʾ but also of *Urim we-ṯummim*.

The Religious Role of Astrology in *Urim we-ṯummim*

Transregional exchange of astrology was also due, in addition to its value for forecasting, to the role astrology acquired in religious thought. In *Urim we-ṯummim*, the merchant of knowledge Moses Cohen Ashkenazi built on the ideas of Ibn Ezra and Kumaṭiano with an unprecedented, systematic integration of astrology and Judaism. In the text, he argued audaciously that astrology was not merely a hermeneutic tool but a way to enhance Jewish observance. Cohen Ashkenazi had commenced writing *Urim we-ṯummim* by 1465 and completed the text on Candia no earlier than 1472.[143] Because both Ibn Ezra and Kumaṭiano's writings were known on Candia at that time, the city was an apt setting for the composition of *Urim we-ṯummim*. Its compelling argument that astrology made one a better Jew also accounts for how the unique manuscript of *Urim we-ṯummim* was most likely first acquired by the merchant of knowledge Moses Galeano/Mūsā Jālīnūs and then by the Fuggers.[144] The contents of *Urim we-ṯummim* were sui generis.

Cohen Ashkenazi framed *Urim we-ṯummim* with his lamentation that, as a Jew of priestly lineage, knowledge about how to use the *urim* and *tummim* had been lost.[145] He aimed to recover his priestly heritage of divination, but because talismans were considered an illicit divinatory tool,[146] astrology was the only remaining path of divination worth pursuing. Referencing textual evidence that even each blade of grass had a star,[147] Cohen Ashkenazi wondered whether the heavens might be an instrument of providence and not of fate. He adduced verses from the Torah in which the heavens acquire an independent role, e.g., Deuteronomy 31:28: "Call heaven and earth to witness against them."[148] Since the Torah obligated one to believe that choice (*beḥirah*) existed, Cohen Ashkenazi reasoned, one could not plausibly be punished for actions that the heavens compelled.[149] Because choice applied only to actions that are possible (*epshari*), but not to impossible (*nimna*ʿ) or necessitated (*meḥuyyaḇ*) actions, God would not want techniques of astrological prediction to be hidden from scholars. Rather, God must have provided Moses, the most important prophet of Judaism, with knowledge of astrology. In turn, Moses would have communicated his knowledge of astrology to Aaron, the first high priest, and to Joshua, Moses's successor as leader of the Israelites. The members of the Sanhedrin, the supreme post-exilic Jewish court, must have known astrology too. Cohen Ashkenazi explained that as Jews had been living among their enemies in the Diaspora for so long, pertinent books containing a full exposition of astrology's religious role had been lost.[150]

Consequently, according to Cohen Ashkenazi, it was imperative for Jews to remedy their ignorance of how astrology could help them differentiate courses of action.[151] One could not, after all, freely choose something that is against nature or choose to prevent something that is fated. One could not choose never to die, for example. One could not freely choose something contingent on future choices that have yet to be made. A natal chart, Cohen Ashkenazi reasoned, allowed the choice of the most appropriate (*raʾui*) path for one's life so that one would not attempt things that would be too difficult. Future bad deeds should not be a surprise because they could be foretold by an astrologer.[152] Cohen Ashkenazi called the second half of the treatise *Tummim*, a word he explained was derived from the Hebrew *tamm* (to be completed),[153] because the use of astrology makes one's knowledge of particulars, including God's judgments, complete.[154]

Cohen Ashkenazi began by acknowledging how astrology functioned as a decision-making tool to help one clarify the particulars of each possible choice.[155] Deciding between going to Damascus or going to Jerusalem was not truly a free choice unless one knew the details of each city, he posited.[156] With knowledge of more particulars, one could optimize Jewish observance. Cohen Ashkenazi asserted that someone furnished with an accurate forecast

> will not perform the commandments and human actions like the blind man searching in the darkness for the "commandment of men learned by rote" (Isaiah 29:13) without intending the purpose from the beginning [of adherence to the commandments]. Rather, everyone who opens his eyes sees the future event, chooses the appropriate action, and actually does it, prevents the evil action. This is the attribute (*middah*) that we reward.[157]

From Cohen Ashkenazi's perspective, astrology was not to cheat the system but to enlighten one to the system within which one fulfilled religious obligations. It is possible that Cohen Ashkenazi advocated astrological forecasting because it freed the soul to focus on higher pursuits such as worship through one's intellect (ʿaḇodat ha-seḵel).[158] However, to attain the highest spiritual stations, one had to appreciate thoroughly God's role as the provider of law and natural order.[159] Astrology was, in this sense, well suited to elevating souls because the telos (taḵlit) of actions was a subject of astrology.[160] Cohen Ashkenazi reframed forecasting as a font of insight rather than as a source of shortcuts. He considered the study of astrology to be part of ʿaḇodat ha-seḵel, not a means to it.

To demonstrate the licitness of astrology, Cohen Ashkenazi identified two earlier perspectives on the topic. The first perspective was that of those who maintained that the varying motions of the heavenly bodies, by moving the sublunar elements (yᵉsodot), were responsible in a general respect for all changes in the lower world.[161] The heavens neither affected nor effected particulars such as whether one was rich or poor, married or unmarried.[162] Philosophers such as Averroës (Ar. Ibn Rushd, d. 1198) maintained this perspective, which restricted the specificity of celestial influences. According to Cohen Ashkenazi, the most recent figure to adopt the philosophers' position was Maimonides, as expressed in his "Letter on Astrology."

Cohen Ashkenazi noted that Maimonides discussed astrology in treatises besides the "Letter," such as in his codification of Jewish law, the magisterial *Mishneh Torah*. Maimonides wrote at the beginning of chapter 1 of *Hilḵot ʿaḇodah zarah* (The Laws of Foreign Worship) that the stars directed the lower world.[163] Cohen Ashkenazi quoted Maimonides's description of the error of the generation of Enosh: "Their error was as follows: 'Since God,' they said, 'created these stars and spheres to guide the world, set them on high and allotted them honour, and since they are ministers who minister before Him, etc.' "[164] The error was honoring the stars and spheres on that basis. From Cohen Ashkenazi's viewpoint, the error was not astrology tout court, but worshipping the stars and divining through astrology without acknowledging the heavens as God's intermediary for control over the lower world. The spheres, sometimes called orbs, were physical bodies that moved the planets and that played a role in the cosmos that was analogous to the heart's role in the body, and the celestial motions provided the vegetative, animal, and rational souls to all living beings.[165] Philosophers believed that God did not know particulars in the way that humans did, because particulars could change, but God, according to the Aristotelian perspective, could not. Since forecasts included particulars, the philosophers were irrevocably opposed to judicial astrology. All the same, Cohen Ashkenazi observed that Maimonides did not attack forecasting in the *Mishneh Torah* as viciously as he did in the "Letter on Astrology." Hence, the position of Maimonides, the most influential Jewish philosopher for the merchants of knowledge, on astrology was nuanced.

The second perspective on astrology that Cohen Ashkenazi identified was ascribed to the astrologers (*ha-ḥozim ba-kok̲ab̲im*). They presumed that God oversaw everything, both universals and particulars, via celestial intermediaries.[166] Cohen Ashkenazi adduced passages from rabbinic literature in support. For instance, in the Babylonian Talmud, Rava (whom Cohen Ashkenazi identified as Rab̲ Ḥanina) held that his lifespan, offspring, and sustenance were not dependent on righteousness (*zᵉk̲uta*) but instead on the heavens.[167] Cohen Ashkenazi acknowledged the existence of a passage in the Talmud where one reads that Israel does not have a star[168] but added that many passages in the midrashic text *Pirqei dᵉ-Rabbi Eliezer* allowed that the stars did influence Jews' actions.[169] He remarked that a number of scholars, including Abraham Bar Ḥiyya (d. 1135), especially in his *Mᵉgillat ha-mᵉgalleh* (The Scroll of the Revealer), sided with the astrologers.[170] Bar Ḥiyya was not a strict astral determinist; the stars only set up the possibility of an event occurring. God's will predominated.[171] Then Cohen Ashkenazi made the crucial point that Maimonides himself was practically in the astrologers' camp (*lo hirḥiq ʿaṣmo mei-ha-daʿat ha-zeh* [!]) because in chapter 11 of book 2 of the *Guide*, Maimonides theorized that the orbs transmit emanations from God down to the lower, sublunar realm.[172] If the orbs were an instrument for effectuating God's providence, then the philosophers, by denying God's knowledge of particulars, eliminated or at least restricted the operation of providence in the sublunar realm.[173] Cohen Ashkenazi was astonished that a scholar as eminent as Maimonides overlooked scriptural reminders of God's involvement with the particulars of nature.[174] Cohen Ashkenazi was also amazed that Maimonides, who asserted in the "Letter" that God was the only judge, managed to hold divergent views of astrology simultaneously.[175]

Though a philosopher would have protested that the astrologers denied human responsibility, Cohen Ashkenazi skillfully portrayed the philosophers' position as the position that posed more of a challenge to Jewish belief.[176] He concluded that the interplay of human responsibility, which the philosophers valorized, with God's power was surely complex, but the reach of God's power was clear from the plain sense of scripture. Thus, the astrologers, by investigating the minutiae of humans' actions, upheld Jewish belief more than the philosophers. Humans' free will might be protected through the denial of astrology, but the astrologers, by acknowledging the heavens' control over particulars in the terrestrial realm, highlighted divine omniscience and providence.[177]

According to Cohen Ashkenazi, the positions of both the astrologers and the philosophers served a purpose. Cohen Ashkenazi drew an analogy with the value of different discourses about anatomy: philosophers such as Aristotle argued that there was a single principal organ, the heart, but physicians, e.g., Galen and Hippocrates, disagreed with the philosophers and each other over whether there were three (Galen) or four (Hippocrates) principal organs.[178] Cohen Ashkenazi clarified that the chief reason for that disagreement was that the philosophers and the physicians had different goals. The philosopher aimed to understand existence in general, whereas a physician aimed to restore health and, therefore, provided a level of specificity that was irrelevant to the philosopher. The existence of different goals did not mean that there was an inherent contradiction between medicine and philosophy.[179] Similarly, astrologers, by investigating particulars, took on a task that the philosophers ceded to them. Cohen Ashkenazi added that because astrology was not fully developed in Aristotle's time, philosophers may yet be able to reconcile philosophy with astrology in Cohen Ashkenazi's time. A merchant of knowledge reading Cohen Ashkenazi's argument would conclude that the forecasting techniques that they and their contacts employed in daily life might enhance Jewish practice.

Cohen Ashkenazi speculated that the contradictions in Maimonides's statements about astrology were intentional; after all, Maimonides mentioned in the *Guide* that he occasionally contradicted himself.[180] Cohen Ashkenazi proposed to resolve the contradictions by consulting a pseudo-Maimonidean text entitled *M^egillat s^etarim* (Scroll of Hidden Things).[181] However, as the passage that Cohen Ashkenazi purported to quote from *M^egillat s^etarim* is absent from the cited text, Cohen Ashkenazi was most likely referring to a no-longer extant text by Maimonides entitled *Iggeret ha-sodot* (The Epistle of Secrets), in which Maimonides explained that the laws of every religion addressed two categories of harmful things.[182] The first category comprised harmful acts that one person did to another. The second category comprised things such as alchemy, talismans, and astrology that most viewed unkindly.[183] The danger was that if most people found out things in either category, they would drop useful pursuits such as agriculture, since astrology, alchemy, and talismans seemed to promise a reward without effort.

Cohen Ashkenazi responded to *M^e^gillat s^e^tarim* (again, actually *Iggeret ha-sodot*) and contended that astrology did not pose a challenge to Judaism by analyzing Ibn Ezra's comment on Exodus 23:26 ("None shall miscarry, nor be barren, in thy land; the number of thy days I will fulfill").[184] Ibn Ezra's interpretation was that if one served God, God empowered the soul over the body. A failure to do so entailed the reverse.[185] Cohen Ashkenazi proposed that the biblical passage meant that serving God could improve upon what the stars decreed.[186] Conversely, a curse from God diminished the power of natural causes. For instance, one might eat and not become satisfied.[187] Thus, divine punishment and reward on earth functioned within the framework of astral causation. In contrast, in an astrology text, Ibn Ezra remarked, "Behold the keeper of the Torah has no need for a physician with God," adding, "For from its light comes [a way] to deviate from the power of the natural causes (*tol^e^dot*)."[188] While Ibn Ezra distinguished between the medicine practiced by physicians, based on intermediate causes, and a cure imparted directly by God, Cohen Ashkenazi naturalized all effects of Jewish observance.[189] Adherence to God's dictates increased the heat and moisture in one's body, thereby lengthening one's life.[190] If someone who desired children devoted himself to God, then God strengthened the kidneys and improved the reproductive function.[191] God's providence operated through the stars.

In his discussion of battle deaths, Cohen Ashkenazi conceptualized a different way in which God might modulate astral causes. In battles, people with different birth times and places, and, thus, different natal charts, died at the same time.[192] Cohen Ashkenazi adduced scriptural evidence showing that God's decrees for battles were not fixed. Isaiah 38:5 reads, "Go, and say to Hezekiah: Thus saith the LORD, the God of David thy father: I have heard thy prayer, I have seen thy tears; behold, I will add unto thy days fifteen years."[193] Hezekiah's original fate was altered on the basis of his right action; if God knew of those actions in advance, they would not be a result of Hezekiah's free choice. Notably, Hezekiah's exercise of free choice did not preclude the operation of astral causes. Rather, God might intervene to save a group of people from a decreed fate without suspending or altering astral causes.[194] Suppose a city was fated to be destroyed by a rising river, but a prophet came and warned the inhabitants to repent, with the warning resulting in the inhabitants' sincere repentance. On the day that the planets' courses were to cause the city's destruction, God sends into the hearts of the inhabitants a directive to pray outside the city.[195] Thus, the people would be saved without God ever having to suspend the causal efficacy of the planets. Because judicial astrology facilitated one's religious observance and because one's practice of judicial astrology benefited from one's piety, judicial astrology became, for Cohen Ashkenazi and his readers, a dimension of the intellectual worship of God. God created the heavens for their own sake, not for human benefit.[196] An argument that began as a defense of the most comprehensive scholarly discipline of the merchants of knowledge and their contacts elevated astrology into a religious philosophy.

Just as some of the technical contents of *Urim w^e-ṯummim* resembled those of an Ottoman almanac, the arguments in favor of astrology as a religious philosophy found in *Urim w^e-ṯummim* would have resonated with a scholar, Galeano/Jālīnūs, familiar with statements about the prophetic history of astrology found in Ottoman almanacs. The author of an 848 AH/ca. 1445 CE almanac asserted that prophets, beginning with Idrīs, the biblical Enoch, were vouchsafed knowledge of forecasting by God: "Enoch, that is, Idrīs, upon him be peace, was the first human to be given prophecy and to write with a pen. . . . Also, the *tafsīr* of the judge [viz. al-Bayḍāwī, d. 1316] says this, that the prophet Idrīs was the first to take a pen and was the first to record and be revealed astrology, and denying it is infidelity."[197] Enoch/Idrīs had long been understood as a prophet of divinatory techniques, and Cohen Ashkenazi observed that Ibn Ezra made similar statements about Enoch/Idrīs.[198] Such parallels between the almanacs and *Urim w^e-ṯummim* account for Galeano/Jālīnūs's acquisition of that text.

In addition, Cohen Ashkenazi's position on astrology meshed with the tenor of discourses of Ottoman Muslim scholars about astrology. Galeano/Jālīnūs's Ottoman Muslim patron, the chief military judge ʿAbd al-Raḥmān Muʾayyadzādah (d. 1516),[199] argued that astrology was licit in a gloss (*ḥāshiya*) on al-Sayyid al-Sharīf al-Jurjānī's (d. 1413) commentary on ʿAḍud al-Dīn al-Ījī's (d. 1355) *al-Mawāqif fī ʿilm al-kalām* (The Stations in the Science of Theology).[200] Jurjānī disputed the existence of any intermediate cause, including the heavens, and asserted that the farcicality of the astrologers' response to the twins paradox was another strike against astrology.[201] Differences in the lived lives of twins were a foundation for an argument against genethliac astrology, since their lives ought to be similar or identical.[202] In response, Muʾayyadzādah denied that astrology made the heavens the sole efficient cause and reminded the reader that astrology contributed to a broader, complex system of natural causes and effects. The heavens could be the true cause or the customary cause (*ʿilla ʿādiyya*). Muʾayyadzādah cautioned that because "the preparation of the receiving subjects (*al-qawābil*) [to receive astral forces] varies as the terrestrial reasons vary, and because the effects of the upper bodies upon them vary, there is a recognition of the lack of the universality of the rules upon which they [viz. astrologers] base their judgments."[203] That is, the qualities that astrologers attributed to parts of the celestial orbs are known only from the effects of their rays on terrestrial matter, and terrestrial matter was variegated for numerous reasons, including God's direct involvement.[204] Muʾayyadzādah, like the author of the ca. 1445 almanac,[205] accepted astrology without asserting that the planets were the sole, independent intermediate causes of terrestrial events.[206] Neither Muʾayyadzādah nor Cohen Ashkenazi adopted a wholly determinist view of astrology. The coherence of Cohen Ashkenazi's explanation of the links between the divine and the quotidian with the perspectives on astrology regnant at the Ottoman court, where astrologers legitimated the sultan's rule and shaped his decisions, helps account for Galeano/Jālīnūs's acquisition of *Urim wᵉ-ṯummim*.

The Integration of Astrology into European Renaissance Thought

Christian contacts of the merchants of knowledge integrated astrology into religious thought. For instance, before composing the *Disputationes*, Pico della Mirandola wrote in his *900 Theses* that "whoever joins astrology to Cabala will see that to sabbatize and rest becomes more appropriate after Christ on the Lord's day than on the day of the Sabbath."[207] Since the cyclical motions of the planets paralleled the cyclical nature of life on earth, Pico concluded, astrology yielded insight into God's role in creation.[208] Pico's correspondent, the humanist philosopher Marsilio Ficino (d. 1499), explored, in the last few decades of the fifteenth century, a similar topic: the relationship between free will, human tendencies governed by the stars, and providence.[209] Ficino met Elijah Delmedigo and may have also been a contact of the merchant of knowledge Saul Cohen Ashkenazi.[210] Ficino theorized how one learned about God through correspondences between the terrestrial and celestial realms.[211] Planets may not be causes but signs,[212] he wrote, that rationalized the natural and political order across religious traditions and helped predict the future.[213] Or, if one understood the planets as instrumental, secondary causes, they then became a way for God to act in response to humans' actions.[214] Ficino's statement that a nativity chart informed one of God's roles in shaping the contours of one's life[215] resembled the thrust of *Urim wᵉ-ṯummim*—namely that astrology was worship through one's intellect (*ʿaḇodat ha-seḵel*). Ficino's source may not have been *Urim wᵉ-ṯummim*, but the common conversation accounts for the acquisition of the manuscript of *Urim wᵉ-ṯummim* by the Fuggers.

There were other parallels between Ficino's ideas and Jewish sources. While explaining why the stars did not necessarily control the mind, a view shared by many Jewish thinkers, Ficino referred to earlier Jewish sources: "Abraham[216] and Samuel,[217] and many Hebrew astrologers, trying to work against the threats of Mars and Saturn, used offerings and sacrifices, their minds elevated to God, clearly confirming that precept of the Chaldaeans; if you set the mind to a serious work of piety, you will keep the body also from falling."[218] Thus, one was not at the mercy of the planets because the spiritus—"the vivifying agent of the soul"—could become a force just like the light and rays of the planets.[219] Self-discipline could be an effective defense and could position one to benefit from astral forces. In the third part of *Liber de vita* (The Book of Life), Ficino urged that if one wanted to benefit from the sun's astral forces, one should find things in nature such as animals (e.g., hens, swans, lions, gold, chrysolite) and plants (e.g., myrrh, incense, musk) that corresponded in some way to the sun.[220] For example, lions ruled the animal kingdom just as the sun ruled the sky. He did not detail how to manipulate these terrestrial objects that corresponded to celestial bodies. Elsewhere in *Liber de vita*, we find a lengthy statement on astrology attributed to the same Samuel explaining that pagans cast statues of their gods at astrologically propitious times.[221] Ficino agreed with Jewish scholars that astral causes were powerful but surmountable through worship. Without pious acts, however, the mind could fall under the influence of the stars. Kumaṭiano, Ibn Ezra, and Cohen Ashkenazi made the same point about the salient effect of religious observance.

Concrete connections between scholars show that these thematic parallels between *Liber de vita* and Jewish texts were no coincidence. A Jewish contact of the merchants of knowledge took note of Ficino's *Liber de vita*. Isaac Abravanel (d. 1508), a wealthy and prominent scholar from Iberia who attempted to get Ferdinand and Isabella to rescind the 1492 order for the expulsion of the Jews, was a correspondent of Saul Cohen Ashkenazi, the son of the author of *Urim wᵉ-ṯummim*. Isaac's son Judah, known as Leone Ebreo, composed a lengthy work entitled *Dialoghi d'amore* around 1511–12.[222] *Dialoghi* bore thematic resemblances to *Liber de vita*[223] and Saul commented critically on *Dialoghi* in a letter to Isaac Abravanel.[224] Hence, Saul had read a text similar to Ficino's *Liber de vita*. According to Judah, the relationship between the sun and the moon was analogous to that between the intellect and the soul: corporeal beauty, he argued, mirrored celestial beauty,[225] and the terrestrial and celestial realms were in love with each other. In Judah's words, "Earth was the proper and regular consort of heaven, whereof the other elements are but paramours."[226] Like Ficino, Judah understood the human body as a microcosm of the intricately functioning universe, though Judah's psychologization of astrology was not as systematic as Ficino's.[227] Leone Ebreo also mediated Hebrew terms to readers of Italian. He explained the Hebrew word for "heavens," *shamayim*, as *eish mayim*, which in Hebrew means "fire [and] water."[228] The fire referred to the planets and the water to the diaphanous part of the heavens. Judah's work subsequently came to the attention of prominent Renaissance scholars such as Giordano Bruno.[229]

The *Dialoghi* is our first evidence that exchange could be multidirectional among the merchants of knowledge and their contacts. First, Cohen Ashkenazi, who wrote from Candia, a Venetian colony, had Ottoman Turkish sources. Second, manuscripts of both *Urim wᵉ-ṯummim* and Kumaṭiano's *Commentary on "Yᵉsod mora᾿"* passed from Jews to Christians. Third, the merchant of knowledge Saul Cohen Ashkenazi was a contact of Ficino and read a text, *Dialoghi d'amore*, that was influenced by Ficino's ideas.[230] Given astrology's outsized intellectual and practical role, exchange of techniques and of constructive criticisms transregionally and across confessional boundaries made sense.

Conclusions

A shared discourse about astrology's role in religious thought accounts for the transaction of *Urim we-ṯummim* from Cohen Ashkenazi to Moses b. Judah (probably Galeano/Jālīnūs) and on to the Fuggers. Merchants of knowledge and their contacts such as Cohen Ashkenazi, Galeano/Jālīnūs, Ficino, and ʿAbd al-Raḥmān Muʾayyadzādah argued that astrology enhanced an individual's religious life. Astrology could be defended as a path to useful knowledge but not as a way to avoid duties and responsibilities. These themes, common to the merchants of knowledge and their Jewish, Christian, and Muslim contacts are why *Urim we-ṯummim* was meaningful to scholars across the Mediterranean. Interest in Abraham Ibn Ezra's integration of astrology into religious thought, the background for *Urim we-ṯummim*, is evidenced by how the merchant of knowledge Michael Balbo owned a manuscript of Kumaṭiano's commentary on Ibn Ezra's *Yesod moraʾ*, a commentary produced in Istanbul. Thus, the composition and exchange of *Urim we-ṯummim* was precedented.

The occurrence of multidirectional exchange reflects the paramount importance of astrology in the worlds of the merchants of knowledge and their contacts. While compensation and the accumulation of social capital mattered, neither would be possible without a premier command of the discipline. Astrology, in turn, would be impossible without up-to-date astronomical information. In the next chapter we will see that the widespread practice of astrology fueled the exchange of texts, tables, and instruments for mathematical astronomy.

THREE

TRANSACTIONS *of* ASTRONOMICAL TABLES *and* INSTRUMENTS

As the intellectual lives of the merchants of knowledge and their contacts were interdisciplinary, exchange in one field—astrology—was intertwined with exchange in another—practical astronomy. For the merchants of knowledge and their contacts, astrology was a comprehensive field that bore upon a great deal of life, including politics, warfare, agriculture, and disease. When the merchants of knowledge and their contacts were active, astrology played a role akin to that played contemporarily by economics.[1] Because forecasts influenced how major decisions were made, astrologers required precise tables to facilitate the computation of the positions of the sun, moon, planets, and constellations at any point in the future, often down to the hour. Planets are visible at different times from place to place, so planetary positions had to be recomputed for new locations, lest forecasts be adversely affected. The production of new astronomical tables was as momentous then as the release of economic data that inform policymakers is today.

Timekeeping was the other principal application of tables in the fifteenth and sixteenth centuries. Jews and Muslims used lunar calendars and disagreed among themselves about whether to base the determination of the new month on observation or calculation. Even if one chose to observe the new moon rather than calculate the beginning of the month, calculations could help one decide when to observe. Lunar calendars were relevant to how the determination of the date of Easter, a Christian holiday, required knowledge of the date of the first full moon after the beginning of spring. There is more than one date for Easter in part because churches disagree over whether to use the Julian or Gregorian calendars. Easter computation was a powerful motive for the study of astronomy in the Byzantine Empire in the fifteenth century.[2] Sometimes new observations motivated the composition of new tables of planetary positions; however, new computations for a new locale or enhanced precision were more frequent rationales for the composition of new tables. Because tables mattered literally every day, we will see that authors of tables curated their contents from a variety of sources for individual tables. Those component sources traveled as well.

The use of tables and astrological forecasting required instruments. For example, as tables informed the user of where a planet is at a certain time, a horary quadrant or a sundial was necessary to know the time. Astrolabes were useful for determining the ascendant at any time at a given latitude and for telling time. Sustained programs of observation were sometimes necessary to improve tables and always called for precision instruments. Because scientific instruments were made by craftsmen, their economic value was evident to the merchants of knowledge and their contacts. Since instruments were used and not read, they are evidence for how scientific practices traversed linguistic boundaries. Recently, scholars found an astrolabe with ornate Arabic labels and supplemented with notes in Hebrew and Latin that were scratched less artistically by the owners of the astrolabe as they learned how to use the astrolabe.[3] Likewise, the merchants of knowledge found that instruments made by Jews were valued by contacts who could not read Hebrew.

Multidirectional Exchange of Scientific Instruments

While scientific instruments did not have to be read like texts, instruments were worth a great deal of money, required specialized knowledge to use, and produced more knowledge. Instruments were exchanged in multiple directions among the merchants of knowledge and their contacts. Instruments yielded information that would help anyone in daily life. The functionality of a GPS-equipped cell phone is a modern, albeit more affordable and ubiquitous point of comparison. A common observation was the measurement of the local altitude of the North Star, as it was equal to the local latitude. One could navigate at sea or travel on land along a latitude line by repeatedly sighting the North Star. Levi b. Gerson (d. 1344), influential in the history of astronomy in Jewish cultures, invented an instrument known as the Jacob Staff with which the user could measure the altitude of an object relative to the observer.[4]

The staff consisted of a ruler with a movable crosspiece perpendicular to the ruler. The user would align the bottom of the crosspiece with the horizon and the top with the object to be measured. The gradations on the ruler provided the angular altitude. Astronomers also used the Jacob (or Cross) Staff to measure the distance between two objects in the sky or the angular diameter of a celestial body.

European astronomers, some of whom credited Levi with its invention, used the staff.[5] Multidirectional exchange of the Jacob Staff occurred on the edges of the network of the merchants of knowledge and their web of contacts. Pico's Hebrew instructor, Yoḥanan Alemanno (d. >1504)—best known for his explorations of the intersection of Platonism, Aristotelianism, and *Qabbalah*—may have learned about the Jacob Staff from someone

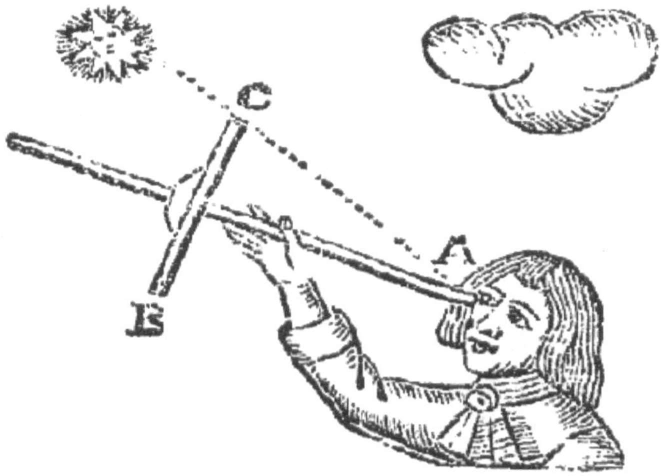

FIGURE 3.1. The Jacob Staff (or Cross Staff). Reproduced from Andrew Wakely, *The Mariner's Compass Rectified*, revised by William Mountaine (London: W. and J. Mount, 1763), 150.

in the circle of Pico della Mirandola (d. 1494), a contact of the merchant of knowledge Elijah Delmedigo (d. 1493).[6] The fact that Alemanno's source for a Jewish instrument was a Christian attests to how highly prized instruments were and how transactional intellectual exchange could be. The teacher (Alemanno) benefited from the more privileged student (Pico).

Since the exchange of an instrument comprised the transfer of knowledge of its operation, repair, and perhaps its construction, exchange of an instrument often entailed multiple steps. In 1466, in Bari, David Kalonymos b. Maestro Jacob translated a text entitled *Marʾit ha-kok̲ab̲im* (The Appearance of the Stars), composed by John of Gmünden (d. 1442) about his equatorium.[7] An equatorium was a gear-driven instrument for displaying and calculating planetary positions, effectively a mechanical *zīj*—an Arabic term for an astronomical handbook with tables. David's account of the acquisition of the instrument indicated that there were several steps in the process. David felt that the accumulated knowledge of the ancients had been lost, so when he saw an equatorium in the hands of a Christian scholar in the Italian city of Taranto, in Apulia, he jumped at the opportunity.[8] As the Christian scholar did not appreciate the instrument, David likened him to a bird who exposed a pearl among rubbish but was unaware of what it had found.[9] Because the tables needed to use the equatorium to find specific positions were not available, David composed new ones himself. And, finally, his translation of John's treatise ensured that readers of Hebrew mastered the instrument. The exchange of instruments was a multistage process linking scholars and documentation with the instruments.

David was on the fringes of the network of the merchants of knowledge. Michael Balbo—a merchant of knowledge—copied David's forecast that the Redemption would begin in 1463/64 and culminate in 1467/68, and a copy of the forecast joined the codex that contained Mordechai Kumaṭiano's (d. 1482) *Commentary on "Yesod Moraʾ"* and Balbo's record of the debate on metempsychosis (see chapter 5).[10] A codex with David's translation of *The Appearance of the Stars* was sold in Venice.[11] The aforementioned exchanges on the periphery of the network confirm that the activities of the merchants of knowledge and their contacts were not an exception.

Indeed, another multistage exchange of instruments occurred when the proficiency of Mordechai Kumaṭiano, a contact of the merchants of knowledge,[12] with astronomical instruments brought him to the attention of an Ottoman military judge.[13] Kumaṭiano possessed a universal astrolabe (*k^eli ṣap̱iḥa*), invented by Ibn al-Zarqāl (d. 1100) by 1066–67.[14] Most astrolabe plates were engraved for use at a specific latitude; only one Islamic universal astrolabe survives today.[15] Once the Ottoman military judge was told (*sipp^eru lo*) about the instrument, Kumaṭiano donated it to him thinking that doing so would enhance his own prestige and believing that there would be another universal plate to be found somewhere in the region. Kumaṭiano's career spanned the Ottoman conquest of Constantinople, and since he had previously assisted a Byzantine minister with an astrological prediction, gifting the instrument to the Ottoman military judge may have helped Kumaṭiano preserve his social connections after the Ottoman conquest.[16] A manuscript of Kumaṭiano's Hebrew commentary on Maimonides's *Guide* is found in the III Ahmet Collection of the Topkapı Library, i.e., in the sultan's personal library. Hence, Kumaṭiano and his students' access to Arabic texts may have indeed accrued from the gifts he made to elite Ottoman Muslim scholars. As this example demonstrates, much of the merchants of knowledge's approach to instruments, due to their economic value and practical applications, was transactional.

Echoing the narrative of the equatorium's rarity, Kumaṭiano reported that he and his colleagues were unable to locate another universal astrolabe like the one he donated. Ultimately, one of Kumaṭiano's students, Rabbi Menaḥem Bashyatchi—the grandfather of another of Kumaṭiano's students, Elijah Bashyatchi (d. 1490)—asked Kumaṭiano to compose a treatise about how to produce the universal astrolabe so that there would always be one such instrument found in the Jewish community. The text Kumaṭiano produced for this purpose is often found in codices with Jacob b. Makir Ibn Tibbon's (d. 1304) translation of Ibn al-Zarqāl's text on how to use the universal astrolabe plate.[17] The knowledge of how to use the universal astrolabe plate was preserved together with the knowledge of its construction.

Kumaṭiano's motivation for writing a commentary on Ibn Ezra's *Treatise on the Astrolabe* (*Seip̱er K^elī ha-n^eḥoshet*; lit. "the brass instrument") was similar to that for composing a text on the universal astrolabe plate: that Jews should be able to repair (*tiqqun*) the instrument themselves and not have to wander around looking for one.[18] As instruments were scarce both inside and outside of the Romaniot Jewish community, knowledge about them was at a premium.

Another student of Kumaṭiano, Caleb Afendopolo, mastered instruments. In 1487, Caleb Afendopolo composed a treatise (*Seip̱er k^eli roḇaʿ ha-*

sha^cot) on the horary quadrant, an instrument that enabled the user to determine the time of day by sighting the altitude of the sun.[19] Though there were many instruments available for telling time, the horary quadrant (*roḇaᶜ ha-shaᶜot*) was, in Afendopolo's view, the instrument best suited to the task. The instrument had a plumb bob to indicate readings on a scale, which could be calibrated in equal hours or unequal hours. Afendopolo fully expected that the reader of his treatise would actually use the instrument; for instance, he provided a finger-based mnemonic for determining when the months of the solar year have thirty days and when they have thirty-one.[20] Afendopolo added that studying astronomy would help one ascend the ladder of knowledge (*kᵉḇar hitḥilu laᶜalot bᵉ-sullam ha-ḥoḵmah*).[21] Increased knowledge of the heavens and the earth, he claimed, augmented one's appreciation of the Creator. Instruments were, in this view, more than equipment: they were physicalized knowledge.

Jewish scholars' earlier involvement in exchanges of instruments and related texts in Istanbul set the stage for the work of the merchant of knowledge Moses Galeano/Mūsā Jālīnūs (d. after 1542). In chapter 1, we learned of the importance of translation for commerce. Half of Galeano/Jālīnūs's translations were of texts about instruments, and one of those translations was the introduction to Ibn al-Zarqāl's *Treatise on How to Use a Universal Astrolabe* (Heb. "Iggeret ha-maᶜaseh ba-luaḥ ha-niqra' ṣapiḥah"). Jacob b. Maḵir Ibn Tibbon had previously translated the rest of the text, and Galeano/Jālīnūs's translation of the introduction is preserved only in a single manuscript of Jacob's translation.[22] Galeano/Jālīnūs's translation is an illuminating record of his activity as a scholarly intermediary.

Galeano/Jālīnūs's word choices in the introduction suggest that he expected the audience of his Hebrew translation to be familiar with Arabic. For instance, he preferred the Arabic words for conical (*makhrūṭī*) and cylindrical (*iṣṭiwānī*) to the Hebrew cognates (*ḥaruṭi* and *iṣṭiwani* [no *alep̱* after the *waw*]).[23] A marginal comment—evidently in the copyist's hand—was in Judaeo-Arabic, meaning that Galeano/Jālīnūs's assumptions about his readership were justified.[24] Although other merchants of knowledge and their contacts interacted with Muslim scholars, Galeano/Jālīnūs and his older relative Moses b. Elijah Galeano were the only merchants of knowledge who translated texts from Arabic into Hebrew. Only five other Arabic-Hebrew translations were performed after 1460.[25] The precise date of Galeano/Jālīnūs's translation of the introduction is unknown, but it had to have been after 1460 because Galeano/Jālīnūs died after 1542. Though he had identified a need for texts on instruments in Hebrew, he, and perhaps his readers, had not mastered Hebrew technical terms.

Galeano/Jālīnūs produced a Hebrew version[26] of Muḥammad b. Muḥammad Sibṭ al-Māridīnī's (d. ca. 1495) *Treatise on the Quadrant.*[27] The Arabic text upon which Galeano/Jālīnūs relied has been attributed to Aḥmad b. Aḥmad ʿAbd al-Ḥaqq al-Sunbāṭī (d. 1582 or 1585), but in the manuscripts I have consulted, the author's (*m*ᵉ*ḥabbeir*) name is given as either Muḥammad b. Muḥammad or Aḥmad b. Muḥammad.[28] There was no separate mention of a commentator. Once again, Galeano/Jālīnūs evidently presumed the reader of his translation would be familiar with technical terms derived from Arabic.[29] He also included matters of Muslim religious timekeeping, such as knowing the altitude of the sun at the beginning of the afternoon (*al-ʿaṣr*; transliterated into Hebrew characters to yield *hal-ʿasr*) prayer, the determination of the dawn prayer,[30] and calculation of the Islamic direction of prayer, the *qibla.*[31] He explained that the latitude circles parallel to the horizon known as *g*ᵉ*sharim* (lit. "bridges") in Hebrew were known as *ham-mūqanṭarāṭ* (more properly, *al-muqanṭarāt*) in Arabic (transliterated into Hebrew characters).[32] In a nod to his Jewish readership, Galeano/Jālīnūs commented on how the letters of both the Hebrew and Arabic alphabets could be used as numbers. He attributed the correlation of the alphanumerics in Arabic with the order of the Hebrew alphabet, rather than the Arabic alphabet, to the divine favor bestowed upon the Hebrew alphabet.[33]

Finally, Galeano/Jālīnūs most likely translated an Arabic treatise by an anonymous author on *zījes*, astronomical handbooks with tables. In the two surviving manuscripts, the title is given as *Peirush ʿal ha-juyub* (Commentary on the Sines) and was bound right after Galeano/Jālīnūs's version of Sibṭ al-Māridīnī's commentary on the quadrant.[34] This translation shares traits with translations that can be attributed with confidence to Galeano/Jālīnūs, not least because of the preference for technical terms derived from Arabic. The Arabic *jayb* (sine) is rendered with a Hebrew transcription rather than a Hebrew equivalent such as *beqaʿ*. Azimuth is rendered as *ha-samṭ*, from the Arabic *al-samt*;[35] the term for observation is *reiseid* from the Turkish *reṣet* (Arabic *raṣad*); and the translator used the Arabic *al-tartīb* instead of the Hebrew *siddur* (order). If Galeano/Jālīnūs did not produce *Peirush ʿal ha-juyub*, then there would have to have been another scholar who shared his interests and approaches. Overall, Galeano/Jālīnūs's translations from Arabic into Hebrew—in the highly transactional subfield of instruments—helped make him the most significant scholarly intermediary in the network.

In the exchange of instruments I have described, we have seen how instruments and texts related to them traversed linguistic and religious borders in a few ways. First, the Jewish scholar Yoḥanan Alemanno learned of the Jacob Staff from his Christian pupil Pico della Mirandola, even though the Jacob Staff originated in a Jewish setting. Second, Kumaṭiano donated his universal astrolabe plate, the work of the Andalusian Muslim Ibn al-Zarqāl, to an Ottoman Muslim elite. In order to recoup lost Jewish knowledge of an originally Islamic instrument, Kumaṭiano composed a treatise on the universal astrolabe plate and Galeano/Jālīnūs translated into Hebrew the introduction to Ibn al-Zarqāl's treatise on the universal astrolabe plate. Third, merchants of knowledge and their contacts translated texts on instruments into Hebrew and authored texts on instruments in Hebrew. This was all because without instruments, the critical activities of timekeeping and astrological forecasting would have been impossible.

Exchanges of Tables

Astronomical tables were the predominant genre of astronomical literature during the era of the merchants of knowledge. Tables, due to their value for timekeeping and forecasting, were probably the first texts to be translated during the rise of Islamic science in the eighth century.[36] Before the time of the merchants of knowledge, there were multilingual exchanges of tables among Jews, Muslims, and Christians. For example, Ibn al-Muthannā's (fl. tenth century) Arabic commentary on al-Khwārizmī's (d. ca. 850) *Zīj al-sindhind*—a handbook of astronomy with tables based on pre-Islamic Indian observations[37]—was translated into Hebrew twice, once by Abraham Ibn Ezra, and once into Latin by Hugo of Santillana in the twelfth century.[38] The next example of multilingual exchange of tables involved the Alfonsine Tables and was foundational for the careers of the merchants of knowledge and their contacts. The Alfonsine Tables, an update of the ca. 1080 Arabic Toledan Tables, were said to be composed by the Jewish astronomers Judah b. Moses ha-Cohen and Isaac b. Sid at the order of King Alfonso (d. 1283) of Castile.[39] Modern scholars speak of an Alfonsine corpus of materials, which comprised tables with flexible contents and other nontabular astronomical and astrological materials.[40] The earliest extant version of the Alfonsine Tables was from 1320, but there is no single canonical version of these tables. The merchants of knowledge, their contacts, Renaissance astronomers such as Nicholas Copernicus (d. 1543),[41] and possibly scholars in Mongol Iran used the Alfonsine Tables.[42] Jewish scholars helped produce and diffuse multiple Latin versions of the Alfonsine Tables in Europe.[43] Mordechai Finzi (fl. 1440–75) of Mantua adapted the Alfonsine Tables to the geographical coordinates of Mantua[44] and translated instructions on their use from Latin into Hebrew.[45] Mediated transregional exchange was in the DNA of the Alfonsine Tables.

But aspects of the Alfonsine Tables had to be reconsidered. Scholars such as Regiomontanus (d. 1476) found that an astronomer relying on them alone could correctly predict neither certain planetary conjunctions nor lunar eclipses, both of which mattered for forecasting.[46] Giovanni Bianchini (d. 1469) knew of these complaints and aimed to simplify the use of the Alfonsine Tables. His ca. 1442 *Tabulae astronomiae* (New Tables of Celestial Motions) became thrice printed tables for planetary motions available in Europe[47] and were used by Copernicus.[48] Some of Bianchini's tables were available in Hebrew[49] and were cited by authors writing in Hebrew,[50] such as Finzi.[51] Bianchini also corresponded on matters of practical astronomy with Regiomontanus. Even before the era of the merchants of knowledge, the Alfonsine Tables traveled across regions and passed from one language to another, in multiple directions. The exchange of improved tables proved to be just as wide-ranging.

In the fourteenth century, Jews authored original tables in Hebrew that facilitated the computation of syzygies: the oppositions and conjunctions of the sun and moon that were critical to calendrical calculations. Because the path of the moon around the earth is inclined by five degrees to the path of the sun about the earth (or, in modern astronomy, the earth's orbit about the sun), syzygies may, but usually do not, result in eclipses of the moon or sun.

As shown in the figure, at the beginning of a lunar month, the sun and the moon are conjoined, though usually not in the same plane. The illuminated side of the moon is not visible from the earth during a conjunction. In the middle of a lunar month, the earth is between the sun and the moon, meaning that the sun, distances, and moon are at opposition. As long as the sun and the moon are not in the same plane, the illuminated side of the

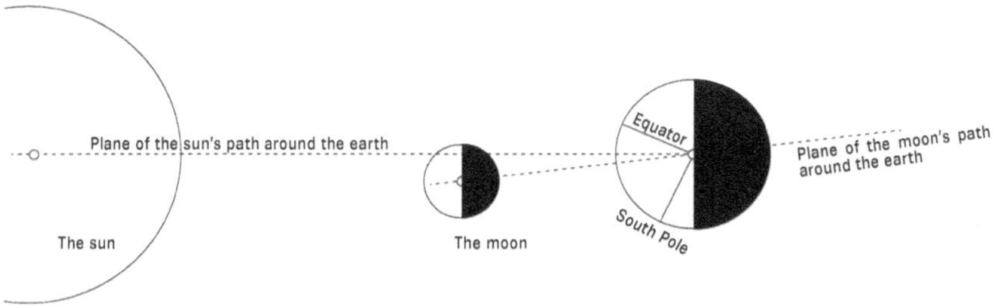

FIGURE 3.2. The sun and the moon at conjunction. The sun, distances, and the moon are not to scale.

moon is fully visible from the earth, creating a full moon. If the sun and moon are in the same plane at conjunction, there is a solar eclipse. If the sun and moon are in the same plane at opposition, there is a lunar eclipse. Because the velocities of the sun and moon vary, and because computations of conjunctions and oppositions depended on knowing the instantaneous velocities of the sun and moon, computing syzygy tables was no mean feat. During the era of the merchants of knowledge, a great deal was at stake in astronomers computing precise tables of syzygies because the time of the true syzygy, e.g., the new moon, could differ from the time of the mean syzygy by as much as several hours. A variation of that magnitude might affect the date on which a month began.

Syzygy tables that had been composed by Jewish scholars between 1350 and 1450 were highly sought after by Christians in the fifteenth century. In *Luḥot ha-Poᶜeil* (The Tables of Jacob ha-Poᶜeil), Jacob b. David Bonjorn (a.k.a. ha-Poᶜeil, fl. ca. 1361) computed a new period for conjunctions and oppositions.[52] Subsequently, Markos Eugenikos (ca. 1444) produced a Greek version of *Luḥot ha-Poᶜeil*,[53] possibly via a Latin intermediary. *Luḥot ha-Poᶜeil* also became the unacknowledged[54] source for Bernat de Granollachs's (d. 1487) *Lunari*, published in 1485 in Latin and Catalan.[55] In the tables entitled *Six Wings* (Heb. *Sheish kᵉnapayim*), Emmanuel b. Jacob (Bonfils; fl. mid-fourteenth century) of Tarascon provided a method for computing the time of the true syzygy.[56] Each titular wing was a section of the text in which the author led the user through a step of the computation. For example, the tables in the first wing facilitated the determination of the mean syzygies, which was the conjunction or opposition of the sun and moon were their velocities never to vary.[57] In the other wings, Bonfils guided the user through the more complex determinations of when the instantaneous velocities of the sun and moon yielded a true conjunction or opposition. Bonfils claimed to rely on Battānī's (d. 929) *zīj* (an astronomical handbook with tables).[58] Since that *zīj* was never translated into Hebrew, Bonfils's source was most likely the tables of Abraham Bar Ḥiyya (d. 1135), a Catalan Jewish scholar whose tables were based on al-Battānī's *zīj*. Hence, Bonfils's sources came from exchange and translation from Arabic into Hebrew.

Due to the convenience of its tables, *Six Wings* garnered the attention of readers of Greek, Latin, and Hebrew (as well as Russian). The text was translated a few times and moved between the Eastern and Western Mediterranean. Johannes Lucae e Camerino translated *Six Wings* into Latin in 1406.[59] Byzantine Christian astronomers shared an interest in syzygy and eclipse calculations,[60] and thus, a Greek version of *Six Wings* by Michael Chrysococcès appeared in 1435.[61] But not all scholars preferred it: in a note appended to a manuscript of *Six Wings*, one of Mordechai Kumaṭiano's (d. 1482) students wrote that Kumaṭiano preferred Bonjorn's tables to those found in *Six Wings*.[62] Kumaṭiano could not abide Bonfils's errors and needless approximations regarding eclipse calculations and lunar positions for the latitude of Constantinople. The number of translations and the existence of *Six Wings*, the tables of Bonjorn, and Kumaṭiano's criticisms are evidence for the indispensability of syzygy calculations.

Despite its wide dissemination, translation, and popularity, *Six Wings* had other shortcomings. Bonfils did not provide a table for the equation of time, which was needed to convert the apparent solar time into mean solar time. Because the sun's velocity varies throughout the year, the time period from high noon on one day to high noon on the next day is often not exactly twenty-four hours as measured by a clock. Approximating the equation of time, as Bonfils did, affected computations of the time period between mean and true syzygies. Bonfils also did not account sufficiently for the effect of the latitudinal motion of the moon, its displacement to the north or south from the path of the sun, on the measurement of the moon's path around the earth, the lunar longitude. Michael Chrysococcès introduced even more errors with his translation, reminding us of the technical expertise necessary to translate tables.[63]

Thus, mid-fifteenth-century Byzantine Christian astronomers obtained other Hebrew sources to optimize syzygy calculations.[64] Isaac Ibn al-Ḥadīb (or al-Aḥdab) ha-Separadi (d. > 1429) composed *Paved Way* (*Oraḥ selulah*) and included in it his tables for eclipse and syzygy computations. The canons—the instructions for using the text, but not the tables—were then adapted into Greek by Matthew Kamariotes (d. 1490/91).[65] In the Italian Peninsula, *Paved Way* was the primary source for Flavius Mithridates's (d. 1489) Latin tables, which he composed after his conversion from Judaism to Christianity.[66] Flavius's tables had a starting date of January 8, 1475, which was before he succeeded the merchant of knowledge Elijah Delmedigo as Pico della Mirandola's teacher.[67] As a former Jew, Flavius realized that by accessing complex Hebrew material inaccessible to most other Christians, he could attract patronage.

While Byzantine Christian scholars translated Hebrew syzygy tables, they also appropriated astronomy from further east. Kumaṭiano and Byzantine Christian authors obtained the contents of *zījes* (astronomical handbooks with tables) composed by scholars associated with the Marāgha Observatory near Tabrīz, which flourished in the second half of the thirteenth century.[68] The Marāgha Observatory is better known for its association with the authors of innovative texts on theoretical astronomy,[69] but a number of influential *zījes* were produced by astronomers who worked there.[70] These Islamic *zījes* contained a method for computing the true syzygy from the mean syzygy that was more precise than the updates of the Ptolemaic method favored by Byzantine Christian astronomers.[71]

Kumaṭiano and his Christian counterparts learned of Islamic tables through Greek intermediaries, like Gregory Chioniades (d. 1320), who was the most crucial conduit of astronomy from the Ilkhanids to the Byzantine Empire.[72] Born in Constantinople, Chioniades traveled to Tabrīz, the seat of the Ilkhanid court, in 1295 and had returned to Trebizond by the late 1290s. Due to the Byzantines' efforts to establish alliances with the Ilkhanid princes, Chioniades was welcomed by the Ilkhanid sultan Ghāzān (d. 1304).[73] Chioniades had returned to Trebizond by the late 1290s and translated some astronomy texts from Persian (and maybe Arabic) into Greek in Constantinople during the first decade of the fourteenth century.[74] These included *al-Zīj al-ʿalāʾī* of al-Fahhād (fl. ca. 1150), a version of *al-Zīj al-sanjarī* by al-Khāzinī (fl. ca. 1120), the *Zīj-i īlkhānī* by Naṣīr al-Dīn al-Ṭūsī (d. 1274), a text known as the *Schemata of the Stars* (based in part on Ṭūsī's Persian text on theoretical astronomy entitled *Risālah-yi muʿīniyya* [The Muʿīnian Epistle]),[75] and maybe a text called the *Persian Syntaxis*.[76] Ṭūsī was the founding director of the Marāgha Observatory, which was built near Tabrīz. Chioniades's informant at Tabrīz was Shams al-Dīn al-Bukhārī (fl. mid- to late thirteenth century), whom Jamil Ragep has identified as Shams al-Dīn b. ʿAlī Khwāja al-Wābkanawī.[77] Shams al-Dīn's marginal position at Marāgha and Tabrīz may explain why Shams al-Dīn, and not, say, students of Ṭūsī, chose to teach Chioniades. As it turned out, Shams al-Dīn (a.k.a. al-Wābkanawī) dedicated a treatise on the astrolabe—an instrument useful for telling time and astrological forecasting—to the Byzantine emperor Andronicus II Palaeologos (r. 1282–1328).[78] Thus, al-Wābkanawī must have believed that he would accrue prestige by welcoming Chioniades. Wābkanawī also had the least to lose given his peripheral status.

Once again, exchange occurred in multiple stages. George Chrysococcès was the next crucial intermediary for the passage of syzygy computation methods from the Ilkhanids to Byzantine scholars. In the first half of the fourteenth century, George traveled to Trebizond in order to learn from a priest named Manuel.[79] By 1347, Chrysococcès had composed a Greek text called the *Persian Syntaxis* (different from any composition by Chioniades): a collection of tables with instructions for how to use them, in the tradition of Islamic *zījes*. Chrysococcès's *Persian Syntaxis* may have been derived solely from Chioniades's translation of Ṭūsī's *Zīj-i īlkhānī*, which was the view of Raymond Mercier.[80] Alternatively, the contents of Chrysococcès's *Persian Syntaxis* may have depended on Chioniades's translations of the older *al-Zīj al-sanjarī* and *al-Zīj al-ʿalāʾī* (and perhaps other texts that Chioniades did not translate) as well. David Pingree and his student Joseph Leichter espoused the latter position.[81] Ragep, by contrast, suggested that Chrysococcès used sources newer than *Zīj-i īlkhānī*, perhaps produced by Shams al-Dīn al-Bukhārī himself.[82] At stake in the scholarly debate over Chrysococcès's sources is the identity of the Islamic sources that Chrysococcès mediated to other Byzantine scholars.

Chrysococcès's *Persian Syntaxis* was translated into Hebrew as *Luḥot Paras* (Persian Tables) by a Jew in Salonika, Solomon b. Elijah Sharbiṭ ha-Zahab (lit. "the golden scepter," a Hebrew calque for Chrysococcès; fl. 1374–86).[83] Sharbiṭ ha-Zahab commented that he translated the *Persian Tables* after determining that Islamic methods of syzygy determination were preferable to the Ptolemaic alternative methods.[84] The unique Paris BnF 1042 MS of Sharbiṭ ha-Zahab's translation of Chrysococcès's *Persian Syntaxis* bears a statement of sale from 1444 with a currency from Candia, which provides additional evidence of how far west his Hebrew translation of the Greek traveled. It also features, in the margins, Latin translations of the Hebrew technical terms used in the tables.[85] A Latin version of the *Persian Syntaxis*, produced on Crete likely in the early 1400s with the Jewish community's assistance, is evidence that scholars in the Latin West were open to appropriating tables from the Islamic East through intermediaries even before the epoch of the merchants of knowledge.[86]

Kumaṭiano and his Byzantine Christian contemporaries had to contend with how George Chrysococcès introduced an error into the method given in Islamic tables for calculating the true syzygy.[87] According to the Islamic *zījes* available to the Byzantines in Greek translation (e.g., *al-Zīj al-ʿalāʾī*), the difference in time between a mean and true syzygy should be calculated as follows: $t = \Delta\lambda/\Delta v$ (i.e., the difference of the lunar and solar positions divided by the difference in their velocities). Ptolemy calculated the difference as follows: $t = 13/12 * \Delta\lambda/v$☾ (i.e., 13/12 multiplied by the difference of the lunar and solar positions divided by the moon's velocity). Chrysococcès created a portmanteau of these two equations, distorting the result by as much as thirty minutes: $t = 13/12\ \Delta\lambda/\Delta v$ (i.e., 13/12 multiplied by the difference in the solar and lunar positions divided by the difference in the solar and lunar velocities). Theoretically, the method found in the Islamic *zījes* should have out-performed updates of Ptolemy's tables because calculating the difference between the solar and lunar velocities was more precise than multiplying by a constant. But Chrysococcès's slip-up, which was reproduced in Sharḇiṭ ha-Zahaḇ's Hebrew translation, clouded matters because multiplication by 13/12 was uncalled for.[88] Understandably, Byzantine Christian astronomers concluded that syzygy calculations performed with Ptolemy's methods and with tables based on Ptolemy's *Handy Tables* were more accurate than the method that Chrysococcès ascribed to Islamic astronomers.[89] Chrysococcès must have misunderstood or transcribed his Persian sources incorrectly.

Further exchange was necessary to adjudicate the dispute between Chrysococcès and his Byzantine opponents. In his ca. 1425[90] *Peirush luḥot Paras* (Commentary on the Persian Tables), Kumaṭiano endorsed and restored the method found in the Persian *zījes*.[91] Since Sharḇiṭ ha-Zahaḇ's Hebrew translation perpetuated Chrysococcès's errors, Kumaṭiano's successful resolution of what he perceived to be a significant weakness in the *Persian Syntaxis* meant that he had pathways to inquire about Islamic astronomy other than via Sharbiṭ ha-Zahab's Hebrew translation. Indeed, Kumaṭiano referred to Chrysococcès's tables as an alternative source.[92] I have not found an independent Hebrew source for the syzygy computation method found in Islamic *zījes*, which means that Kumaṭiano had independent, more direct access to Islamic texts, perhaps through oral intermediaries. There is additional evidence that Kumaṭiano had sources for Islamic astronomy other than Sharbiṭ ha-Zahab's Hebrew translation of the *Persian Syntaxis*. For example, Kumaṭiano's value for the obliquity of the ecliptic—the angle between the celestial equator and the path that the sun traces (the ecliptic) against the background of the zodiacal constellations—23;35°, is the same as that found in the *Persian Syntaxis*. Yet Kumaṭiano noted (correctly) that his parameter was found in many Islamic *zījes*.[93] Such a remark is absent in the Hebrew translation of the *Persian Syntaxis*, indicating that in order to reach this conclusion, Kumaṭiano must have had multiple ways to learn about the contents of Islamic tables. Not all exchange was traceable through translation.

Eclipse prediction also depended on measurements of the apparent diameters of the sun and moon, which Kumaṭiano revised from those found in Sharbiṭ ha-Zahab's translation. Since one should not gaze into the sun, measurement of the solar diameter is challenging. Because the sun's speed was believed to vary throughout the year and because these variations in speed were explained by positing that the earth was removed from the center of the sun's path, it followed that the apparent size of the sun ought to vary throughout the year. Ptolemy remarked that he could not observe changes in the sun's visible diameter,[94] but subsequent astronomers did.[95] Kumaṭiano reported in his commentary that the solar diameter was 0;31,30° at apogee—when it was farthest from the earth—and 0;35,20° at perigee—when it was closest.[96] Values for the visible diameters of the moon and sun did not appear in Sharbiṭ ha-Zahab's Hebrew translation of the *Persian Syntaxis*, nor in the *Paradosis*, a Greek commentary on and instruction manual for Islamic astronomical tables produced ca. 1352. Nor did Kumaṭiano borrow these parameters from Ṭūsī[97] or from the *zījes* that had been translated into Greek.[98] Chioniades, in the Greek version of *al-Zīj al-sanjarī*, recommended deriving the apparent diameters from the sun and moon's speed but did not provide specific values.[99] Kumaṭiano did not mention performing his own observations, so his sources for the sizes of the sun and moon have yet to be uncovered.

Kumaṭiano was frustrated by what he learned of Byzantine astronomy. He alleged that Byzantine Christian scholars, singling out Isaac Argyros (fl. mid-fourteenth century), claimed that "these [viz. Persian] tables are not correct according to astronomy, rather that the truth found in them in calculating the paths is accidental."[100] He added, exasperated, that Byzantine astronomers nevertheless still used Islamic tables. Indeed, in the late fourteenth century, the Byzantines debated whether Islamic *zījes* might, in fact, have better parameters, despite the perceived shortcomings in the instructions.[101] The measure of the motion of the solar apogee, the point in the sun's path when it is most distant from the earth, was one important parameter. Astronomers in Islamic societies discovered that the solar apogee had an independent motion. The anonymous Byzantine author of the *Paradosis* reported that the solar apogee moved fifty-one or fifty-two seconds per year.[102] Kumaṭiano provided, in his *Commentary on the Persian Tables*, the figure of fifty-one seconds per year, adding that this equated to one degree about every seventy years.[103] The tables from Chioniades's version of *al-Zīj al-ʿalāʾī* yielded 54.53° seconds per year.[104] The shared preference for the figure of fifty-one seconds per year for movement of the solar apogee in light of alternatives is evidence that Christian scholars in the Byzantine Empire, despite any reservations they had about Islamic astronomy, joined Kumaṭiano in adjudicating among competing Islamic parameters for the motion of the solar apogee.

Kumaṭiano's familiarity with Byzantine Christian astronomy was even deeper, suggesting that he had access to the text of the *Paradosis* and not just the parameters. His frustration with the Byzantines' vacillation on the reliability of Islamic astronomy was informed. In one section of his *Commentary on the Persian Tables*, Kumaṭiano paraphrased portions of a text that he attributed to Argyros.[105] The paraphrasing came in the context of Kumaṭiano's expression of concern that one might be misled in determining the latitudes of the planets by a dispute between his views and those of Isaac Argyros over the use of the tables to determine longitude. First, he quoted his critic's method and, in so doing, paraphrased the *Paradosis*:[106] "For when you find the two latitudes of the sought planet, if both are in one direction, that is to say two to the north or two to the south, combine them. If one is northern and the other southern, subtract the lesser from the greater and there shall remain the latitude of the planet from the greater and its side, according to whether the greater was northern or southern."[107] After skipping a paragraph of the *Paradosis*, Kumaṭiano continued his paraphrasing:[108] "After that, look at Venus, and if its latitude is northern, subtract from it what you find in the table of the direct [viz. true] motion, and what remains is the true latitude. If its latitude is southern, add to it what you find in the table of the direct motion, and that is the true latitude. With Mercury, do the opposite, so if its latitude is northern, add to it what you find in the table of the direct motion, and if it is southern, subtract from it what you find in the table of the direct motion, and that is the true latitude. Up to here is his commentary."[109] Kumaṭiano may not have been the only Jewish scholar who knew the *Paradosis*; the translator who produced a Latin version of the *Paradosis* may have received the Byzantine original from Jews on Crete.[110]

Kumaṭiano's other claims about Byzantine astronomy are puzzling. First, modern scholars have not concurred with his ascription of the material resembling the contents of the *Paradosis* to Argyros. Second, the author of the *Paradosis* did not reject Islamic astronomy to the extent that Kumaṭiano alleged.[111] Thus, even if we presumed that Argyros authored the text, it would still be unclear why Kumaṭiano classified him as an opponent of Islamic astronomy. And if Argyros did criticize the method for the determination of true syzygies found in the *Persian Syntaxis*, then he was correcting a mistake found in Greek versions of Islamic tables. Argyros's known activities, i.e., the copying of an MS of the *Paradosis*, suggested an affinity for Islamic astronomy. Though the merchants of knowledge did not have any contacts who were Byzantine Christians, we have found one contact, Kumaṭiano, who did.

Kumaṭiano's intervention in the debate with Byzantine astronomers over syzygy calculation resonated with his students. Kumaṭiano had students not only from the Rabbanite sect of Judaism but also from the Qaraite sect.[112] Traditionally, Rabbanites held that God revealed an oral law to Moses on Mount Sinai along with the written law recorded in the Bible. The Qaraites, however, deny the authority of the Talmud, the Rabbanite legal code that is believed to record discussions of the oral revelation from Sinai.[113] The authors of the Talmud also commanded computing the Jewish calendar; consequently, Qaraites, in contrast to the Rabbanites, held that the new moon had to be observed. Things changed in the late fifteenth century when many Qaraites moved from Edirne (Adrianople), a city where Kumaṭiano lived for a time, to Istanbul as the city was repopulated after the 1453 Ottoman conquest.[114] One Qaraite student of Kumaṭiano, Elijah Bashyatchi, composed *Adderet Eliyahu* (Elijah's Mantle), a renowned compendium of Qaraite law. Bashyatchi's student and brother-in-law, Caleb Afendopolo (d. 1525), completed the text after Bashyatchi's death. *Adderet Eliyahu* was a comprehensive work on Qaraite law that contained a set of tables and instructions for calculating the visibility of the new moon.[115] With Bashyatchi's work, "the great divide between the calculated Rabbanite calendar and the observed Karaite calendar was finally breached."[116] A later Qaraite holdout, Samuel b. Benjamin, who preferred observation to calculation, conceded that one should know about astronomy in order to rationalize why observation was more precise than calculation.[117] Rabbanites and Qaraites continued to disagree over other calendrical matters, such as the determination of Passover and whether there should be two days of certain holidays. Bashyatchi and Afendopolo's willingness to calculate the calendar led to additional exchange.

Adderet Eliyahu contains evidence that Bashyatchi and Afendopolo learned of the contents of Islamic tables. Based on my analysis of Bashyatchi and Afendopolo's tables, I concluded that they or an intermediary had read Battānī's *zīj* in Arabic.[118] More important, they propounded the method ($t = \Delta\lambda/\Delta v$) for calculating the time from the mean syzygy to the true syzygy gleaned from the Islamic *zījes* in Greek translation.[119] In that key respect, they followed their teacher Kumaṭiano, who had defended Islamic methods of calculating the true syzygy against opposition from Byzantine Christian scholars. Their method of calculating lunar crescent visibility used a process of approximation close to the method described in the *Zīj-i īlkhānī*, but Bashyatchi and Afendopolo modified their approximation for the latitude of Istanbul (41°), as opposed to Marāgha (37°).[120] The precision of their method compared favorably to that found in the *Zīj-i īlkhānī* and the ca. 1430 *zīj* of Ulugh Beg (d. 1447).[121] Moritz Steinschneider (d. 1907) suggested that Bashyatchi and Afendopolo may have accessed Ulugh Beg's tables directly.[122]

And they were not alone in that respect. Romaniot and Italian Jewish scholars contemporary with Kumaṭiano but outside his circle also encountered Islamic tables. Paris MS BnF hébreu 1091 contains a Hebrew translation of the planetary tables from Ulugh Beg's *zīj*. The paper of the manuscript was produced in Venice between 1477 and 1508 and the numbering scheme is typical of Italian manuscripts, indicating that Ulugh Beg's tables were available to a Jewish scholar in Venice at the end of the fifteenth century.[123] Abraham b. Yom Ṭoḇ Yerushalmi, who was a rabbi in Istanbul in

1510,[124] was quoted in a prayerbook (*siddur*) that he had composed as saying that he verified the information in the prayerbook with the Hebrew version of Ulugh Beg's tables in the early 1500s.[125] Thus, the availability of Ulugh Beg's tables to Jewish scholars led to access.

Jewish scholars did not just appropriate knowledge; they shared their own methods and tables as Hebrew texts continued to be sources for Greek astronomy well into the fifteenth century. Kumaṭiano's *Commentary on the Persian Tables*, along with *Six Wings* and the work of Sharb̲iṭ ha-Zahab̲, were sources for Gemistus Plethon's (d. 1452) *Manual of Astronomy*.[126] Plethon also relied on Jewish sources for his 1433 tables. George Scholarios (d. 1473), an adversary of Plethon and patriarch of Constantinople, claimed that Plethon studied with a Jew.[127] That charge was not a compliment and serves to remind us that the Eastern Mediterranean was not a utopia for Jews in the fifteenth century, despite the remarkable exchange with which they were involved.

Christians in Europe consulted the *Paradosis*. Cardinal Bessarion (d. 1472), originally from Trebizond, produced a copy of the *Paradosis* in which he modified the structure of the text and its wording in order to make mathematical astronomy of Islamic origin even more accessible.[128] In the early 1460s, the renowned Renaissance astronomer Johannes Regiomontanus (d. 1476) became part of Bessarion's circle while concluding work on the *Epitome of the "Almagest,"* which his teacher Georg Peuerbach (d. 1461) had left incomplete.[129] The *Epitome* was the most sophisticated fifteenth-century astronomy text produced during the Renaissance before Copernicus wrote. Regiomontanus gained access to Bessarion's library, which contained four copies of the *Paradosis*.[130] Later, Pico della Mirandola (1463–94) borrowed—from the Medicis—and annotated a manuscript that included a text entitled *Persian Tables*.[131] While there is no evidence that Pico, a contact of the merchant of knowledge Elijah Delmedigo, learned anything about Islamic tables from the merchants of knowledge, Pico was aware, though his gaze was mediated by Greek sources, of the same Islamic tables as Kumaṭiano and his Byzantine Christian counterparts.

The calculation of eclipses and the lunar calendar was a challenging problem that brought together Muslim, Byzantine Christian, and Jewish astronomers, including Kumaṭiano, a contact of the merchants of knowledge. Sometimes complete tables with instructions for their use were translated. In these cases, the originals and translations were concrete objects of value. In other cases, discrete methods and parameters moved across regional and linguistic boundaries and found their way into original compositions and commentaries. With the annotation of the *Paradosis* by Pico della Mirandola, we have evidence that another contact of the merchants of knowledge explored mathematical astronomy that originated in Iran. The most technical problems linked a wide swath of scholars.

Ottoman Scholars' Interest in Latin Tables

Given the prominence of astrology in the Ottoman Empire, it is not surprising to learn that Ottoman Muslim scholars devoured *zījes*. The authors of the Ottoman almanacs that we encountered in the previous chapter relied on many of the sources (e.g., *Zīj-i īlkhānī* and *Zīj-i Ulugh Beg*) available to Kumaṭiano, his students, and Byzantine Christian scholars.[132] After the Ottoman conquest of Constantinople, a new vector of exchange opened up as the Ottomans looked west. One fascinating transaction depended on the mediation of the merchant of knowledge Moses Galeano/Mūsā Jālīnūs. An elite Ottoman scholar patronized his translation of a Latin text entitled *Almanach perpetuum* into Arabic. The *Almanach perpetuum* was based on the work of the Jewish scholar Abraham Zacut (d. 1515), whose career was a synecdoche for the passage of his writings from the Iberian Peninsula to the Ottoman Empire. Zacut was born in 1452 in Salamanca and likely completed *ha-Ḥibbur ha-gadol* (The Grand Composition) there in 1478.[133] After leaving Salamanca, he was patronized by the knight, nobleman, and future bishop Juan de Zuñiga in Extremadura.[134] Astrology motivated Zacut's interest in astronomy, and he composed a treatise in Castilian entitled *Tratado en las ynfluencias del cielo* (Treatise on the Influences of the Heavens) at Juan's request in 1486.[135] Following the Expulsion from Spain, Zacut served the King of Portugal as royal astronomer until 1497. Zacut composed *Seiḇer ha-yuḥasin* (The Book of Genealogy), a history of the Jews, in Tunis in 1504[136] and died in 1515 in the Ottoman Empire, in Damascus.[137] Zacut's peregrinations matched the geographic influence of his astronomy.

Zacut's range of sources is striking. For instance, his 1513 tables for Jerusalem modified Nicholas de Heybech of Erfurt's (fl. ca. 1400) tables for syzygy computations.[138] Zacut, in his most famous set of tables with canons, *ha-Ḥibbur ha-gadol* (The Grand Composition), drew on earlier material, including the Alfonsine corpus, and he cited the tables produced by Emmanuel Bonfils, Jacob b. David Bonjorn, Judah b. Asher II (d. 1391), and Judah Ben Verga (fl. 1455–80).[139] Zacut's discernment was evident throughout *ha-Ḥibbur.* For example, one Hebrew manuscript of *ha-Ḥibbur* contained a gloss in which Zacut observed an occultation of Venus by the moon and realized that such an event could not result from the latitudes of Venus and the moon predicted from his tables, nor from the version of the Alfonsine Tables produced by John of Lignères (d. ca. 1335), nor from the explanation found in Averroës's *Compendium of the "Almagest."*[140] In other words, he checked his sources, as well as his own computations, through observations. Zacut's sophisticated development of the Alfonsine corpus garnered a multilingual readership. Zacut cooperated with Juan de Salaya, the occupant of the chair of astronomy/astrology at Salamanca, to translate the canons of *ha-Ḥibbur* into Castilian in 1481[141] and may have assisted with the translation of the tables of *ha-Ḥibbur* into Latin.[142] In any case, the result was that Zacut's astronomy attained its greatest diffusion through the *Almanach perpetuum*, a text that was not his original composition.

The relationship between the *Almanach perpetuum* and *ha-Ḥibbur ha-gadol* is complex. The canons of the *Almanach perpetuum* are distinct from those of *ha-Ḥibbur*, but the tables of the *Almanach perpetuum* depend on the tables of *ha-Ḥibbur.*[143] The *Almanach perpetuum*'s canons were produced by Zacut's Jewish student José Vizinho (or Vizinus) in Castilian and Latin and first printed in 1496.[144] Besides Vizinho, Zacut had a Christian student, Augustinus Ricius, who wrote a book on the motion of the eighth sphere and cited a number of Hebrew texts.[145] Zacut likely participated neither in the printing of the *Almanach perpetuum* in 1496 nor in the edition of the work,[146] but printing made his work more widely available.[147] The tables from the *Almanach perpetuum* were used for navigation,[148] and in some sources Zacut is credited with educating Portuguese explorers.[149] Even though the *Almanach perpetuum* was not Zacut's independent work, it was the source of his renown.

A merchant of knowledge brought the *Almanach perpetuum* to the attention of an elite Ottoman Muslim scholar soon after it was printed. Galeano/Jālīnūs dedicated his Arabic translation of some of the canons of the *Almanach perpetuum* to the chief military judge ʿAbd al-Raḥmān Muʾayyadzādah (d. 1516).[150] Not only was Muʾayyadzādah drawn to astrology, as we learned in the previous chapter, but he also he studied astronomy with Jalāl al-Dīn al-Dawānī (d. 1502) and acquired a number of texts on astronomy.[151] Bayezit

II (r. 1481–1512), the sultan during Muʾayyadzādah's career, learned astronomy from Mīrim Çelebī (d. 1525), a prominent Ottoman author of practical and theoretical astronomy. Two astrolabes were made for Bayezit II, indicating the significance of instrument production when Galeano/Jālīnūs was present at the court.[152] The existence of a second set of Arabic canons, along with a separate colophon and six additional tables about lunar crescent visibility, in the unique manuscript of Galeano/Jālīnūs's translation, led Parra to conclude that he was not the sole translator.[153] Galeano/Jālīnūs translated the canons principally from the Hebrew but relied also on the Latin.[154] The tables came mostly from the Latin version of the *Almanach perpetuum*,[155] and technical terms in the translation are transliterations of the Latin.[156] The table of contents of the Arabic version followed neither the contents of the Castilian version nor the Latin version of the tables that Goldstein and Chabás have studied.[157] As the canons were printed in Portugal only in 1496, and in Venice in 1502, I speculate that Galeano/Jālīnūs secured a copy of the *Almanach perpetuum* while abroad, perhaps with the aid of the printer Gershom Soncino (d. 1534), whom Galeano/Jālīnūs met in Venice between 1497 and 1503, because Galeano/Jālīnūs anticipated the potential value of this knowledge in the Ottoman Empire.[158] Venetian officials' disinclination to let the 1499–1503 war with the Ottoman Empire disrupt trade rendered Galeano/Jālīnūs, as an Ottoman subject, less subject to harassment while traveling.[159] More important, Galeano/Jālīnūs, because he was a physician, may have been able to evade the 1496 requirement that Jews wait a year between visits to Venice.[160] That and his knowledge of Latin and Arabic made him an ideal intermediary to bring European knowledge back to the Ottoman Empire.

In the introduction of this book, we learned of Mehmed the Conqueror's interest in Western goods. Other scholars attempted to attract his attention with Greek and Latin texts. For example, the Byzantine Christian George of Trebizond, who was born on Crete and spent a lot of time in Venice, tried to dedicate his Latin translation of the *Almagest* to Mehmed. George perceived the Byzantine heritage as a potential commonality.[161] Galeano/Jālīnūs's translation of the *Almanach perpetuum* was evidence that a market for European science in Istanbul endured under Mehmed's son Bayezit II (d. 1512). A translation of Latin tables and instructions for their use into Arabic was an asset for Galeano/Jālīnūs to hawk as he jockeyed for influence at the Ottoman court.

His task was to mold an astronomy text that took shape in Europe to fit the intellectual context of the Ottoman Muslim patron, Muʾayyadzādah, whom Galeano/Jālīnūs eventually located. His introduction to the Arabic translation of the canons reflected his Islamic setting. He wrote as someone aiming for a station at the Ottoman court rather than as a visiting scholar from afar. The text begins, "In the name of God the Merciful and Compassionate. Praise to God who created the earth and the heavens, and who adorned the lower heavens with the lamps of the radiant, especially through the two most luminous, dazzling luminaries. And who made its light the cause of the development of creations, whether animals, minerals, and plants. And who apportioned the lunar mansions by constellations and degrees. Praise and peace be upon our lord [Muḥammad], the most noble of created beings and upon his family and his companions, possessors of grace and dignities. Now then . . ."[162] Many Islamic scientific texts begin with similar language. By choosing to add these sentences, Galeano/Jālīnūs demonstrated to his patron that the *Almanach perpetuum*, once translated into Arabic, could also be translated to an Islamic intellectual milieu.

After demonstrating the general fit of the translation, Galeano/Jālīnūs provided specific, compelling evidence for why Muʾayyadzādah patronized the translation. New tables could yield deeper and better insights into historical astrology and improve decision-making. Galeano/Jālīnūs explained:[163]

> The reason that prompted me to compose this wondrous, pleasant, welcome, and desired epistle written [originally] in the Frankish tongue is that there came to pass a day, on a certain date in 911 AH [1505 CE] in which I, the poor man in need of the wealthy God, Mūsā Jālīnūs al-Tīrawī [of Tire],[164] mentioned in the council of the most eminent of the penetrating scholars, the most perfect of the virtuous ones who investigate, the one who combines the perfections of the ancients and the moderns, the inheritor of the sciences of the prophets and the apostles, the one protected by the gaze of the liberating king, I mean ʿAbd al-Raḥmān, the chief judge for the victorious army for the exalted threshold and the Sublime Porte, may God support[165] its columns and level its edifice, the transfers (*taḥwīlāt*) and aspects (*anẓār*) of the seven planets in a summary fashion without going on at length and dispensing with the re-

> newed extraction [of their positions] and their longitude and latitude as long as the planets are moving and revolving. So the most magnificent lord [ʿAbd al-Raḥmān] commanded me to translate it from the Frankish tongue into Arabic.[166]

The references to the planets' motions, astrological terminology (e.g., transfers and aspects), and the *qāḍī*'s council (*majlis*) confirm how the widespread interest in astrology at the Ottoman court occasioned the commissioning by Muʾayyadzādah of a translation of the entire treatise from the "Frankish tongue" to Arabic. Through social contact, Galeano/Jālīnūs became attuned to Muʾayyadzādah's and other Ottoman elites' astrological interests. When in Italy, he would have jumped at the opportunity to bring back a text, the *Almanach perpetuum*, that could instruct those whose patronage he coveted.

Interestingly, Galeano/Jālīnūs seems not to have updated the technical contents of the *Almanach perpetuum* in his translation. In the Arabic canons, the starting point for computing cycles of leap years was 1472,[167] and in the *Ḥibbur*, the table for finding the day of the week commenced with 1473.[168] That was also the beginning date for the table of the motions of the sun in the Arabic translation.[169] One might expect that the syzygy tables would have been updated in a manner similar to how Zacut modified Bonjorn's syzygy tables for the longitude of Salamanca from the latitude of Perpignan where Bonjorn worked.[170] But the first syzygy in the Arabic translation is listed as being on March 4, 1478 at 6:00 a.m.[171] Not only is 1478 almost thirty years before the translation, but Galeano/Jālīnūs also recorded when the syzygy occurred in Salamanca, not in Istanbul![172] Hence, he produced a translation in the strict sense, not a new, updated version of the *Almanach perpetuum*.

Still, the Ottoman Turkish marginalia in the unique MS of the Arabic translation provide evidence that Muʾayyadzādah or a colleague wasted no time in using the translation. Galeano/Jālīnūs translated in 1505, and a reader corrected a calculation for the longitude of the sun on August 14, 1505.[173] A long marginal comment treated the effect of the quadrennial cycle of leap years for computing the longitude of the sun in its annual path about the earth. Because a year is close to 365 and a quarter days, but is counted as only 365 calendar days, the position of the sun at the beginning of each successive year in a cycle lags about one-quarter of a degree behind its position at the beginning of the previous year. Then, the position of the sun at the beginning of the year after a leap year advances just over a degree. Hence, one needs four different sets of tables to find the precise position of the sun, one for each year of a four-year cycle. In a marginal comment in the Arabic version, the user is instructed to subtract 1472, the most recent leap year before the beginning of the solar motion tables in 1473, from the given year and divide by four. The quotient is the number of quadrennial cycles (*adwār*) of the sun, comprising one leap year and three ordinary years. In the main text, we find that if the remainder is one, the first of four tables of the sun's position is used. The marginal comment, written in Turkish, continued: "Look at the cycles [viz. of the sun]; if the remainder is two, [the position of the sun] is entered in the second table." [174] If the remainder was three, the third table was used. The reader was directed to the table of the sun's position appropriate for the year of the quadrennial cycle. A remainder of four was equivalent to no remainder, meaning that the table of solar motions particular to leap years should be used.[175] If Galeano/Jālīnūs's aim was to provide Muʾayyadzādah with a usable text, Galeano/Jālīnūs seems to have succeeded without having to update the tables. Galeano/Jālīnūs's awareness of the Ottoman market for knowledge from Latin texts motivated him to search for Christian contacts when in the Veneto.

Conclusions

Much like the exchanges narrated previously in the field of astrology, the exchange of tables, instruments, and texts about them among the merchants of knowledge and their contacts was multidirectional. For example, the universal astrolabe plate originated in eleventh-century al-Andalus. The Jewish scholar Kumaṭiano, a contact of the merchants of knowledge, gave away his own universal astrolabe plate without thought to securing a replacement. The plate was mobile between Jewish and Muslim cultures. To take another example, Islamic mathematical astronomy from Marāgha appeared in the work of Kumaṭiano and his students probably through Islamic intermediaries. Marāgha astronomy also reached Pico della Mirandola, another contact of the merchants of knowledge via a different route, through the *Paradosis*. Finally, the Latin *Almanach perpetuum* that Galeano/Jālīnūs translated into Arabic was based on Abraham Zacut's Hebrew *ha-Ḥibbur ha-gadol* (The Grand Composition). While Ottoman court astrologers drew on *zījes* (astronomical handbooks with tables) from Marāgha and Samarqand, Galeano/Jālīnūs correctly gauged ʿAbd al-Raḥmān Muʾayyadzādah's openness to something new.

The production of tables and the construction of instruments were remarkable computational and three-dimensional (respectively) solutions to complex problems. Yet the exchange of instruments and tables was also highly transactional. Kumaṭiano mentioned how scarce the universal astrolabe plate was among the Jews. Yet he was so eager to curry favor with an Ottoman Muslim judge that Kumaṭiano seems to have given insufficient thought in the moment to the scientific value of the instrument. Only after the exchange did Kumaṭiano realize that Jews should never lack knowledge of how to engrave such a plate. Conversely, the privileged position of the Ottoman scholar who received the astrolabe plate aided his acquisition. In this chapter, we have learned how the historical actors profoundly appreciated how intellectual exchange could enhance or benefit from their social capital.

The findings of this chapter corroborate recent findings in Ottoman studies that the sciences were cultivated during the reign of Bayezit II.[176] In addition, the exchange facilitated by Galeano/Jālīnūs and in which Kumaṭiano's students participated is evidence that Ottoman Muslim scholars under Bayezit II remained open to contact with non-Muslim cultures. The transregional contacts of Ottoman scholars mattered for other fields. The exchange of tables could coincide with the exchange of theoretical astronomy (chapter 6), since scholarly intermediaries with access to the best tables were also aware of theoretical innovations. For instance, Finzi, the author of a commentary on the Alfonsine Tables, was aware of a significant innovation in theoretical astronomy that originated at the Marāgha Observatory in Northwest Iran in the thirteenth century.[177] Gregory Chioniades, the translator of Islamic *zījes* into Greek, also composed the *Schemata of the Stars*, a Greek précis of Islamic theoretical astronomy containing some of the theories that also appeared in the work of Copernicus. Scholars interested in the best tables also sought innovative theories, and we will see that Galeano/Jālīnūs was no exception.

In order to procure the copy of the *Almanach perpetuum* that he subsequently translated into Arabic, Galeano/Jālīnūs must have been in contact with scholars of astronomy who were conversant in Latin scientific literature. Yet better tables did not require better models. In addition, reflections on the structure of the cosmos also occurred in other fields: Aristotelian philosophy and *Qabbalah*. His connections with European Christian scholars were the outgrowth of connections forged by other merchants of knowledge in the fields of Aristotelian philosophy and *Qabbalah*. These are the topics of chapter 4 and chapter 5, respectively.

FOUR
ARISTOTELIANISM *across* BORDERS

The practices of astrology and astronomy covered in the previous two chapters did not differ markedly from region to region; however, Aristotelian (capaciously defined) philosophical traditions did. There was no single philosopher who was equally influential for the merchants of knowledge and their Christian and Muslim contacts. For example, Avicenna (d. 1037; Ibn Sīnā in Arabic) blended Aristotelianism and Platonism to wield the greatest influence on subsequent Islamic philosophy, particularly in the Islamic East (*mashriq*), which included the Ottoman Empire. While Avicenna was known to Jews and Christians, particularly through his summa of medicine entitled *The Canon of Medicine*, the Andalusian arch-Aristotelian Averroës (Ar. Ibn Rushd; d. 1198) was the single most compelling philosopher for the merchants of knowledge and their Jewish and Christian contacts. Translations of his works from Arabic to Hebrew and Latin first appeared in the early thirteenth century. Both philosophers were renowned for their explorations of how revealed texts cohered with human reason, but Averroës has retained more of an image as a rationalist in modernity. The French Orientalist Ernest Renan (d. 1892) remarked that Averroës's thought contrasted sharply and positively with that of his Muslim contemporaries, meaning that Averroës's critical faculties were not dulled by religion in the same way that those of his Muslim contemporaries were.[1] In the thirteenth-century Latin West, Averroism, a philosophical school based on the texts of

Averroës, was also proscribed for its contradictions of Christian theology. Indeed, Averroism was condemned as heretical at the University of Paris in 1277. To Renan, not all religious opposition to Averroës was equal, as Renan had a less jaundiced view of condemnations of Averroism in the Latin West.[2] By treating Averroës as an avatar of a modern, critical perspective, Renan avoided serious exploration of why Averroism was criticized by the church.

In the 1997 film *Destiny* (Ar. *al-Masīr*), the renowned Egyptian director Youssef Chahine used Averroës as an allegory for freethinking intellectuals in the modern Islamic world who were threatened by Islamists to portray a certain vision of modernity. In the film, Averroës appeared widely respected by his Muslim adversaries. Unlike Renan, Chahine avoided reducing matters to a dichotomy between an open-minded West and a restrictive Islamic world; as the film opens, a Christian Averroist is burned at the stake. Averroës's life was shaped more by his political struggles with Muslim rulers and ostensibly by the frustration of officials with his rationalist interpretation of Islam, but the relevance of Averroism to Christian theology was more important to the merchants of knowledge and their Christian contacts. Why was Averroës so controversial for so many Christians? The answer lies in the theological implications of how Averroists attempted to attribute the multivarious complexity of the world to a unitary God. Because astrology was widespread and provided a different answer to that question, much was at stake for Averroist (and Avicennan)[3] philosophers during the careers of the merchants of knowledge and their contacts.

Through translation, composition, and instruction, the merchants of knowledge facilitated a philosophical conversation that linked Averroism and Avicennism, two schools of Aristotelian philosophy. The merchants of knowledge situated themselves in a long tradition of applying Hellenistic philosophy (*falsafa*) and philosophical theology (*kalām*) from Islamic societies to Jewish questions. The most famous early Jewish thinker to engage with Islamic rationalist traditions was Saadia Gaon (d. 942) from Fayyūm in Egypt. Saadia was influenced by *mutakallimūn* (philosophical theologians) and, like them, argued that God was inherently just, that God was beyond anthropomorphization, and that humans had free will.[4] Isaac Israeli (d. ca. 955), who also spent part of his life in Egypt, was the first Jewish philosopher in the Neoplatonic tradition.[5] Neoplatonists synthesized many currents in Hellenistic philosophy, especially Aristotelianism and Platonism. Isaac knew the work of al-Kindī (d. 873 or 877), the first Muslim to compose original works of philosophy. Solomon Ibn Gabirol (d. ca. 1058) was a later Jewish Neoplatonist, and Abraham Ibn Ezra (d. 1164 or 67) brought Neoplatonic philosophy to bear on the interpretation of Jewish texts. Averroës has been understood to be the most faithful medieval interpreter of Aristotle, but Romaniot Jews encountered Neoplatonism, through the corpus of Ibn Ezra's writing, first.

The Philosophy of Abraham Ibn Ezra among Romaniot Jews

Though nothing in Ibn Ezra's voluminous oeuvre was formally in the genre of philosophy, he was a thinker widely read by Romaniot Jews.[6] Ibn Ezra's writings garnered attention more for their range of sources than for their analytic depth.[7] Because Ibn Ezra also composed in Latin (with assistance) and because other texts of his were translated into Latin, he acquired a Christian readership.[8] His Hebrew works circulated throughout the Mediterranean and reached Constantinople, attracting admirers from two sects of Judaism: the Qaraites and the Rabbanites.[9]

Traditionally, Rabbanites held that God revealed an oral law to Moses on Mount Sinai along with the written law recorded in the Bible. The modern movements of Reform, Conservative, Orthodox, and Hasidic Judaism are developments of Rabbanite Judaism because the debates among these modern movements are about the applicability and interpretation of the oral and written law. The Qaraites, however, denied the authority of the Talmud, the Rabbanite legal code that is believed to record discussions of the oral revelation from Sinai.[10] Qaraites purported to reject figurative (*d^e^rash*) and homiletic readings of scripture because they believed that any interpretations that departed from the plain sense (*p^e^shaṭ*) were the foundation of the oral law of the Rabbanites. The Qaraite approach to Jewish law, based on what they took to be the plain sense of scripture, sometimes yielded more stringent rulings. For a time, Qaraites did not eat chicken because nowhere in the Bible was chicken named as a permitted food. The breach between the two approaches to Jewish law, and hence between the two communities, could be bridged because from the Rabbanite perspective, a principal goal of Talmud study has been to discover the divine reason embedded in the oral law. If Jewish law was founded upon reason as well as the double revelation (written and oral) at Sinai, outreach to Qaraites by Rabbanites through the study of science and philosophy was possible.[11] Given the presence of significant Qaraite communities in Asia Minor, Ibn Ezra's incorporation of science and philosophy into Jewish religious thought met the moment.

One of Ibn Ezra's goals in composing his biblical commentary was to engage Qaraites on their own intellectual turf by focusing on the plain sense of the text rather than the figurative (*d^erash*) interpretations that they decried. Mordechai Kumaṭiano (d. 1482), who was a famous scholar in Istanbul[12] during the lifetime of the merchant of knowledge Moses Capsali (d. 1495), authored a biblical commentary in which he drew on Ibn Ezra's biblical commentary in content and approach. Kumaṭiano was born in Istanbul but fled[13] to escape the plague in the 1450s to Edirne, the Ottoman capital before 1453, where he studied with Ḥanok̲ Sap̲orta, a refugee from Catalonia who gained a reputation for instructing Qaraites.[14] Capsali knew of Kumaṭiano because Capsali tried unsuccessfully to forbid Sap̲orta and other Rabbanites from teaching the oral law to Qaraites.[15] Kumaṭiano certainly appealed to Qaraites and perhaps to non-Jews as well.[16] Kumaṭiano was the teacher of Elijah Mizraḥi (d. 1526), who, in turn, instructed the merchant of knowledge Moses Galeano/Mūsā Jālīnūs in astronomy. Mizraḥi was also noted for his willingness to have Qaraite students.[17] Mizraḥi succeeded Capsali as a rabbinical authority in Istanbul and issued a legal opinion negating Capsali's proscription of instructing Qaraites.[18] The intellectual encounter between Rabbanites and Qaraties in Asia Minor accounted for the rationalist tenor of Romaniot intellectual life.

In chapter 3, we saw that Kumaṭiano's most famous Qaraite student, Elijah Bashyatchi (d. 1490), along with Caleb Afendopolo (d. 1525),[19] another of Kumaṭiano's Qaraite students and Bashyatchi's brother-in-law, embraced the Rabbanites' technique of calculating the new moon.[20] Bashyatchi and Afendopolo also developed an interest in Averroës, an interest which linked Afendopolo indirectly to the Fuggers. In 1497, Afendopolo purchased a copy of Averroës's *Long Commentary on Aristotle's "Metaphysics"* and, at the same time, a copy of Thomas Aquinas's (d. 1275) commentary on Aristotle's *Ethics*.[21] An inscription in the manuscript informs us that after Afendopolo's death, his widow sold the manuscript along with his manuscript of the Hebrew translation of al-Ghazālī's (d. 1111) *The Intentions of the Philosophers* (perhaps the future Vatican MS Ebr. 346) to Solomon b. Solomon. *The Intentions of the Philosophers* was a summary of the views of Avicenna. This may have been the same Salamone of Canea who transacted another manuscript to the agents of the Fuggers in 1542.[22] In any case, the agents of the Fuggers eventually acquired Afendopolo's manuscript of Averroës's *Long Commentary on Aristotle's "Metaphysics"* and probably Afendopolo's manuscript of Ghazālī's *The Intentions of the Philosophers.* The intellectual exchanges between Kumaṭiano and his students were paralleled by significant exchanges of manuscripts.

Kumaṭiano's *Commentary on "Yᵉsod Moraʾ"* (The Foundation of Awe)

Kumaṭiano's *Commentary on "Yᵉsod Moraʾ"*—Ibn Ezra's treatise on the rationales for the commandments—contains details of how Kumaṭiano persuaded Qaraites to accept the Rabbanite position on calendrical calculations. He completed the commentary in Edirne around 1450.[23] Ibn Ezra commenced *Yᵉsod moraʾ* by cataloging three types of intellectuals, all of whom were concerned with scripture. Scholars of the first type, he specified, knew much about the scribal practices of the Masoretes (early medieval Jewish scholarly scribes), but he likened them to someone given a book of cures who then decided just to count the pages—in other words, readers who could not have been less thoughtful. Scholars in the second category were experts in biblical grammar, knowledge which is of great use for Judaism. Scholars in the third category meditated day and night on scripture, had recourse to the Aramaic translations, and truly mastered the meanings of scripture.

Ibn Ezra cautioned, however, that scholars who fell into these three categories—scholars with skills that the Qaraites possessed in spades—did not know a single commandment without the benefit of the oral law.[24] Without the benefit of the oral law, he wrote, they would be unable to determine something as fundamental as the date of the new moon. To that end, Kumaṭiano recommended Maimonides's treatment of crescent visibility in the *Mishneh Torah*, a topical rearrangement of the Talmud.[25] For Kumaṭiano, the mathematical precision of Maimonides's method epitomized the rationality of Jewish law, written and oral.

Then, Kumaṭiano reversed direction and admonished Rabbanites for what they would lose as Jews were they to forsake the study of science and philosophy.[26] Yes, Qaraites skilled in astronomy came to recognize how rational and, therefore, correct the Talmudic position was.[27] But Kumaṭiano reminded readers that no Rabbanite could understand the Talmudic position on calculating the calendar *without* astronomy. Hence, Kumaṭiano concluded that his work on mathematical astronomy provided information essential for appreciating and fulfilling God's law.[28] Other sciences mattered too. Avicenna's foundational *The Canon of Medicine* (*al-Qānūn fī al-ṭibb*) helped Kumaṭiano rationalize the indisputable biblical prohibition of the consumption of blood: "Only be stedfast in not eating the blood; for the blood is the life (*ha-nep̱esh*)" (Deut. 12:23). Kumaṭiano explained that since blood that is eaten and digested remained a carrier of life, the prohibition existed so that we do not eat another human being. In support, he mentioned how Avicenna wrote, in the first book of his *Canon*, that whatever is consumed leaves in the body an imprint of its heat, in the case of garlic, or coldness, in the case of lettuce.[29] The consumer of blood is the consumer of souls, the carriers of life. Souls rightfully belong to God. Though Rabbanites did not need science to accept the authority of the Talmud, they could not fully grasp the import of the Talmud without science and philosophy.

Kumaṭiano brought psychology (which was considered part of philosophy) to bear in his *Commentary on "Yᵉsod mora'"* upon the interpretation of Psalm 73:26 ("My flesh and my heart faileth; but God is the rock of my heart and my portion forever"). He produced a long excursus about the definition of the human intellect.[30] Like all Aristotelians, Kumaṭiano believed that the process of understanding transpired in the intellect, which was part of the soul. He followed earlier philosophers in describing the part of the intellect where understanding occurred as "material" (or hylic), not because the intellect was material but because the material intellect shared matter's potential for being acted upon.[31] Something changed in the hylic intellect as one gained knowledge.

Definitions of the intellect and of the process of understanding had religious stakes because, as Kumaṭiano explained in his comments on Exodus 3:13, the acquisition of rational knowledge could lead to the conjunction of the soul with God and the soul's separation from the body.[32] This was a different variation on the definition of *'aḇodat ha-seḵel*, worship through one's intellect, than Moses Cohen Ashkenazi espoused in *Urim wᵉ-ṯummim*. Kumaṭiano did not provide more details about the operation of the hylic intellect in his *Commentary on "Yᵉsod mora',"* but he seemed to side with the position that Averroës articulated in the middle commentary on *De anima* (On the Soul).[33] There, Averroës wrote that the material intellect dwelled in the body but was connected to an eternal intellect.[34] Hence, in order to explicate fully Ibn Ezra's statement about the intellect, Kumaṭiano consulted Averroës. Ibn Ezra had led Kumaṭiano to Averroës!

Kumaṭiano's integration of science and philosophy into religious thought resonated with the Candiote merchant of knowledge Michael Balbo. Balbo acquired a manuscript (the future Vat. MS Ebr. 105) of Kumaṭiano's *Commentary on "Yᵉsod mora᾽."* By the time he took possession of the manuscript, Balbo had been a prominent figure on Candia for at least three decades.[35] How the manuscript traveled from Istanbul to Candia is not known, though a 1458 decree from Moses Capsali about the illegitimacy of a cantor appointed by the Venetian authorities addressed to Balbo and others also arrived in Candia from Istanbul.[36] However the manuscript traversed the Eastern Mediterranean, its passage is not surprising given the centrality of Ibn Ezra's oeuvre to Romaniot intellectual life. Likewise, the eventual acquisition of Vat. MS Ebr. 105 by the Fuggers from the leaders of the Candiote community was the result of the integration of science and philosophy into religious thought in the conversations between the merchants of knowledge and their Christian contacts. The itineraries of manuscripts mapped the use and combination of knowledges, such as this transregional conversation about Aristotelian philosophy, described in this book.

Kumaṭiano and Islamic Philosophy

Another text of Kumaṭiano's, a commentary on Maimonides's (d. 1204) *The Guide of the Perplexed*, attracted attention from elite Ottoman Muslim scholars. The *Guide* was completed in Judeo-Arabic in 1190 and translated into Hebrew at the end of Maimonides's life. As an attempt to reconcile Aristotelian philosophy and science with Judaism, the *Guide* was a staple of intellectual life for Candiote scholars.[37] Ottoman Muslim scholars copied the *Guide* in Arabic script.[38] In addition, there is a 1480 manuscript of Kumaṭiano's Hebrew commentary on the *Guide* in the III Ahmet Collection of the Topkapı Palace Library (no. 53).[39] Shabbetai b. Moses, the copyist,[40] was from Candia, and the Topkapı manuscript of the *Guide* shares physical characteristics with other Greek manuscripts that were copied during Sultan Mehmed II's reign, such as Kristoboulos's Greek *History of Mehmed the Conqueror.*[41] The presence of the manuscript in the Topkapı Library suggests that Kumaṭiano had access to Muslim elites.

Kumaṭiano was not as interested in fifteenth-century Ottoman Islamic thought as Ottoman Muslim scholars were in his. Consider Kumaṭiano's commentary on *Guide* I.73, on the foundational premises of Islamic philosophical theology (*kalām*), a passage that reflected Maimonides's imbrication in his Islamic context. In the base text we find a number of premises of *kalām*, the tenth of which is that "they [the theologians (*mutakallimūn*)] are of the opinion that everything that may be imagined is an admissible notion for the intellect."[42] Then, Maimonides added that the *mutakallimūn* held that the conjunction of opposites, though conceivable, is intellectually inadmissible.[43] In a sense, the *mutakallimūn*'s rejection of the conjunction of opposites seemed the preferable position; something cannot be both alive and dead at the same instant. But *mutakallimūn* in the fourteenth and fifteenth centuries, who were influenced by Avicennism, reconsidered the inadmissibility of the conjunction of opposites. Though conjoined opposites were not thought to exist externally, *mutakallimūn* influential for Ottoman Muslim intellectual life in Kumaṭiano's time, such as al-Sayyid al-Sharīf al-Jurjānī (d. 1413) and Saʿd al-Dīn al-Taftāzānī (d. 1390),[44] allowed that the conjunction of opposites may have had an objective existence in the fact of the matter (*fī nafs al-amr*).[45] That is, the conjunction of opposites was verifiable according to an objective standard in the same way that mathematical equations can be verified without existing externally. Kumaṭiano did not recount these or any recent developments in *kalām* anywhere in his commentary. The most likely explanation was that he was disinterested.

One might conclude that Kumaṭiano perceived his task as a commentator to be confined to analyzing only the positions of the Muslim thinkers to whom Maimonides referred. However, Kumaṭiano relied on Islamic sources unnamed by Maimonides that were available in Hebrew.[46] Among these texts were Averroës's commentaries on the Arabic translations of Aristotle's corpus, Ibn Sīnā's *Canon*, Averroës's *Epistle on the Possibility of Conjunction* and *De substantia orbis* (On the Substance of the Celestial Orb), and Ghazālī's *The Intentions of the Philosophers*. Kumaṭiano studied some of these texts with the commentaries of Narboni (d. > 1362).[47] While Kumaṭiano adduced Islamic sources to which Maimonides had not referred, none were from the fifteenth century.

More important, Kumaṭiano cited a source that was unnamed in the base text *and* unavailable in Hebrew. Maimonides and Kumaṭiano both investigated how mathematics was suited to demonstrating the existence of things that were difficult or impossible to imagine, such as two lines that approach each other but never meet.[48] Such lines are conceivable on a curved surface. Kumaṭiano commented that while Apollonios's *Conics*, to which Maimonides referred, was not available, he came across a book by Heron of Alexandria (d. 70 CE), either in Arabic or Greek, with demonstrations that Kumaṭiano believed to be similar.[49] Other texts by Heron were sources for Kumaṭiano's work on geometry.[50] Thus, Kumaṭiano's sources were not wholly constrained by Maimonides's library or by the availability of Hebrew translations.

Rather, Kumaṭiano's inattention to fifteenth-century Ottoman Islamic thought was intentional, as evidenced by how frequently he took Averroës's position in debates with Avicenna, whose views predominated in fifteenth-century Islamic theology and philosophy. One such debate was over God's relationship to the outermost, nested orb of the heavens. Was that orb moved directly by God, which was Averroës's position, or by a separate celestial intellect, which was Avicenna's position?[51] While the question was relevant to astronomy, the answer lay in the field of natural philosophy. The locus for Kumaṭiano's intervention was Maimonides's statement in *Guide* I.70 that God resided upon the heavens and not in the heavens. Kumaṭiano argued that any discrete body could not be the source of the totality (*keʹlal*) of existence. Hence, God must be the direct mover of the outermost orb.[52] Kumaṭiano's citation of and concurrence with Averroës in the course of his comments indicates that Kumaṭiano's philosophy was, unlike much of Islamic thought in the fifteenth century in Istanbul, oriented away from Avicenna.[53] Rather, questions Kumaṭiano had about Ibn Ezra and Maimonides's texts led him to Averroës's philosophy, which was not in vogue with Muslim scholars in the Ottoman Empire.

The Unicity of the Intellect

Averroism was attractive to the merchants of knowledge and their contacts because, according to Averroists, knowledge brought the soul ever closer to God in a novel way. Averroist epistemology depended on the theory that there is a single human intellect, a position that Kumaṭiano began to explore in his *Commentary on "Yesod mora᾽."* The concept of the unicity of the intellect is challenging for modern readers to grasp, and the urgency of the debate during the time of the merchants of knowledge and their contacts is difficult for modern readers to appreciate. For the merchants of knowledge and their contacts, an acceptance of the unicity of the intellect was the most telling indicator of an overall commitment to Averroism.[54] The doctrine of the unicity of the intellect meant that all true human knowledge was shared.

The unicity of the intellect was understood by Jewish and Christian Averroists in the Renaissance in the following way. Some philosophers located human thought in the material intellect (Ar. *al-ʿaql al-hayūlānī*; also hylic intellect). This intellect was called "material" (or hylic) not because it was thought of as matter but because it was understood to share with matter the potential for being acted upon. The material intellect was thought to change as one gained understanding in the subsequent way: as individuals cogitated on particulars, the divine agent intellect (Ar. *al-ʿaql al-faʿʿāl*) provided universal abstractions to the hylic intellect.[55] For instance, to explain how, after viewing and sampling a variety of bagels, one acquired a universal knowledge of bagels, an Averroist would comment that the agent intellect was the source of that universal truth about bagels. Universal truths were believed to be immaterial, immutable abstractions unsuitable for corporeal beings. They could dwell only in the intellect. If there were multiple hylic intellects, say one for each individual, there would have to be multiple identical universals in the agent intellect. Also, were there multiple hylic intellects, there would have to be an infinite regress of causes, which Aristotelians believed was impossible. Averroists concluded that there was a single, eternal human hylic intellect. For an Averroist, the universal truth of bagels remained in the single, universal intellect. Averroës came to another surprising conclusion: if the single human intellect was eternal, then the human species was too.[56]

The unicity of the intellect was contested because much Jewish, Christian, and Islamic doctrine entailed that God created an eternal soul for each human at birth that is judged at the end of time.[57] Those who believed that each human has an individual soul that outlasts death rejected the doctrine of the unicity of the intellect and maintained that the individual souls grasped universals. Lived experience yielded other counterexamples to the doctrine of the unicity of the intellect: With a single intellect, how would forgetting be possible? When one person learns, why do not others cogitating on the same (or even slightly different) particulars always learn the same thing? If individual brains are laptops, Averroists failed to explain how some people have a better Wi-Fi connection to the cloud, i.e., the unified material intellect, than others. This question was compelling for the merchants of knowledge and their Christian contacts because gaining understanding meant contact with the agent intellect, hence beginning to think and know in the way God did.[58] In this way of thinking, humans had no independent capacity to reason without a connection to the cloud.

The doctrine of the unicity of the intellect developed out of an interpretation of a curious passage in *De anima* where Aristotle wrote (429a22–24) that "that part of the soul which is called the intellect (and I call the intellect that part by which we discern and cogitate) is not one of the beings in act before it understands."[59] Hence, one had no intellect before one grasped universals. Because the hylic intellect received universal forms from the agent intellect on the basis of abstraction from particulars, Averroës commented in his *Long Commentary on "De anima"* (*LCDA* hereafter),[60] which contained what Averroists believed was his most definitive statement, that "it is apparent that this nature is not a determinate particular nor a body."[61] Thus, if the hylic intellect was neither matter, nor form, nor something composite, then the hylic intellect was not subject to generation and corruption.[62] Hence, the hylic intellect was not a component of individual human bodies. If understanding was the acquisition of universals, which cannot be individuated at all, then the intellect had to be unitary and shared.[63] The puzzle for Averroës was how humans could be one in terms of final actuality by the intellect, since all educated people would know the same things, but multiple in terms of the first actuality of the hylic intellect, since when one person learned something, others did not.[64]

While cogitation occurred in the cloud, late antique and medieval philosophers wondered whether there was any processing capacity in individuals. Averroës responded to that question in *LCDA* by examining the views of Theophrastus (d. 287 BCE), Alexander of Aphrodisias (fl. ca. 200 CE), and Themistius (d. ca. 387 CE). Alexander of Aphrodisias[65] argued that the hylic intellect was a disposition in the human soul, while Themistius and Theophrastus contended that since intellection depended on sensations which themselves could not be eternal, the hylic intellect was an actual nonmaterial substance that entered humans at birth.[66] Averroës disagreed with his predecessors and concluded in *LCDA* that "there are three parts of the intellect in the soul, one is the receptive intellect, the second is that which makes [things],[67] and the third is the product [of these]. Two of these three are eternal, namely, the active[68] and the recipient; the third is generable and corruptible in one way, eternal in another way."[69] The "active" intellect in the preceding quote was the agent intellect, the eternal recipient intellect was the hylic intellect, and the third part, which he called the "theoretical intellect," was both eternal because it was part of the hylic intellect *and* corruptible because it depended on human sense perception.[70] The theoretical intellect accounted for individual differences. Herbert Davidson remarked that there was little Aristotelian about the doctrine despite Averroës's reputation as an arch-Aristotelian,[71] but in any case, Averroës's compromise was not the last word.

Paul of Venice (d. 1429), who taught at the University of Padua, was the first prominent Renaissance defender and expositor of the doctrine of the unicity of the intellect. Due to the controversial theological implications of the doctrine, while most Averroists believed that the hylic intellect and individuals were connected only during the process of understanding, Paul allowed that the hylic intellect contributed to an individual human's identity.[72] While Paul's modification of the doctrine of the unicity of the intellect was advantageous with regard to Christian faith and individuals' motivation to learn, Paul's successors continued to wonder what, exactly, Averroës thought. The debate over the role of the agent intellect was abstract, but the stakes were immense: the purpose of learning was conjunction with the divine agent intellect, which was a form of salvation.[73] Bland explained that "conjunction implies not only that we know differently, but that we exist differently when the objects of thought are no longer objects in the physical world but the self-reflective thoughts about those thoughts residing in the mind. . . . In this epistemic and ontological state of transformed human consciousness, humans begin to think and know in the way that the Active[74] Intellect thinks and knows. In turn, the Active Intellect resembles in an inferior way God's thinking which is understood to be a divine mode of knowing that transcends the distinction between particulars and universals."[75] If conjunction with the agent ("active" in Bland's words) intellect meant thinking in a way that approaches how God thinks, then only a unified human intellect conjoined with the agent intellect would be closer to the way God thinks. The doctrine of the unicity of the intellect was both tantalizing and an endless source of complexities and theological contradictions. Thus, Christians sought out scholars like the merchant of knowledge Elijah Delmedigo, an influential teacher, translator, and author of Averroist texts.

Elijah Delmedigo and Averroism

The merchant of knowledge Elijah Delmedigo's (ca. 1460–97) ability to translate the Hebrew versions of Averroës's texts into Latin, and to compose in Latin, accounted for why he was in demand among Renaissance luminaries such as Pico della Mirandola (d. 1494). Delmedigo had traveled from Candia to Venice by 1480, and by the end of 1480, he had completed his three Latin *Quaestiones* on Averroës's *Physics*. These questions were eventually printed with Jean de Jandun's (d. 1328) *Quaestiones* on the same text in 1488.[76] The print diffusion of some of Delmedigo's original compositions evidenced and enhanced his reputation.

Despite his talents and renown, Delmedigo was but a private docent or a *ripetitore* in Padua, someone who assisted students in studying, rather than an official professor.[77] Delmedigo's association with Pico began in 1481 in Padua, perhaps through the mediation of the Christian Cretan scholar Manuel Adramitteno.[78] Delmedigo instructed other Christian students besides Pico, but Pico was the dedicatee of many of Delmedigo's works.[79] At Pico's request, Delmedigo composed a Latin text on psychology entitled *Investigations in Accordance with the Principles of the Philosophers* in 1482. The text survives only in Delmedigo's Hebrew version (*D^{e}rushim k^{e}-p̱i shor^{e}shei ha-pilosop̱im*). This was Delmedigo's most sophisticated composition.

In *Investigations*, Delmedigo defended the Averroist doctrine of the unicity of the intellect against Thomas Aquinas's critiques[80] and contested Jean of Jandun's interpretations. Though Delmedigo agreed with Jean that the hylic and active intellects were two aspects of a single substance,[81] they disagreed regarding the details of how humans came to know.[82] Jean de Jandun believed that intelligible species, mental representations of the objects of cognition, were the first step by which cognition occurred.[83] Delmedigo countered that the intellect, with the involvement of the agent intellect, conceptualized the objects of cognition directly. Hence, the disagreement was over whether any conceptualization occurred without the agent intellect.

By 1485, Delmedigo relocated to Florence, and there is no evidence that he occupied an official professorial position there either, but Delmedigo's relationship with Pico continued. Marsilio Ficino (d. 1499), a renowned Renaissance Platonist, recorded Delmedigo's presence at Pico's house in a letter, and Delmedigo attempted to accommodate Pico's increasing interest in Plato.[84] Averroës's epitome of Plato's *Republic*, which Delmedigo translated from Hebrew into Latin in 1485 in Florence, corroborated the presentation of the unicity of the intellect found in the *LCDA*.[85] Delmedigo's views on the unicity of the intellect were mirrored in Pico's *900 Theses*, published in 1486, when Pico was twenty-three and still in contact with Delmedigo.[86] In the chapter on Averroës's conclusions, Pico stated:

> 7.2. The intellective soul is one in all men.
>
> 7.3. Man's greatest happiness is achieved when the active intellect is conjoined to the possible intellect as its form. This conjunction has been perversely and incorrectly understood by the other Latins whom I have read, and especially by John of Jandun, who not only in this, but in almost all questions in philosophy, totally corrupted and twisted the doctrine of Averroës.
>
> 7.4. It is possible, upholding the unity of the intellect, that my soul, so particularly mine that it is not shared by me with all, remains after death.[87]

Pico's attack on Jean de Jandun echoed Delmedigo's exasperated language,[88] and Pico highlighted in section 7.4 the potential for human felicity through the conjunction of the hylic and active aspects of the intellect.[89] Pico proposed to debate the theses publicly with anyone willing to embrace the challenge, meaning that he was willing to uphold philosophical positions that he adopted from his Jewish teacher.

Delmedigo's disagreement with Jean de Jandun and the influence of Delmedigo's Latin compositions on Christian readers meant that he participated in the late fifteenth-century phenomenon of Hebrew scholasticism, in which Jewish scholars were embroiled in the dialectical debates of medieval Christian philosophy. Hebrew scholasticism was a means by which Aristotelianism traversed borders.[90] Because a few medieval Hebrew technical terms were calques for the Latin, Jews must have been drawing on Latin sources by 1300. [91] Still, most fourteenth-century Jewish philosophers did not know the relevant Latin texts directly as Delmedigo did in the fifteenth century. Though he was a distinguished, influential scholar, Hebrew scholasticism remained robust after Delmedigo's death.[92] For example, translations of Averroës's texts from Hebrew into Latin by Abraham de Balmes (ca. 1460–

1523), a Jewish correspondent of Delmedigo's,[93] were published at Venice between 1522 and 1523.[94] De Balmes was the physician of Delmedigo's student Domenico Grimani (d. 1523), a Venetian patrician who became a cardinal. De Balmes, like Delmedigo, authored original works in Latin and Hebrew.[95] As the only merchant of knowledge who got published, Delmedigo had an exceptional reach. But he had peers among other Hebrew scholastics.

Hebrew scholasticism was not confined by Averroism. Outside of the network of merchants of knowledge and their contacts, David b. Samuel b. Shushan, perhaps the David b. Shushan esteemed by Muslims for his knowledge,[96] translated a Latin digest of Aristotle's natural philosophy composed by Thomas Bricot (d. 1516) into Hebrew as *Tol^{e}dot Adam* (Genealogy of Man).[97] The opening of the Bodleian manuscript of *Tol^{e}dot Adam* contains David b. Shushan's first-person account of the expulsion from Iberia.[98] On that basis we can deduce that David's original translation was produced in the 1490s.[99] Hebrew scholastics outside of the network of the merchants of knowledge and their contacts composed remarkable work.

Details of Delmedigo's participation in Hebrew scholasticism distinguished him as a merchant of knowledge. He spotlighted his philological skills in his published Latin works. His 1480–81 "Question on the Prime Mover" ("De primo motore"), along with other short texts, appeared in the 1488 printed edition of Jean de Jandun's *Quaestiones in libros physicorum Aristotelis*. Delmedigo, in his annotations on Averroës's *Physics* printed in that volume, pointed out that the turn of phrase *simplicium apprehensio* followed the Hebrew (*ha-koleil ha-muskal*) and (now lost) Arabic versions. [100] In "Question on the Prime Mover," Delmedigo drew attention to what he believed were helpful alternative expressions found in the Hebrew version of Averroës's *Physics*: "We speak also of the divine substance. Truly the origin and the first being, it is moved neither essentially nor accidentally. Rather, it sets in motion and produces the first motion as well. Moreover, it is, however, true that in the Hebrew[101] translation it is found thusly: 'We should say in this way, by what is set forth and defined, that the principle and the first of all beings moves neither essentially nor accidentally.' And that it moves with the first, eternal motion and it is not put (in motion)."[102] The Latin translation of Averroës's *Long Commentary on Aristotle's "Physics"* has been found to be less readable than the Hebrew translation.[103] Delmedigo's ability to refer to the Hebrew translation was a selling point of his work.

Through these printed Latin texts, Delmedigo's influence on Christians exceeded his web of personal contacts. Delmedigo transacted knowledge to Jews as well. He translated his Latin works, such as his *Investigations*, into Hebrew to highlight to Jews the weaknesses of Christian theology and to illustrate the areas of overlap between Averroism and Judaism.[104] As a merchant of knowledge he was attuned to the benefits of an exchange of philosophy for both his Christian patrons and the Jewish community. Hebrew scholastics like Delmedigo created a hybridized philosophy in order to mediate the exchange of Aristotelianism.

The Reverberations of Averroism for Astronomy

Because premodern philosophy had a wider ambit, encompassing more topics than modern philosophy, scholars in the Aristotelian tradition investigated how the celestial orbs move, questions that were not always explored in equal depth in theoretical astronomy. Just as the doctrine of the unicity of the intellect meant that humans knew in a way that resembled God's knowledge, the merchants of knowledge and their contacts probed God's most proximate relationship to the heavens.[105] Delmedigo composed his "Question on the Prime Mover" ("De primo motore") because he hoped to attract Pico's attention with a brief treatise on a lively topic of discussion at Padua.[106] In that text, Delmedigo investigated whether God moved the outermost celestial orb (or sphere) that was thought to contain the fixed stars—which make up the constellations and which were theorized to be in a single orb—with more than one motion. Though the question arose from natural philosophy, i.e., physics, explanations for how two motions proceed from a unitary God implicated theology. The wide-ranging Aristotelianism of the merchants of knowledge and their contacts was interdisciplinary.

In determining the mover of the outermost orb, Delmedigo took into account observations that showed that the fixed stars move with two motions. One observed motion was the daily twenty-four-hour motion from east to west, causing day and night, and the other was a slow motion of one degree every sixty-six to seventy years, depending on the source, from west to east. These observations led to an additional dilemma. If the stars were fixed in a single orb (the outermost orb), how could one mover produce two motions about two different axes? The daily motion is about the North Pole of the heavens while the slower motion from west to east is about a pole 24°

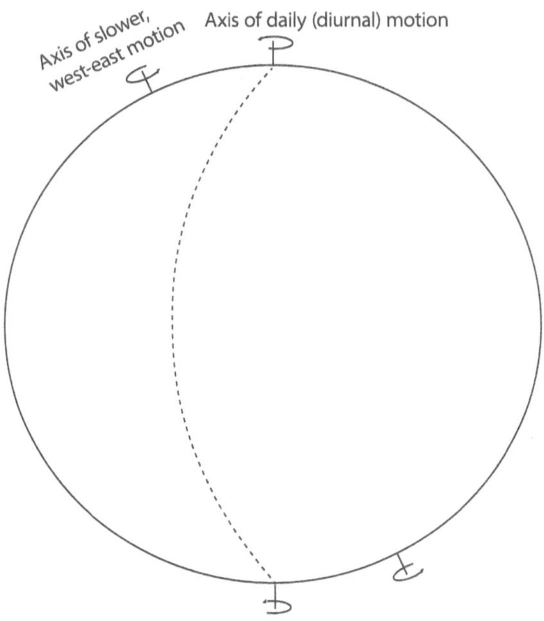

FIGURE 4.1. The diurnal and annual motions of the orb of the fixed stars.

away. The easiest answer, which most astronomers accepted, was that there must be separate movers (i.e., orbs) for those two motions of the fixed stars. But then, how could a single God be the cause of these multiple motions?

Delmedigo commenced his answer by stating that the inquiry into the first principle (*principio*) of all things was one about which many thinkers were uncertain (*dubitarunt*) and added that recent Latins had said that God did not move the first body, i.e., the outermost celestial orb.[107] Then, he elucidated Averroës's contrasting position that each orb had a separate soul that moved the orb by desiring an intellect associated with that orb. Because desires are often motivations, Averroists held that desire could be a source of motion. As God, the objection of intellection and desire, was "the unmediated mover of the first celestial sphere," i.e., the outermost orb,[108] the soul of the outermost orb, like the unified human intellect, desired God's thought.

Delmedigo entertained Maimonides's position, which was at odds with that of Averroës:

> Rabbi Moses[109] of Egypt argued in this manner, saying these words: "And it is not suitable that the intellect moving the first mobile should be the first cause that is self-necessary because, moreover, it should accord with the other intellects in a disposition, which is to move bodies." Like this therefore he argued, that just as there are different substances each of which one is the cause of the other, a single effect does not come from them. But the substance of the first principle differs very much from the substance of other intellects in terms of a first perfection and rank. It is the self-necessary cause of them. Therefore, it does not share with the remaining intellects in moving [bodies].[110]

To Maimonides, since the first mover was simple in a way that the other intellects were not, then, contra Averroës, the prime mover could not be the mover of the outermost orb. Rather, there was a first intellect caused by God; this first intellect moved the outermost orb. From Delmedigo's perspective, Maimonides's solution compromised God's unmediated relationship with the cosmos.

Delmedigo next turned to the question of the placement of these two orbs. The orb responsible for the diurnal motion must be the outermost orb because all the orbs, not just the orb of the fixed stars, participated in the diurnal motion. Though the orb with the stars would be beneath the starless orb responsible for the daily motion, the starred orb would nevertheless be nobler due to the presence of stars. [111]

Thus, according to Delmedigo, the motion of the starless orb was to be on account of (*propter*) the motion of the lower, starred orb. Hence, the mover of the outermost orb was not the prime mover. Delmedigo, who wanted to believe that God had an unmediated relationship with the outermost orb, found himself in a quagmire: God could not be the direct mover of the outer, unstarred orb if the motion of the unstarred orb was on account of the orb of the fixed stars nested beneath the outermost orb.

Later in "Question on the Prime Mover," Delmedigo returned to the question of which orb was the outermost and how God moved it. He wondered whether Averroës's position might be modified: "If, however, there should be proposed a starless ninth orb that is moved only with the daily motion, and if [its existence] is demonstrated, then it will be necessary to say that Averroës did not perceive it in his time to be for the motion of the fixed stars. But [it is found] above them."[112] Instead, Delmedigo revisited

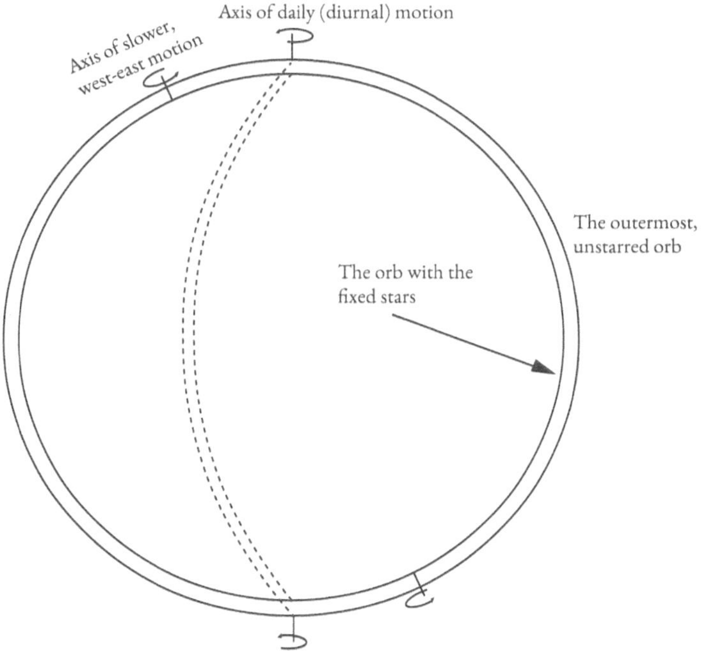

FIGURE 4.2. Separate orbs for the diurnal and annual motions of the orb of the fixed stars.

the principle that entailed that the heaven, i.e., the orb, existed on account of the stars: "But then it is proper to restrict the proposition saying that the heaven exists on account of the stars. For this reason, the motion of the heaven in which there is a star is on account of the star. These are my views." Because not all heavens (i.e., orbs) had to have stars, Delmedigo did not foreclose the possibility of the existence of a starless outermost orb responsible for the daily motion were it concentric with the other orbs. The uppermost, starless orb could be noblest due to its size and velocity. Thus, it was eligible to have an unmediated relationship with God. In this way, Averroës was reinterpreted to accommodate observations without jeopardizing God's unitary relationship to the heavens, which paralleled God's unitary relationship to the human intellect. Delmedigo's investigation illustrated the breadth of the transregional Aristotelian tradition, comprising topics that today would be considered scientific and theological, mediated by the merchants of knowledge.

Another matter of theoretical astronomy presented Delmedigo with an additional opportunity to instruct Pico in Averroist natural philosophy. A concern for mathematical precision affected how some scholars theorized the orbs that were believed to move the planets. Instead of attributing a planet's motions to a single orb, astronomers in the Ptolemaic tradition instead divided an orb centered on the earth into partial orbs called epicycles and eccentrics so that theories accounted more precisely for observations.

Delmedigo, in his commentary on Averroës's *De substantia orbis*, which he authored for Pico in 1485, elucidated physical contradictions that followed from theorizing epicycles and eccentrics. Epicycles and eccentrics did not share the center of the celestial orb, which meant that their existence jeopardized Averroës's distinction between the celestial realm, in which change did not occur, and the terrestrial realm, where it did. Additionally, the positions of partial orbs changed. Delmedigo elaborated: "It

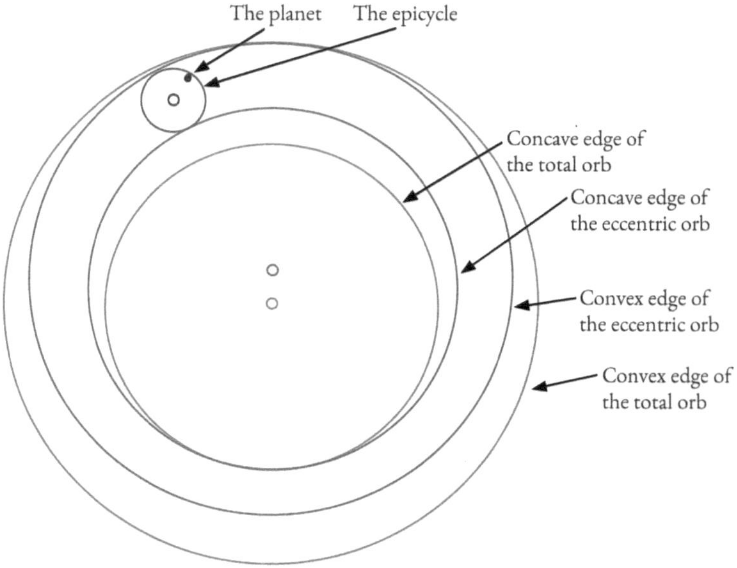

FIGURE 4.3. Partial orbs embedded in the total orb.

followed also that if they had different centers, then they would truly be of different species and, thus, subject to generation and corruption."[113] Pico echoed Delmedigo's concerns about eccentrics and epicycles.[114] Their mathematical value aside, epicycles and eccentrics jeopardized the theoretical elegance of the Averroist cosmos.

The Averroist debate over God's relation to the cosmos in the field of natural philosophy continued after Delmedigo's career ended. After Delmedigo's death, Agostino Nifo (Lat. Augustinus Niphus; 1473–1538) took up a question related to determining what caused the motion of the outermost orb.

Nifo studied at the University of Padua with Nicoletto Vernia, another of Pico's teachers at Padua, and later taught at Padua himself. In *De substantia orbis*, Averroës had asked whether, in a cosmos without epicycles and eccentrics, all of the orbs had to move in the same direction.[115] Recall that observations implied that the fixed stars moved with two motions in different directions about different poles. In his commentary on *De substantia*,

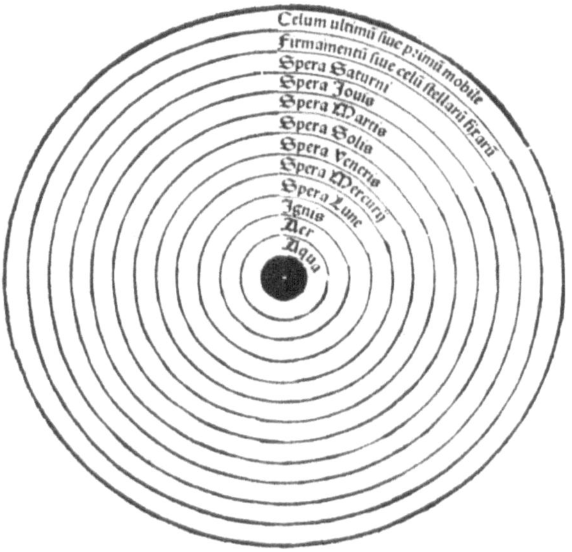

FIGURE 4.4. A cosmos of geocentric, homocentric orbs. Reproduced from Johannes de Sacrobosco, *Spera mundi*. Image courtesy of the Linda Hall Library of Science, Engineering, and Technology, Kansas City, MS.

first printed in 1508, Nifo agreed with Averroës that God, as first intellect, was both the motive cause of the heavens, i.e., the source of the heavens' existence and motions, and the agent cause, i.e., the heavens' proximate mover.[116] Presuming the heavens moved eternally, then the intellect that moved the heavens was eternal and, hence, divine. Because there could not be multiplicity in that divine intellect, Nifo agreed with Delmedigo that God could not cause two primary motions. Averroës's position in *De substantia* needed to be corrected or reinterpreted.

Still, Nifo considered the same interpretation that Delmedigo proffered in "Question on the Prime Mover": that the diurnal motion was the first motion and that the starless, outermost orb was responsible for the diurnal motion. According to Nifo, Averroës's response would be that "the motion of the first heaven and of all the orbs is one, that is to say diurnal. The rest are by accident and through secondary [causes]. It is absolutely not to be denied that different motions produce different specifics, thus in one animal there can be a plurality of motion in species (as he says himself) of which one is from the right to the left. The other is, on the other hand, opposite."[117] Hence, Nifo took a different tack and resolved that the first, diurnal motion of the heavens could be in the opposite direction of the other motions of the orbs because the other motions were accidental. As was true for Delmedigo, Nifo's Averroism[118] did not lead him to deny observations that suggested two motions for the fixed stars about different axes. The disagreements between Nifo and Delmedigo confirmed that a dynamic conversation about Averroist astronomy, and its agreement with observations, continued at Padua into the first decades of the sixteenth century. Delmedigo's influence broadened due to the printing of his texts. The interdisciplinary reach of these technical questions meant that Delmedigo influenced exchanges not only in physics and theology but also in astronomy and, as we shall see now, in astrology.

God's Knowledge of Particulars and the Averroist Denial of Astrology

Though the merchants of knowledge and their contacts lived in a late medieval and early Renaissance world in which God's existence was not in doubt, many of them questioned the way God interacted with creation. A significant debate among scholars, as we saw in chapter 2, was about the extent of God's control over the details of human lives. Frameworks of astral causation implied that God controlled, and thus knew, particulars, such as the identity and actions of individual humans. Averroist philosophers denied that God would have to know the details in order to control them. The doctrine of the unicity of the intellect was a starting point for investigations of how God's knowledge was different from humans' knowledge. God could not arrive at knowledge of universals through inference from particulars due to the concomitant that God would have learned and changed. And particulars were not present in the shared human intellect. Thus, the breadth of transregional exchange of philosophy mediated by the merchants of knowledge cannot be appreciated without further discussion of the relevance of the denial of God's knowledge of particulars. While Avicenna also denied God's knowledge of particulars, Averroists emphasized God's unicity. If God's knowledge comprised particulars, salvation through conjunction with the agent intellect might be jeopardized.

Delmedigo's discussion of the possibility of God's knowledge of particulars appeared in his 1482 *Investigations in Accordance with the Principles of the Philosophers*, dedicated to Pico. Delmedigo reported that Christian theologians posed the question as follows: If God is the prime mover, and if there are no other eternal intellects, then how could it be that God does not know particulars?[119] After all, the prime mover caused everything, and God ought to know that which God caused. But if God did know particulars, then would God's knowledge depend on corruptible, inferior beings?

Delmedigo selected an eclipse as a case study.[120] Because the sun and the moon, which cause eclipses, move eternally, an eclipse is a particular sort of particular. At the beginning of an eclipse, nothing has come to be, and at its end, nothing has passed away. Delmedigo first asked whether the knowledge that the eclipse would happen, is happening, and, then, has happened constituted a change in the knower because the knowledge itself changed. Because eclipses result from changes in the positions of the luminaries, Delmedigo concluded that the man who is to the right of someone and to the left of someone else is, nevertheless, the same man. Of course, if you did not know that the eclipse was going to happen, and then you see it happen, your knowledge would have truly changed, even though the change in knowledge was not due to a change in the essence of the luminaries but only in their relationship (Heb. *ṣeirup̱*) to each other. But neither the essence of a human who understands eclipses nor the essence of knowledge changed.[121] Because eclipses are perfectly predictable if one perfectly understands the motions of the sun and the moon, Delmedigo concluded that eclipses are an example of how God's knowledge was unitary, not even divided by universals.[122] Still, God's knowledge of particulars could not be likened to human knowledge.[123]

Though God did not know particulars as humans do, might God still cause particulars? In societies in which astrological forecasting was widespread, this controversy was unavoidable because connections between the heavens and the earth could be a way for God to cause events on earth. In a lengthy remark in his commentary on Averroës's *De substantia orbis*, Delmedigo explored the connections between the celestial and terrestrial realms. To explain the panoply of variations observed on earth, astrologers theorized that different planets had different natures. But according to Aristotle and Averroës, the celestial realm (i.e., the orbs, planets, and fixed stars) was made of the same matter. Since the planets were visible but not the orbs, Delmedigo determined that more explanation was necessary.

In the base text, Averroës rationalized how the planets differed from yet shared the body of the celestial orb that carried them: "For it seems that density and rarity are the cause of transparency and nontransparency . . . the cause of the luminosity of the parts of the celestial bodies, that is the stars, is the density of that particular part of the celestial sphere occupied by the star that is transparent in actuality."[124] Despite appearances, Averroës did not differentiate the planets from the orbs in matter or form. Likewise, in his commentary on the *Meteorology*, Averroës mentioned parts of the heavens with so many stars so close together that the orb illuminated the vapors just below.[125] Luminosity was due to greater density, not to a difference between the matter of the orb and the planet.

Delmedigo in his commentary on *De substantia orbis*, which survived in Hebrew and Latin versions from 1484, refined Averroës's explanation of how the matter of the celestial realm was uniform.[126] The orbs, he wrote, had different levels of rarity (Lat. *raritas*; Heb. *sᵉp̱ogiyyut*), which correlated with increased transparency (Lat. *diaphanitas*; Heb. *sappiriyyut*).[127] The opposite of rarity was density (Lat. *densitas*; Heb. *moqshiyyut*), which correlated with being luminous (Lat. *lucidus*; Heb. *meiʾir*). Averroës proposed that both density and rarity were different accidents in the heavens, citing the different colors of the Milky Way as evidence that luminous bodies were not all equally dense.[128] Delmedigo countered that since an eclipse was caused by one planet shading the light of another, shading could come only as a result of density, not rarity. Hence, not every dense body was luminous.[129] Rather, a body that was naturally luminous was divided between density and rarity. Delmedigo's influence on Pico extended to this position on the composition of the orbs, which Pico adopted in his "Philosophical Conclusions."[130] Yet Averroist philosophical principles might not account for the differences between the planets' natures proposed by astrologers. Hence, astrology might not be a viable rationalization of God's knowledge of particulars.

Merchants of knowledge had a special role to play in resolving these high stakes questions. Delmedigo argued that defining the cause of the planets' luminosity might have been complicated by others' errors in translation. According to Averroës, Aristotle asserted that the moon was more like the earth than other planets because the moon was not at all self-luminous.[131] Delmedigo opposed Averroës on this point because, since the moon is visible during an eclipse when it does not receive light from other planets, he determined that the moon must be self-luminous like the other planets.[132] That is, it must be self-luminous even if it is not as self-luminous as the other planets. Delmedigo wondered whether Averroës's failure to consider that the moon is visible while eclipsed was due to an error in the Latin translation and stated that if he were to obtain a Hebrew translation of *De substantia*, he would consult it.[133] Delmedigo was aware of what could be lost in translation.

Once Delmedigo pinpointed why differences in luminosity could not arise from differences in the orbs' matter, he turned to the question of whether variations in calefaction (i.e., the heating of terrestrial bodies by celestial ones) were an alternative way for a unitary prime mover to bring about multiple terrestrial events. Delmedigo identified two reasons for the variations in the strength of the heat generated by the celestial orbs for the terrestrial realm: "The first is that the heat of the sun, the rest of the planets in motion, and the proximity and distance of the sun from us are the reasons for the strength and weakness of the heat. The other [reason] is the light and the sparks. The lines of the sparks, when they are straighter or closer to being straight, are a reason for the strength of the heat."[134] Still, variations in calefaction accounted for some terrestrial phenomena but not the minutiae of humans' decisions. For example, one might experience extreme thirst due to the heat of the sun, but the decision about whether to drink remained one's own. One could also choose to stand in water that the sun does or does not heat. According to Delmedigo, the heavens might give someone the aptitude and disposition to excel in philosophy, but the heavens did not and could not give that person philosophy itself.[135] Things were organized by the heavens but not necessitated by them.[136]

Pico della Mirandola's attacks on astrology (chapter 2) depended on his engagement with the Averroist philosophy in which Elijah Delmedigo instructed him. Without Delmedigo's tutelage, Pico's religious opposition to astrology would not have been buttressed with physical arguments.[137] Pico agreed that terrestrial events could not be explained by differences in calefaction[138] and even cited Averroës's critique of astrology at the beginning of his posthumous *Disputationes adversus astrologiam divinatricem*.[139] He argued that calefaction resulting from the velocity of celestial motions was not part of important causal chains in the terrestrial realm: "The stone would not be moved downward, nor fire upward, by a resting heavens; and it will not descend there or ascend here more rapidly because the heavens be revolved more rapidly."[140] Heavenly control over particulars, he claimed, detracted from the significance of God's all-encompassing role as the mover of the heavens.[141] According to Pico, celestial heat served only the good, vivifying the terrestrial realm.[142] Although astrology did much work in the world of the merchants of knowledge and their contemporaries, the concomitant of astrology that God would know particulars contradicted the logic of the doctrine of the unicity of the intellect. Pico and Delmedigo respected the implications of the denial of God's knowledge of particulars for natural philosophy.

In addition, Delmedigo's interpretation of the physics of calefaction, which Pico accepted, militated against the existence of epicycles and eccentrics. If epicycles and eccentrics existed, the sun would be farther from the earth in the northern hemisphere's summer, suggesting that the summer should not be as hot as it actually has been observed to be.[143] Alessandro Achillini (d. 1526), a later Averroist philosopher at Padua, concurred that variations in the distances of the sun from the earth were not necessary for variations in calefaction.[144] He held that Averroës's earlier acceptance of epicycles and eccentrics was instrumentalist. Thus, the Averroist denial of God's causation of particulars, a consequence of the doctrine of the unicity of the intellect, became implicated in their cosmology. The unicity of the intellect was the flagship doctrine of Renaissance Averroists, but Delmedigo engaged his contacts and readers on a wide variety of related philosophical questions.

Galeano/Jālīnūs's Novel Islamic Sources on God's Knowledge of Particulars

Galeano/Jālīnūs was the final merchant of knowledge to enter the debate about God's knowledge of particulars. With him, the conversation shifted to the Ottoman Empire and away from the Averroism espoused by Delmedigo, his contacts, and readers. Like earlier merchants of knowledge and their contacts, Galeano/Jālīnūs recognized the stakes of the debate. Averroists presented compelling arguments for why knowledge of particulars compromised God's unicity. Because particulars come to be and pass away, God's knowledge of particulars could not be unchangeable and eternal. Because Galeano/Jālīnūs could not escape the tug of a belief in a personal God cognizant of the particulars of one's life, we saw in chapter 2 how Galeano/Jālīnūs honed techniques of astrological forecasting in *Puzzles of Wisdom* (Heb. *Taʿalumot ḥoḵmah*). Galeano/Jālīnūs mirrored the eclecticism of Abraham Ibn Ezra in his pursuit of fascinating material in order to intervene in the debate over whether God knew particulars.

The longest discussion of philosophy in *Puzzles of Wisdom* covers four folios and addresses the question of God's knowledge of particulars. Galeano/Jālīnūs observed that

> the error occurring with respect to the universal and the particular is arguing that the absence of God's oversight of particulars is due to a change that would follow in His, may He be exalted, knowledge. For if someone of righteous intention were saved from death, though before his righteous action he had been ordained to die then His, may He be exalted, knowledge would have changed from what it had been. The error in this is that this scholarly questioner thought that our knowledge of things resembles the knowledge of God, may He be exalted.
>
> It is not like this for His ways are elevated over ours and His thoughts over ours. God's way of knowing things and God's oversight of them is according to how they emerge from him. He, may He be exalted, is their first cause just as the heavens are a cause of lower things. Thus, the thoughts of God, may He be exalted, differ from our thoughts since His knowledge is the reason for the actualization of things, not that His knowledge comes from things, as is the case with our knowledge. As God, may He be exalted, is, with total unicity, the reason for all exist-

> ing things and the particulars of their particulars, though by means of many intermediaries, then all existents are subject to God's order. God's providence is separate from the potential for multiplicities. That is to say, multiplicity is according to the recipients and is due to the different dispositions of the recipients owing to their ranks. The action of God, may He be exalted, is unified and total and it perfects and positively influences everyone according to His convention. Through this it is said that He knows all, but only in a general way, not in a particular way, as if he caused them through intermediaries.[145]

Galeano/Jālīnūs, like Delmedigo, his Christian contacts, and readers of his printed works, had accepted the Averroist denial of God's knowledge of particulars and affirmation of God's total unicity. For Galeano/Jālīnūs, as for Delmedigo, the explanation was that multiplicity on earth arose from multiplicity in the recipients, not from any multiplicity of the celestial causes.

A few folios later, Galeano/Jālīnūs returned to the question of how humans could possibly learn more about God's knowledge if God knew in a different way than humans. That Averroists were not the only Aristotelians who denied God's knowledge of particulars was relevant to Galeano/Jālīnūs's intervention. Galeano/Jālīnūs responded to Maimonides's urging to be satisfied with ignorance of the answer[146] by introducing an Islamic source that he believed to be new to his readers. Galeano/Jālīnūs wrote:

> I always used to find difficult the response of Maimonides to this problem in his commentary on *Pirqei Aḇot*.[147] When I reflected, it appeared to me that the aforementioned concerned our response that preceded, and Fārābī said in his commentary on *De interpretatione*[148] that the true answer[149] for this is saying that something following necessarily from something else does not mean that the thing that follows is necessary in itself. If they believed what is said, the thing that followed followed from it by the necessity of the thing's existence. It does not follow from this that the thing necessarily exists on its own, but that its following [from something else] is due to the truth of what is said to be necessary. Its following on its own is not necessary.[150]

Though God caused all particulars, Galeano/Jālīnūs explained, only God's existence was necessary, so God did not know contingent particulars. Fārābī was an important source for Maimonides and Avicenna, and the long quotation from Fārābī's *De interpretatione* concludes: "This opinion is more useful for religions (*datot*) than the opinion of one who sees otherwise."[151] According to Fārābī, a particular meteorological and terrestrial event might follow from God's knowledge, but one should not conclude that God's knowledge comprises or necessitates those events. This quotation from Fārābī helped Galeano/Jālīnūs solve the problem of God's knowledge of particulars.

What was Galeano/Jālīnūs's source? Comparison of the Hebrew text of *Puzzles* with Arabic texts by Fārābī indicates that Galeano/Jālīnūs must have accessed Fārābī's *Long Commentary on "De interpretatione."* The introduction and some quotations from the *Long Commentary on "De interpretatione"* were translated into Hebrew,[152] but even those quotations were not necessarily available to Galeano/Jālīnūs and were, in any case, too brief to be the source for the long passage he reproduced in *Puzzles*. Little suggests that a Latin version of the abridgment of Aristotle's logical corpus was Galeano/Jālīnūs's source.[153] It is possible that there was a now lost Hebrew complete version of the long commentary available to Galeano/Jālīnūs, but it is most likely that he translated from Arabic. As we saw in the previous chapter, translation was part of the commerce in knowledge just as translation mattered for commerce, as we learned in chapter 1. His presentation of a relevant text unavailable to readers of Hebrew exemplifies how Galeano/Jālīnūs melded the roles of translator and interpreter as a merchant of knowledge. To Muslim scholars, Fārābī's logic would have been old hat, but given Kumaṭiano's lack of awareness of Islamic philosophy and *kalām* beyond what was available in Hebrew, Galeano/Jālīnūs's translation from Fārābī's *Long Commentary on "De interpretatione"* must have been novel for readers of Hebrew. In chapter 1, we learned that his family was less wealthy and prominent than other merchant-scholar families; perhaps his more marginal, interstitial position was what motivated him to search harder for new sources.

Additional evidence also indicates that Galeano/Jālīnūs was more motivated than other Jews with connections to the Islamic world to read widely in logic. He produced an original text on logic, entitled *Mel'eḵet ha-higgayon* (The Art of Logic).[154] Jews from the Islamic world like Galeano/Jālīnūs barely wrote on logic, so while we do not know where and when he composed this text, its mere existence is significant.[155] In *Mel'eḵet ha-higgayon*, Galeano/ Jālīnūs explained that the study of logic was the key to other sciences (*ḥoḵmot*).[156] Logic for the intellect was akin to grammar for language. Galeano/Jālīnūs mentioned in *Melʾeḵet ha-higgayon*, in the course of a critique of the followers of Thomas Aquinas and Duns Scotus,[157] a certain אמאם מולכס (Imām Mulakhkhaṣ). This figure was Fakhr al-Dīn al-Rāzī (d. 1210), a scholar who drew on the work of Avicenna, and the author of *al-Mulakhkhaṣ fī al-ḥikma* (A Concise Exposition of Philosophy). In addition, *Melʾeḵet ha-higgayon* contained translations of short passages from the section on logic from *al-Mulakhkhaṣ* (a.k.a. *Manṭiq al-Mulakhkhaṣ*; The Logic of "The Mulakhkhaṣ") into Hebrew.[158] The inventory of the Ottoman Sultan Bayezit II's (d. 1512) books include that volume, so it is plausible that Galeano/Jālīnūs learned of that text at the Ottoman court.[159] Other Jewish scholars knew of Fakhr al-Dīn Rāzī, but Galeano/Jālīnūs was the only one known to refer to and quote from Rāzī's *Mulakhkhaṣ*.[160]

In *Melʾeḵet ha-higgayon*, Galeano/Jālīnūs cited Walter Burley and Marsilio of Inghen,[161] other scholars who wrote originally in Latin.[162] Mordechai Kumaṭiano, in his commentary on Maimonides's *Millot ha-higgayon* (The Terms of Logic), the Hebrew translation of *Maqāla fī Ṣināʿat al-manṭiq*, also cited scholastic Latin texts.[163] But Galeano/Jālīnūs innovated by integrating his quotes from Rāzī's *Mulakhkhaṣ* into debates about Latin logic texts.[164] That is, Galeano/Jālīnūs was likely motivated by Hebrew scholasticism to seek out sources that would have been novel for Jewish readers. One might see a parallel in the Byzantine Christian Gennadios Scholarios's (d. after 1472) project to translate Latin Scholastic logic texts into Greek, which he integrated into his defense of Aristotelianism against Plethon's attacks.[165] Scholarios was most likely Romaniot Jews' intermediary with the sultan before 1453–55.[166] The discipline of logic influenced Galeano/Jālīnūs in other ways too. *Puzzles* was not organized according to discipline or topic but according to errors in reasoning. Langermann observed that the organization of *Puzzles* around logical fallacies means that logic exemplified to Galeano/Jālīnūs "the unity of all disciplines."[167] For Galeano/Jālīnūs, new sources enhanced both his intellectual capital and his reputation as a merchant of knowledge.

Conclusions

The merchants of knowledge mediated an expansive conversation about Aristotelian philosophy that spanned the logic of Fakhr al-Dīn Rāzī to the philosophy of Pico della Mirandola. The exchange of Aristotelianism among the merchants of knowledge and their contacts was always multidirectional and multilayered. The oeuvre of Abraham Ibn Ezra, the most influential thinker for Romaniot Jews, was facilitated by the influx of Islamic philosophy and science into Jewish cultures. The interest of the students of Mordechai Kumaṭiano in Ibn Ezra's integration of science and philosophy into Judaism was paralleled by the exchange of manuscripts. And the acquisition of Kumaṭiano's commentary on Maimonides's *Guide of the Perplexed* by elite Ottoman Muslim scholars paralleled Kumaṭiano's donation of a universal astrolabe plate that I described in the previous chapter. Yet, because Avicenna exerted a greater influence on Ottoman Islamic philosophy than on the merchants of knowledge, Kumaṭiano did not engage in a reciprocal manner with fifteenth-century Ottoman Islamic philosophy.

Because Aristotelianism was multivocal, the merchant of knowledge Elijah Delmedigo and his Christian contacts were absorbed in the Averroist school, concerned with the doctrine of the unicity of the intellect and its reverberations. Delmedigo's Latin Averroist treatises on the intellect and cosmology attracted a wide readership, and he mentioned his Christian adversaries in the Hebrew versions he produced for Jewish readers. Delmedigo's linguistic skills made him a valuable intermediary in both directions.

So far in this recap, the Averroist layer that occupied Jews and Christians has barely intersected with the Avicennan layer that shaped fifteenth-century Islamic philosophy in the Ottoman Empire. But in the final part of this chapter, we see that Galeano/Jālīnūs translated from Arabic into Hebrew an Islamic philosophical text on God's knowledge of particulars, a question that preoccupied Delmedigo and his Christian contacts. In fact, Galeano/Jālīnūs distinguished himself among the merchants of knowledge through his knowledge of Arabic and Latin logic. His ambition and command of Arabic and Hebrew enabled him to link the nonintersecting Averroist and Avicennan layers of Aristotelianism. The stakes of these philosophical questions spurred this remarkably capacious and multidirectional exchange.

Looking forward, the exchange of Aristotelianism between Jewish and Christian scholars bled into conversations about *Qabbalah*. In the next chapter, we will see how the possibility of harmonizing Aristotelianism with *Qabbalah* was an axis of exchange and a zone of contention between the merchants of knowledge and their Christian and Jewish contacts.

FIVE

ARISTOTELIANISM *and* QABBALAH

The exchange of philosophy mediated by the merchants of knowledge was geographically wide-ranging because it was disciplinarily capacious. Some of the philosophy that was exchanged intersected with Judaism's esoteric teachings, often called *Qabbalah* after the twelfth century. Jews interested in *Qabbalah* sought meanings and knowledge not available to all, as well as a religious experience unattainable through the intellect alone. *Qabbalah* appeared in Iberia and Provence in the twelfth century and spread quickly.[1] *Qabbalah* has been disseminated in the modern world, outside the Jewish community, as a New Age religion. In modern progressive Jewish communities, as traditional barriers to the study of *Qabbalah* have fallen away, it has been invoked as an alternative progressive path within Judaism and opened to study by all. But the word *Qabbalah* means "tradition," and its foundational texts, despite their medieval origins, were communicated to the reader as the received wisdom of revered Jewish sages from Late Antiquity. The merchants of knowledge recognized that *Qabbalah* was in conversation with other fields of Jewish learning.

Though there are connections between *Qabbalah* and Christian mysticism and Sufism, all three traditions should be studied as discrete historical phenomena, not as instantiations of a universal mysticism. Apparent commonalities obscure important differences, such as how many Qabbalists did not hold that self-effacement and self-surrender were necessary to gain esoteric knowledge.

Esoteric speculation in Judaism began by the dawn of the Common Era. The terms *ma'asei b^e^reishit* (the acts of creation)—questions of cosmogony unanswered in Genesis 1 (e.g., Did God create heaven or earth first?)—and *ma'asei merkaḇah* (the acts of the chariot; a reference to Ezekiel 1)—esoteric teachings about the visible manifestations of God based on Ezekiel's visions—appeared in the Talmud, a postbiblical compendium of Jewish law and lore that is foundational for Jewish practice and study. The presence of those terms, among others, is decisive evidence that esoteric teachings were found in the Judaism of Late Antiquity and that esoteric ideas and practices were not a mistaken divergence from proper Judaism. Rather, esoteric speculation was sparked by biblical passages and was integrated into a text focused on the exoteric. Likewise, exchange of *Qabbalah* blended with exchange in other fields.

The first of *Qabbalah*'s foundational texts was *Seiḇer Yᵉṣirah* (Book of Creation), a text of unknown authorship probably composed before or during the ninth century CE. The contents have been attributed to the patriarch Abraham, and in one manuscript copy, the text was referred to as the laws of creation, evoking the authority of biblical law.[2] The *Zohar* (The Book of Splendor), another foundational work of *Qabbalah*, was composed in the thirteenth century most likely by Moses de Léon, in Aramaic, the language of the Talmud. Because the author wanted *Qabbalah* to be perceived as grounded in tradition, biblical and Talmudic, he attributed the contents of the *Zohar* to the second-century rabbi Simon bar Yoḥay and his disciples, figures cited in the Talmud.

Just as the authors of foundational texts of *Qabbalah* integrated the Bible and the Talmud into their work, the merchants of knowledge harmonized *Qabbalah* and Aristotelianism. Steven Katz explained that "from Crescas to Mendelssohn, no Jew dared venture into the field of metaphysical speculation without treading the approved pathways of Qabbalah."[3] Along these lines, the merchant of knowledge Elijah Delmedigo presented *Qabbalah* as a mode of contemplation connected to Aristotelianism and a hermeneutic for deepening one's appreciation of Jewish ritual. His Christian student Pico della Mirandola demurred and held that *Qabbalah* was a discrete dimension of Judaism that confirmed Christianity.

Qabbalah was multifarious, and it is difficult to identify a central belief or practice. But two concepts from *Qabbalah* frequently surfaced in exchanges among the merchants of knowledge and their contacts. The first was *ein sop̱* (lit. "the infinite"), a term that denoted the aspect of God that was transcendent and unknowable. Isaac of Acre (fl. thirteenth to fourteenth century), in an influential commentary on *Seip̱er Y*^{e}*ṣirah*, defined *ein sop̱* as divinity that was beyond human contemplation.[4] The second concept, introduced by the author of *Seip̱er Y*^{e}*ṣirah*, was that there are ten divine forces immanent in creation called *s*^{e}*p̱irot* (sing. *s*^{e}*p̱irah*). In the *Zohar*, the *s*^{e}*p̱irot* symbolized how God emerged from hidden nothingness. In a comment on the first chapter of *Genesis*, we read:

> A spark of impenetrable darkness flashed within the concealed of the concealed, from the head of Infinity—a cluster of vapor forming in formlessness, thrust in a rung, not white, not black, not red, not green, no color at all. As a cord surveyed, it yielded radiant colors. Deep within the spark gushed a flow, splaying colors below, concealed within the concealed of the mystery of *En Sof*. It split and did not split its aura, was not known at all, until under the impact of splitting, a single, concealed, supernal point shone. Beyond that point, nothing is known, so it is called ראשית (*Reshit*), *Beginning*, first command of all.[5]

In this passage, the *s*^{e}*p̱irot* were the colors, and *ein sop̱* was beyond that supernal point from which the *s*^{e}*p̱irot* flowed. The *s*^{e}*p̱irot* could be immanent in creation in more than one way.

Some Qabbalists interpreted the *s*^{e}*p̱irot* as powers that mediated the divine in creation, as the divine essence, or as an instrument of the divine.[6] In this view, God created and maintained order through the *s*^{e}*p̱irot*, which could be arranged in the form of a man or a tree.

Qabbalists disagreed about whether the *s*^{e}*p̱irot* were hypostases, entities through which one may know and affect God. A belief in hypostatic *s*^{e}*p̱irot* was heretical to some because the existence of hypostatic *s*^{e}*p̱irot* entailed multiple divine entities. Others denied that the *s*^{e}*p̱irot* were hypostases and thought of the *s*^{e}*p̱irot* as primordial numbers and letters.[7] As an example of a nonhypostatic interpretation of the *s*^{e}*p̱irot*, in which they were letters, the author of the *Zohar* told of God creating through language.[8] The *s*^{e}*p̱irot* could have concrete applications. The Qabbalist Joseph Gikatilla (d. 1305) explained the protective power of the mezuzah, parchment with verses from the Torah affixed to the doors of the houses of Jews, in terms of the

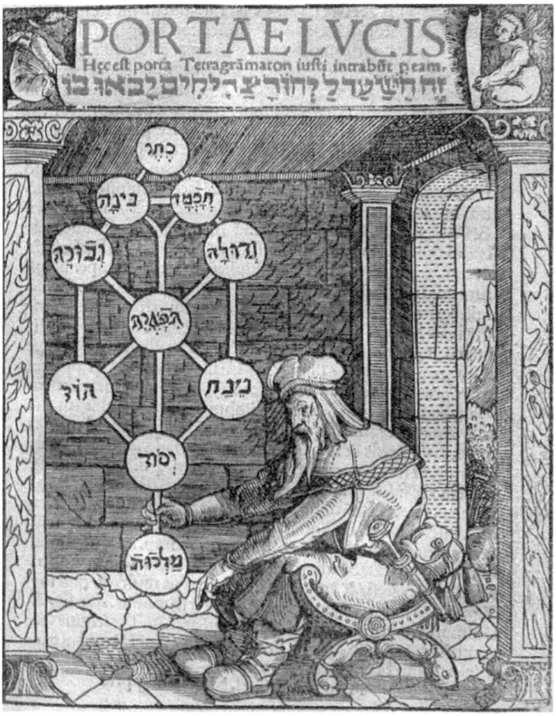

FIGURE 5.1. *s*^{e}*p̱irot* on the frontispiece of the Latin translation of *Sha'arei Orah* (The Gates of Light). Joseph ben Abraham Gikatilla, Paulus Riccius, Leonhard Beck. Public domain via Wikimedia Commons.

s^{e}*p̱irot.* [9] According to the author of the *Zohar*, the mezuzah drove away evil entities. Qabbalists contested the definition and function of the *s*^{e}*p̱irot.*

Despite *Qabbalah*'s Jewish origins, some Christians interpreted it to accommodate Christianity. After Christians took an interest in *Qabbalah* in the thirteenth century, *Qabbalah* impelled Christian scholars to seek out Jewish or formerly Jewish informants. Renaissance Christian scholars of *Qabbalah* believed that it represented the *prisca theologia*, an ancient theology that was the root of Christianity. They used *Qabbalah* to attract Jews to Christianity and to bolster the Christian perspective that Jesus superseded Jewish law.[10] The merchants of knowledge grappled with these entreaties and arguments. They also had to reckon with the potential challenge *Qabbalah* posed to other intellectual approaches to Judaism when *Qabbalah* was imported to the Eastern Mediterranean by Sephardic Jews fleeing the Iberian Peninsula. [11] For instance, the authors of *Qabbalah* texts entered debates over the rationales for the commandments (*ṭaʿamei ha-miṣwot*), much as Ibn Ezra did in *Yᵉsod Moraʾ*.[12] The question for the merchants of knowledge was whether the answers to the questions found in *Qabbalah* texts complemented or contradicted those of Ibn Ezra. Ultimately, any qualms that merchants of knowledge had about *Qabbalah* were balanced by the opportunities that became available for patronage from interested Christians.

The result was that the merchants of knowledge put *Qabbalah* in conversation with Aristotelian philosophy. By tracking the contours of this conversation between the merchants of knowledge and their contacts, we gain an unprecedented, granular perspective on patronage relationships. The chapter begins with an analysis of a debate in which Averroist (an Aristotelian school) philosophy was adduced in support of the qabbalistic doctrine of metempsychosis and against astrology.

Metempsychosis, *Qabbalah*, and Averroism in the Debate about Metempsychosis on Candia

In the second half of the fifteenth century, two merchants of knowledge, Moses Cohen Ashkenazi (the author of the astrology summa entitled *Urim wᵉ-ṯummim*) and Michael Balbo, drew on Aristotelian philosophy when they debated the qabbalistic doctrine of metempsychosis. Metempsychosis (Heb. *gilgul*) was the doctrine that souls can pass from a deceased person into another body and, for some scholars, even into the bodies of animals.[13] Some Jewish thinkers advocated metempsychosis before the rise of *Qabbalah*.[14] Qabbalists used the doctrine of metempsychosis to make broad sense of the apparent lack of justice in the world. Events that an astrologer attributed to the stars were instead interpreted through the doctrine of metempsychosis to be the result of one's deeds in a past life.[15] This belief that souls were shared facilitated new interpretations of scripture and human experience.[16] Because the doctrine of metempsychosis was presumed without argument in major sources of *Qabbalah*—beginning with the early thirteenth-century *Seiper ha-Bahir* (The Book of Brightness)—either the doctrine is older than the opposition to it or the authors of those *Qabbalah* texts had nothing to do with the scholars who opposed it.[17] Since Aristotelian philosophy neither confirmed nor denied this aspect of *Qabbalah*, the two intellectual fields were not necessarily opposed.

The debate about metempsychosis between Cohen Ashkenazi and Balbo in the 1460s on Candia was notable for its interdisciplinary character. Both scholars reflected on the practice of *yibbum* (levirate marriage), in which the surviving unmarried brother of a man who died childless should marry the deceased's widow so that the deceased brother's name and, according to Balbo,[18] the deceased brother's soul could live on through their offspring.[19] Moses Cohen Ashkenazi initiated the debate by denying metempsychosis. He opposed the doctrine because, from his perspective, the encouragement to practice *yibbum* rendered metempsychosis impossible.[20] If metempsychosis occurred, then the deceased's soul could pass to *anyone* else's body, not just the offspring of his widow and brother. Since *yibbum* was necessary to keep the deceased's soul from being erased, metempsychosis did not occur.

By contrast, Michael Balbo defended the doctrine of metempsychosis, but his commitment to *Qabbalah* was not clear at first blush. Though his father Shabbetai Balbo copied an important manuscript of the *Zohar*,[21] there was a paucity of references to the *Zohar* and other texts of *Qabbalah* in Balbo's defense of metempsychosis.[22] However, Michael Balbo entertained a broad range of justifications for metempsychosis in his writings and, in so doing, elucidated the multiple relationships of *Qabbalah* to other disciplines. For example, in a brief treatise written for a student of *Qabbalah* in which he defended metempsychosis, Balbo explained that terrestrial cycles of generation and corruption were caused by the motions of the celestial orbs (*gilgulei ha-galgalim*), and that an understanding of the philosophers' ideas about the human soul—e.g., that it receives universals from the agent intellect—were necessary to understand his arguments.[23] Hence, it should not be surprising to find that Balbo, in the controversy over metempsychosis, relied on reasoned proofs more than on citations from classics of *Qabbalah*.[24] Philosophical arguments were required to counter Cohen Ashkenazi's assertions, and since Balbo's father also copied several classics of philosophy, philosophy must have been part of Balbo's formation.[25]

In addition, Balbo was partial to the *Qabbalah* of Abraham Abulafia (d. after 1291),[26] who allowed "the union of the actualized human intellect and the divine intellect"—a position close to the Averroist doctrine of the unicity of the intellect.[27] Abulafia held that the *s^e^pirot* might represent divine or even human attributes but not hypostases.[28] Still, while *Qabbalah* could be harmonized with Averroism,[29] and even though Balbo's arguments were not replete with references to texts and concepts from *Qabbalah*, *Qabbalah* was superior as far as Balbo was concerned.[30]

The results of the debate emanated from Crete. The Istanbul-based merchant of knowledge Rabbi Moses Capsali learned of the debate as Balbo sent him separate booklets, one for each scholar's position.[31] The dispatch of the results to Istanbul was no coincidence as Balbo was connected to other scholars in Istanbul, such as Shalom ʿAnab̲i, a mathematician.[32] Balbo also composed a dirge on the Ottoman conquest of Constantinople in 1453. Christians were interested in the debate too. The Fuggers acquired both Cohen Ashkenazi's record, now preserved in Vatican MS 254, and Balbo's record, now found in Vatican MS 105.[33] Both codices include other texts. The reverberations of the debate were not confined to an island.

In the milieux of the merchants of knowledge and their contacts, Balbo was not the first to blend esoteric thought with other disciplines. Rabbi Mordechai Kumaṭiano (d. 1482), a leading scholar in Istanbul during Balbo's lifetime, referenced texts of *Qabbalah* in some of his own writings. For instance, Kumaṭiano discussed *Seip̲er Yᵉṣirah* in his commentary on Ibn Ezra's *Yᵉsod Moraʾ*, a treatise on the rationales for the commandments. In chapter 12 of *Yᵉsod Moraʾ*, where Ibn Ezra attributed changes on the earth to the motions in the celestial realm, on the basis of a quote from *Seip̲er Yᵉṣirah* (4.12; "so that the mouth cannot speak and the ear cannot hear"), he determined that a certain celestial configuration would never recur.[34] In his interpretation of Ibn Ezra's text, Kumaṭiano agreed with the thrust of Ibn Ezra's argument. But Kumaṭiano counseled the reader that Ibn Ezra ought to have cited other parts of the passage from *Seip̲er Yᵉṣirah* to explain how a few or several components yielded many combinations. *Seip̲er Yᵉṣirah* reads: "Two letters build two houses," with Kumaṭiano glossing "houses" as "words." The verse continues: "Three stones build six houses," i.e., three letters build six words—meaning that the celestial bodies are numerous enough so that an effectively limitless number of combinations is possible. Hence, a single celestial configuration would not repeat itself. Ibn Ezra's texts, which were more popular among Romaniot Jews than those of Averroës, beckoned readers like Kumaṭiano to explore the esoteric in the course of studying other disciplines.

Many Jewish scholars thought that Maimonides's *Guide for the Perplexed*—another key text of Jewish Aristotelianism that Kumaṭiano commented upon in 1480—included an esoteric dimension.[35] Qabbalists produced and studied commentaries on the *Guide* between 1270 and 1290,[36] and sometimes the *Guide* was even cited in *Qabbalah* texts. For example, in *Seiper ha-Meliṣ* (Book of the Interpreter),[37] the Qabbalist Abraham Abulafia interpreted *Guide* III.7.[38] Kumaṭiano concurred with Qabbalists that the *Guide* had an esoteric meaning. For example, he cited *Seiper ha-Bahir* in his own commentary on the *Guide* as a source for the esoteric meaning of the divine name.[39] Esoteric texts turned out to be relevant even to the most famous harmonization of Judaism with Aristotelianism!

Developments in Christianity were also relevant to Kumaṭiano's discussion of the esoteric in his commentary on the *Guide*. While analyzing Maimonides's argument that it was possible to interpret the expression *k*ᵉ*ḇod adonai* (glory of the Lord), in Exodus 40:35 to mean "created, yet divine light," Kumaṭiano remarked that Maimonides "explained that the divine presence [*sh*ᵉ*ḵinah*; also the lowest of the *s*ᵉ*p̱irot*] is a created light existing outside the soul of the prophet."[40] This statement by Kumaṭiano was likely informed by hesychasm, a mystical movement in Eastern Christianity in the fourteenth century.[41] Gregory Palamas (d. 1359), who defended and practiced the hesychasts' techniques, contended that although humans could not participate in God's essence, they could participate in God's light, which was manifested in the world, through solitary prayer and contemplation.[42] Gregory's opponent in the debate, Barlaam of Calabria (d. 1348), argued that the human intellect grasped neither the essence of God nor God's light; God was exalted beyond human experience. This debate continued during Kumaṭiano's lifetime.

Kumaṭiano's interpretation of the thirty-fifth chapter of book 1 of the *Guide* regarding the denial of God's corporeality may also have been influenced by hesychasm. The base text reads: "It behooves that they [the masses] should be made to accept on traditional authority the belief that God is not a body."[43] Kumaṭiano commented that Maimonides's decision to teach the denial of God's corporeality to the masses "separated him from the ancients, even though some of the later philosophers say that it is not appropriate to spread this [viz. the denial of corporeality] among the masses."[44] Eisenmann and Schwartz conceded that Kumaṭiano's remark that God is light depended partially on Moses Narboni's (d. after 1362) commentary on the *Guide* but concluded that hesychasm also informed Kumaṭiano.[45] For contacts of the merchants of knowledge such as Kumaṭiano, esoteric

Jewish thought was neither hermetically sealed off from other disciplines nor from other religions. The porosity of all these boundaries facilitated the pervasiveness of *Qabbalah* in intellectual exchange.

Cohen Ashkenazi's Interdisciplinary Twentieth Argument against Metempsychosis and Balbo's Rebuttal

Cohen Ashkenazi composed twenty arguments against metempsychosis. Each of the first nineteen arguments he called *reʾayah* (evidence), while the twentieth and longest argument was called a *mopet* (proof or demonstration). Though the whole debate has been studied by Gottlieb, Ravitzky, and Ogren, the interdisciplinary dimension of the twentieth proof has not yet been fully appreciated. In the twentieth argument, Cohen Ashkenazi had recourse to Aristotelian texts and modes of reasoning to explain how one soul could be distinguished from another. Bodies could be differentiated through accidents, which were adjectives such as tall or short that distinguished one material object from another. On this point, Cohen Ashkenazi quoted a passage from the first chapter of Maimonides's *Book of Knowledge* where Maimonides stated that "entities, that can be enumerated and are equal in their essence, are only distinguishable from each other by the accidents that happen to physical bodies."[46] Cohen Ashkenazi reasoned that since the soul was not mixed with a body, then the soul was not material and was, hence, bereft of accidents. Cohen Ashkenazi cited *Guide* I.74 to support his contention that since souls could not be part of the chains of causes and effects that account for differences, then there could not be multiple, distinct eternal souls.[47] In other words, multiple eternal souls could not exist.

Next, Cohen Ashkenazi entertained the Averroist doctrine of the unicity of the intellect, which entailed that there was a single eternal human intellect, the unique part of the human soul. He, however, noted a problematic ramification of the doctrine: such unity implied that "when I study something, then you know it too, and when I forget it, then you forget it as well."[48] He also noted that the doctrine of reward and punishment in Jewish law failed if there was but a single soul. Because the soul must be rewarded or punished along with the body, individual accountability was impossible. He concluded that if there were neither single nor multiple eternal souls, then there was no rational basis for maintaining the doctrine of metempsychosis. Even if the doctrine of the unicity of the intellect were correct, the concept of the transmigration of souls would still be impossible because motion, e.g., of a soul from one body to another, occurred with respect to time, yet the multiple eternal souls required for metempsychosis existed outside of time.[49] In support, Cohen Ashkenazi quoted Ghazālī's[50] argument in *Maqāṣid al-falāsifa* (The Intentions of the Philosophers) that universals residing in the agent intellect, the source of intellectual abstractions, could not reside in a separate body because bodies are divisible and universals are not.[51] If there was not a single human intellect, there must be multiple human souls. Multiple human souls could exist only if they were created individually.

Balbo's response had ramifications for astronomy and astrology as well as psychology. For Aristotle and Aristotelians (as well as Plato and Platonists), the seemingly perfect and eternal motions of the stars and planets rendered the cosmos a locus for metaphysical reflection. Balbo, as he searched for evidence for metempsychosis in his rebuttal to Cohen Ashkenazi, trod the same path, in that he drew on the teachings of Aristotelian philosophy about souls and intellects. Scholars theorized multiple celestial orbs to move the planets, and these orbs revolved through their desire for the perfection of the celestial intellects. In other words, the orbs were alive and ensouled, and their motions were motivated by thoughts and desires, similar to how humans' motions were motivated by the desire to be somewhere else. Balbo explored whether the multiple celestial intellects were self-subsisting, i.e., created directly by God and dependent on nothing else for existence, or whether one intellect emanated from another, meaning that one intellect was the cause of the next. Emanation (Ar. *fayḍ*; Heb. *aṣilut*) was a philosophical metaphor that accounted for how God was responsible for multiplicity in the cosmos without multiplicity being imputed to God. According to Avicenna, an intellect issued forth (i.e., emanated) from God, the first cause. From that intellect necessarily emanated a second caused intellect and from it a soul and the outermost orb. Because the first intellect emanated necessarily from God, as light necessarily emanates from the sun, Islamic and Jewish philosophers used the myth of emanation to accommodate an eternal God who created an eternal cosmos. Emanation was a type of causality. Orbs and intellects emanated one from the next until the agent intellect, the cause of human souls, emanated from the intellect of the orb of the moon.[52] The doctrine of metempsychosis depended, in this sense, on the existence of multiple uncreated, incorruptible, and incorporeal souls and, hence, on the denial of emanation.[53] In the Averroist cosmos, emanation was not required because multiplicity proceeded directly, without emanation, from a unitary God.[54] One separate intellect was not the cause of another.

Thus, an assertion of the Averroist rejection of emanation helped Balbo's cause. Specifically, an Averroist attack on Ptolemaic astronomy was a key plank in Balbo's rejection of emanation and his argument for metempsychosis. We learned in the previous chapter that Averroists denied the existence of epicycles and eccentrics and theorized a cosmos of homocentric orbs (see chapter 4, fig. 4.3 for partial orbs and fig. 4.4 for homocentric orbs). Epicycles and eccentrics enabled astronomers such as Ptolemy (fl. 125–50 CE) to predict and retrodict observations more precisely than astronomers who theorized only homocentric orbs. That mathematical precision improved astrological forecasts. The unitary God theorized by Averroists could not be the proximate mover of epicycles and eccentrics revolving in multiple directions with multiple velocities about multiple centers.[55] Scholars who proposed epicycles and eccentrics also presumed that each epicycle and eccentric had to have its own intellect that was produced through emanation. Thus, according to Averroism, the denial of emanation correlated with the denial of the epicycles and eccentrics of Ptolemaic astronomy.

Balbo undermined Ptolemaic astronomy by mentioning numerous doubts (*sᵉp̱aqot*). As Maimonides had in the *Guide*, Balbo inquired as to whether Ptolemy knew of a proof for the existence of the epicycle.[56] Balbo also questioned the reliability of humans' overall knowledge of astronomy and remarked that no one knew the number of the stars and whether they were perfectly spherical (*kadduri*). There was no decisive answer, he wrote, to the question of whether the matter of the orbs was the same as that of the planets. Balbo remarked that the most trenchant of his many critiques was of Ptolemy's inability to demonstrate whether Jupiter and Venus were above or below the sun. Supporters of Ptolemy might have claimed that systems with epicycles and eccentrics afforded more precise predictions of the planets' positions, but how could that be true if Ptolemy's observations did not even establish the true locations of Jupiter and Venus?[57] If Ptolemaic astronomy was not as precise as advertised, Balbo contended, then the justification for preferring the Ptolemaic cosmos to an Averroist, homocentric cosmos weakened or even disintegrated. If there was no need, from the disciplinary perspective of astronomy, for epicycles and eccentrics, then there was no philosophical need for the metaphor of emanation to explain the origins of the multiple souls and intellects that epicycles and eccentrics required.

Having dispensed with emanation, Balbo had to explain how a unitary unmoved mover moved all the orbs. He provided the example of the lodestone and iron, with the iron playing the role of the orb and the lodestone the unmoved mover, as an analogy for how God moved orbs. The lodestone (God) attracted the iron (the orb) without the lodestone being moved in the least by the iron.[58] Yet there was an important difference between the way a magnetized stone attracted iron and the way the unmoved mover moved the outermost orb: a piece of iron would stop moving once it reached the stone, but because the outermost orb's motion was rotational, not rectilinear, the motion of the outermost orb never ceased. Thus, Balbo modified the explanation of the iron and lodestone as follows:

> Each part of the orb has in it a nature to be attracted opposite the body that moves it. When this [first] part arrives at it, it is not possible for it to rest because the nature of the part coming after it pushes it, necessitating for this first part a motion without rest, in two respects. The first is its being spherical, and this first part does not touch directly the mover, which would prevent it from moving. And the second respect is that this second part cannot reach the body moving it except by the movement of the part that is before it. Thus, each part is pushed by the part coming after it. The orb will be found to be moving in an eternal motion without rest according to this opinion.[59]

The analogy of pushing and pulling meant that if the orb's distance to God ever changed, then the orb's speed would change because the object of its desire, God, would be off center. But since all the orbs moved uniformly with the daily motion through their desire for God, all orbs had to be equidistant from the unmoved mover. Thus, Balbo's explanation functioned *only* if there were no epicycles and eccentrics.

For additional understanding of how the orbs in an Averroist cosmos were thought to move, we turn to Averroës's *Long Commentary on Aristotle's "Metaphysics."* There, Averroës reflected on the uniform daily motion of the cosmos to account for the other motions of the orbs of the planets: "We ought to understand that the motions of the rest of the orbs are led by the motion of the starred orb itself and that the perfection of every single orb, I mean the first mover for each orb, is perfected by the first mover of the universe. Thus, it all is led by this motion, I mean the diurnal."[60] The quoted passage helps us differentiate between the intellects of the orbs in the Averroist cosmos and the intellects of the epicycles and eccentrics in the Ptolemaic cosmos. Averroës commented that the intellect associated with each orb gained its form, i.e., its perfection, by contemplating the first cause, i.e., God, who is responsible for the diurnal motion. Then, the orb's desire for the associated intellect caused the orb's motion. Astronomers who hypothesized epicycles and eccentrics termed them partial orbs because they were components of the total orb for a planet. So, while the first mover of the orb of a planet in the Ptolemaic system may be led and perfected by the diurnal motion, the partial orbs and the partial intellects that moved them were not responsible for imparting the diurnal motion to the planet. That was because, in a Ptolemaic cosmos, the epicycles and eccentrics did not share the center of the diurnal motion. Hence, the relationship of the epicycle and eccentric with the first mover was mediated by the total orb and its intellect through emanation. In contrast, because emanation was not necessary in an Averroist cosmos, there could be multiple, self-subsisting intellects. The existence of these multiple, self-subsisting intellects paralleled the possible existence of multiple eternal human souls. Thus, for Balbo, evidence for the qabbalistic doctrine of metempsychosis was to be found at the intersection of Averroist philosophy and astronomy.

Cohen Ashkenazi, in his response to Balbo's rebuttal, agreed that astronomy was germane to discussions of emanation and metempsychosis. He, however, protested that the followers of Averroës took positions, including the denial of partial orbs, that Averroës never adopted. In the margins of his record of the debate, Cohen Ashkenazi retorted that R. Albalag (second half of the thirteenth century), a Hebrew translator of Ghazālī's *The Intentions of the Philosophers*, averred that Averroës believed

> that the many motions do not come from a single mover, only [that] each individual motion needs an individual (*m*[e]*yuḥad*) perception and desire and every particular perception and desire depends on a particular object of desire. Then the number of the movers follows from the number of the motions. For him, therefore, the eternal intellects are at least thirty-eight. If there are more, the philosophers do not have a

> proof or denial of their existence. Therefore, the philosophers do not prohibit that there be more. What is impossible for them is that there be an eternal intellect that does not have an agent in the orbs or in the realm beneath them.[61]

Thus, an alternative interpretation of Averroës's cosmology entailed, first, that partial orbs were possible and, second, that one intellect was the cause of the next.

Cohen Ashkenazi's response is evidence that Balbo's general recourse to Averroism in the preface to his rebuttal was not controversial. After the preface, Balbo moved to his direct rebuttal of Cohen Ashkenazi's twentieth proof. First, he noted that some of Cohen Ashkenazi's conclusions did not meet the standards of a proof because the premises were not reliable. For instance, against Cohen Ashkenazi's contention that one could not speak of multiple separate essences except in terms of cause and effect, Balbo adduced Narboni's commentary on the *Guide* and Levi b. Gerson's *Wars of the Lord*, texts in which the authors held that the intellect of each orb was not a cause of another orb's intellect.[62] Evidence for the existence of multiple eternal entities that did not cause each other, as required by the doctrine of metempsychosis, accumulated. Balbo argued that the different velocities of homocentric orbs could be ascribed to the orbs' different dispositions (*haḵanah*, pl. *haḵanot*), just as humans, according to the doctrine of the unicity of the intellect, had different dispositions for knowledge.[63] These subsequent arguments of Balbo's were more powerful in light of his dismissal of the astronomical evidence for a cosmology in which one orb and intellect emanated from the next.

Second, Balbo provided critical evidence from rabbinic texts for why there were multiple, eternal self-subsisting entities in the heavens. Balbo imported the interpretation of Ezekiel 1:6 ("And every one had four faces, and every one of them had four wings") from the fourth chapter of the eighth- to ninth-century CE midrashic text *Pirqei d^{e}-Rabbi Eliezer* (The Chapters of Rabbi Eliezer) into his comment. Balbo analogized the angels to celestial intellects, writing: "The camp (*maḥaneh*) of Michael is to his [God's] right and the camp of Gabriel is on his left. The camp of Uriel is in front of him and the camp of Rafael is behind him.[64] They are the four camps of the heavens, and the divine presence (*sh^{e}ḵinah*) is in their midst. In the earth are four standards[65] and the tent of meeting is among the camps.[66] The intellects that move [the orbs] are most numerous and all are effects[67] in a single way as preceded. One is not the cause of the other."[68] To the author of *Pirqei d^e-Rabbi Eliezer*, the angels were eternal entities analogous to the celestial intellects, and each angel was associated with a different orb. But one angel was not the cause of another. If there were multiple eternal celestial intellects, then there were multiple, eternal human souls.[69] Because tradition taught that there were far more angels than celestial orbs, Balbo contended that there were more orbs than astronomers realized.[70] Thus, despite the role of astronomy in Balbo's arguments, the cosmology of astronomers turned out to be subalternate in his arguments to the cosmology of *Qabbalah*.[71] And in determining that there were multiple human souls, Balbo refuted the doctrine of the unicity of the intellect.[72] For Balbo, Averroism became just a means to the end of refuting emanation.

The culmination of Balbo's rebuttal with revealed evidence is a significant indication of his view of the superiority of *Qabbalah*. Because Balbo demonstrated that Cohen Ashkenazi was wrong about astronomy and that the Averroists were wrong about psychology, alternatives to the qabbalistic doctrine of metempsychosis were flawed. The Averroist doctrine of the unicity of the intellect, because it was false, did not bring one closer to God.[73] Astrology, conceptualized by Cohen Ashkenazi as a religious philosophy, had no grounding in astronomy. Yet the evidence for metempsychosis, this central doctrine of *Qabbalah*, was contained only in revealed texts; the truth was not available to all. Still, because Balbo's demonstration of the truth of metempsychosis depended on his engagement with other fields in which the merchants of knowledge and their contacts exchanged knowledge, it is not surprising that the codex with Balbo's record of the debate contains his copy of Kumaṭiano's *Commentary on "Y^esod Mora*ʾ." The material relationship between texts is evidence for the interdisciplinarity of debates in *Qabbalah*. Celestial dynamics were intimately relevant to Balbo's defense of the qabbalistic doctrine of metempsychosis. He could not have dispatched Cohen Ashkenazi's position without arguing against the existence of epicycles and eccentrics.

Pico, Averroism, and *Qabbalah* in Delmedigo's *Commentary on "De substantia orbis"*

The imbrication of *Qabbalah* and other disciplines affected relationships between the merchants of knowledge and their contacts. Notably, some modern scholars have attributed Delmedigo's loss of the prominent Renaissance scholar Pico della Mirandola (d. 1494) as a student and patron to disagreement over the Jewish authenticity of *Qabbalah*.[74] Christian scholars of *Qabbalah* contested the Jewish origins of *Qabbalah* because they interpreted aspects of *Qabbalah*, such as the *s*^e^*p̱irot*, in a way that destabilized the Jewish emphasis on God's unicity. Qabbalists who believed that humans' actions affected the *s*^e^*p̱irot* implied that God relied on humans, a stance that troubled the beliefs of Averroists in hierarchy in creation and of many Jews in God's transcendence. Delmedigo's reservations about *Qabbalah* expressed in his last known work, *B*^e^*ḥinat ha-dat* (The Examination of Religion), could be understood as a response to Pico's enthusiasm for *Qabbalah* but not to Delmedigo's repudiation of *Qabbalah*. Modern scholars have reassessed Delmedigo's misgivings about *Qabbalah* in the *Examination* and have found them to be less pronounced than previously believed. Though the *Examination* was not modeled on Averroës's *Decisive Treatise* (Ar. *Faṣl al-maqāl*), in which Averroës did not identify an inherent conflict between religion and philosophy, Delmedigo may not have rejected *Qabbalah* outright in favor of Averroism in the *Examination*.[75] We should reconsider whether the end of the relationship between Pico and Delmedigo was due to Delmedigo's rejection of *Qabbalah*.

One objection that could be leveled against *Qabbalah* is that it was unmentioned in the Bible and Talmud, and thus not permitted. In the *Examination*, Delmedigo noted the novelty of *Qabbalah* by reminding the reader that even later Jewish luminaries such as Rashi (d. 1105) did not know of *Qabbalah* and, thus, did not judge it.[76] Though novel did not mean wrong or that all of *Qabbalah* was problematic, Delmedigo cautioned that *Qabbalah* should not supersede rabbinic tradition.[77] After all, he reasoned, if the concepts found in the *Zohar* were, in fact, grounded in rabbinic tradition as the author of the *Zohar* implied, then those concepts should not be controversial.[78] Though Delmedigo denied that legal practices should conform to certain esoteric interpretations,[79] he allowed that *Qabbalah* and observances enjoined by rabbinic law might occupy separate realms.[80] Delmedigo did disapprove of the notion that God and the angels could be restored only through human intervention when he remarked that the belief that humans' actions might restore the upper realm was, at best, a false explanation that motivated praiseworthy actions.[81] But that disapproval was not a fulsome rejection.

Thus, the *Examination* was far from a head-on critique of *Qabbalah*. Instead, *Qabbalah* was part of the intellectual world that Delmedigo inhabited. He must have been aware of the debate about metempsychosis, because Cohen Ashkenazi's son, the merchant of knowledge Saul Cohen Ashkenazi (ca. 1470–1523), was Delmedigo's student and the addressee of the *Examination*.[82] Like Balbo, Delmedigo believed that Averroism and *Qabbalah* did not contradict each other, to an extent.[83] More important, Pico's patronage of Delmedigo led Delmedigo to address the connections of Averroism to *Qabbalah*. Delmedigo's message to Pico was that hypostatic *s^e^p̱irot* were a misinterpretation of *Qabbalah* that ought to be eschewed—a conclusion he had reached through his assessment of the connections of Averroism to *Qabbalah*.

For at least a year, Delmedigo had reservations about furnishing Pico with a complete assessment of *Qabbalah*. There is textual evidence for how Delmedigo managed his qualms. In 1485, Delmedigo composed for Pico a Latin commentary on *De substantia orbis*, but Delmedigo, in the Hebrew version of the same commentary,[84] which he produced later in 1485, added comments on *Qabbalah* not found in the Latin original. For example, a complete explanation of the connections between providential order in nature and *Qabbalah* was found *only* in a long passage from the Hebrew version of the commentary on *De substantia*.[85] Subsequently, Delmedigo's wariness diminished, and he decided to share some thoughts on *Qabbalah* with Pico in a Latin letter in 1486.

In that letter, before commenting directly on *Qabbalah*, Delmedigo ruminated on Averroist texts. In the course of those remarks, he unraveled a difficulty in Averroës's *Meteorology*: "Your lordship said he queries the cause by which east winds should be warmer, for it is seen that the reason that Averroës ascribes[86] is null, because just as before the sun appears at the eastern point, it heats that part indirectly for one or two hours. Thus, also before it arrives at the western origin (*principium*), it heated it previously for one or two hours."[87] Averroës's *Meteorology* was important to a debate over *Qabbalah* because the passage at hand implicated God's relationship to creation. The immediate problem was that if God were equidistant from all points on earth, as required by Averroist cosmology, and if God's effect on earth were unitary, why would east winds be warmer?[88] Delmedigo proffered a solution:

> Yet perhaps we can respond to this objection in the way that he [viz. Averroës] himself does because the sun is not of altogether similar aspect when it is above the eastern horizon, that is, the eastern part of the earth, as in each western part, with it receding from the middle of the heavens. And by it being given, that should the appearance be similar, as is mostly true, the rest is not equal. Since if the sun faces an eastern part, it heats it, and coldness has no effect in return. Thus, the part that is below the eastern horizon is already heated, with the sun near at hand over there. It is not however thus in the western part, since it [viz. heating] is contrary to the effect of the cold part of the earth that is beneath the western horizon with this part being for some time cold, as there not having been sun for the same long time. This is seen to me as a good response.[89]

According to Averroës, east winds were warmer than west winds because the sun, when in the east, heated parts of the earth that were still below the horizon. The converse did not hold when the sun was in the west because the part of the earth below the horizon was cooling off all day. Thus, God's effect on the different regions was static, as Averroists believed. In Delmedigo's analysis, meteorological evidence buttressed the Averroist position that God was the unitary, unmediated cause of all celestial motions, which was a crucial argument of his attempt to steer Pico away from a belief in hypostatic *s^e^pirot.*[90]

To that end, Delmedigo wrote: "Because I see that your lordship exerts much effort in that blessed *Qabbalah*, I want to mention to you those things, that I already was noting in my commentary on *De substantia orbis*, speaking in Hebrew about spiritual powers, that I never wished to say to you. And because of this I should say, truly it is so occult, that no one in this time, who devotes themselves to this, comes to know this of them."[91] Delmedigo was skeptical that many Qabbalists, Pico included, properly understood the *s^e^pirot*. To that end, in the letter Delmedigo supplied Pico with a Latin version of his remarks on the *s^e^pirot* found in his Hebrew commentary on *De substantia orbis*. The following passage was notably absent from the earlier Latin version of the commentary, meaning that Delmedigo had initially intended not to divulge this information to Pico. Delmedigo wrote:

These scholars[92] suppose there are certain entities, whose grade is inferior to the grade of the glorious God, whom they call infinite. They are emanations that I say are neither made nor produced (*fluxa non dico facta, nec producta* [absent from the Hebrew]), from that which they call infinite. These [entities] themselves have different grades, and the highest grade of these is the mover of the heavens and the sensible celestial bodies. And the order through which the created entities are produced and subsist is according to the order through these, that is to say through the *çephirotb*, i.e., numbers. Indeed, they name them emanations from the infinite.

For they believe that in the infinite there occurs neither cogitation nor apprehension nor some definition or determination, nor, furthermore, an intellectual disposition. Nor can there be said of it a will, nor an intention, nor cogitation, and altogether no disposition. It is impossible that anything originate or emanate from it, that is to say from the infinite.[93] This universe would therefore be diminished next to this, or would fall short from its perfection.

But the first emanation from it are these entities, which they call *çephirotb*, according to their grade (*secundum gradus eorum*), as we say, and they themselves act through the power of God, whom they call infinity, and through emanations that come from Him. Thus, all is through that power. Therefore, they, namely the *çephirotb*, depend on Him and are emanations from Him, namely infinity. Hence, according to this, that order found in the universe is through these *çephirotb*.

First, there is no disposition or positive attribute to be said of that which they call the infinite. Neither, in no way, do they want to call it an intellect, as Averroës likewise says in the book *Incoherence of the Incoherence*, speaking of attributes or properties. Plato or certain Platonists want to say that God is an intellect or to affirm of him that he is understanding. Moreover, these very *çephirotb*, they make a proper noun, and an emanation [from God] or something dependent [on God]. According to their opinion, they grant that they are a cause whereby those things are obliged to exist [. . .][94] neither more nor less and in this they produce books and volumes.

Everything else, however, I have appended in my commentary on *De substantia orbis*, but it is not necessary for you. And they all, or most of them, who make this a doctrine for themselves, are entirely ignorant of all. All they say are words and they do not understand.[95]

In this passage, Delmedigo has counseled Pico that hypostatic *sᵉp̱irot* were a Neoplatonist misconstrual of *Qabbalah*. That intent is clear through one of the few departures of the text of the Latin letter from the Hebrew version of the commentary on *De substantia*. In this passage from the Latin letter, though not in either version of the commentary on *De substantia orbis*, Delmedigo glossed the *sᵉp̱irot* as numbers, not as entities through which one knows God.[96] As for *ein sop̱*, the transcendent aspect of divinity beyond human contemplation which Delmedigo called "infinite," that entity lacked volition and thought; the intellectual conjunction with the divine sought by Averroists was not possible with *ein sop̱*. Delmedigo's message for Pico in this letter was that the theosophic possibilities of the Averroist doctrine of the unicity of the intellect were richer than those of *Qabbalah*. But there is no evidence that Delmedigo dismissed *Qabbalah* in toto.

A full understanding of this debate over the relative merits of Averroism and *Qabbalah* is crucial for unpacking the granular dynamics of the patronage relationship between Delmedigo and Pico. Crucially, Delmedigo's nonhypostatic interpretation of the *sᵉp̱irot*, which was more consonant with Averroism and plausible for some Qabbalists, did not land with Pico. Consider Pico's *900 Theses*, written the same year as Delmedigo's letter, in which Pico declared that the *sᵉp̱irot* represented positive attributes of God. Pico's sixty-sixth thesis reads:

> I adapt our soul to the ten *sefirot* thus: so through its unity it is with the first, through intellect with the second, through reason with the third, through superior sensual passion with the fourth, through superior irascible passion with the fifth, through free choice with the sixth, through all these as it converts to superior things with the seventh, through all these as it converts to inferior things with the eighth, through a mixture of both of these—more through indifferent or alternate adhesion than simultaneous inclusion—with the ninth, and through the power by which it inhabits the first habitation with the tenth.[97]

Delmedigo's tactical decision to divert Pico from the doctrine of hypostatic *sᵉp̱irot* clearly failed. But there is no evidence that Delmedigo aimed to debunk *Qabbalah*. In that light, the absence in the Latin letter to Pico of Delmedigo's affirmation of the authenticity of the *Qabbalah* found in the Hebrew commentary on *De substantia orbis* is curious. What was going on?

The answer lay in the dynamics of the competition for Pico's patronage between Delmedigo and a scholar from outside the network of merchants of knowledge. In 1485–86, the convert to Christianity Flavius Mithridates (born Samuel b. Nissim Abū al-Farāj; fl. late fifteenth century) was making a play to become Pico's next teacher. Pico's relationships with Flavius and Delmedigo overlapped. Marsilio Ficino (d. 1499), a renowned Renaissance Platonist, testified to the complicated relationship when he recounted his favorable impression of Delmedigo's defense of Judaism in a 1485 debate at Pico's house in Florence. Ficino recalled that "in these discussions, Elias[98] and Abraham, Hebrew physicians and Aristotelians, argue against Guglielmo of Sicily [that is, Flavius Mithridates].[99] They contend that the divine words of the prophets do not in any way apply to Jesus but were spoken in some other sense. Thus, they turn them all in another direction and wrench them from our hands with all the strength they can muster. Nor does it seem possible to refute these men easily, unless divine Plato were to come forward into the judgement hall as the invincible defender of holy religion."[100] Though Ficino did not elaborate on the relationship between Aristotelianism and the defense of Judaism, recall that, around the year of the debate, Delmedigo wielded Averroism against Pico's Christian interpretation of the *s^e^pirot*. In any case, even from the Platonist Ficino's perspective, Delmedigo's performance brought him social capital. One can imagine that the experience of being shown up by Delmedigo motivated Flavius to tempt a student from Delmedigo.

Flavius Mithridates succeeded in that endeavor. His skillful accommodation of all Pico's interests[101] was the principal reason why Flavius supplanted Delmedigo as Pico's instructor. Flavius strategically treated everything he taught Pico as if it were the esoteric material Pico desperately sought.[102] Consider the case of Flavius Mithridates's Latin translation of Gersonides's commentary on *Song of Songs*.[103] Although Gersonides's commentary on *Song of Songs* was not a qabbalistic text, Flavius's translation contained Latin interpolations relating Jewish law and philosophy to *Qabbalah*. Yoḥanan Alemanno (d. 1504), who instructed Pico in Hebrew and possibly *Qabbalah*, also dedicated a commentary on *Song of Songs* to Pico,[104] but the evidence that Flavius's translation pleased Pico the most is clear.[105]

Flavius and Delmedigo concurred that instruction in *Qabbalah* was critical for retaining Pico's patronage.

This contest for Pico's patronage affected how Delmedigo presented *Qabbalah* in the Latin letter. From Delmedigo's perspective, Averroist arguments against hypostatic *s*ᵉ*pirot* had the potential to steer Pico away from the *Qabbalah* that Flavius Mithridates was peddling. Acknowledging in the Hebrew version of the commentary on *De substantia orbis* that *Qabbalah* was valid if properly understood would not have helped Delmedigo's case because he would just be vouching for the authenticity of Flavius Mithridates's wares if not Flavius's interpretation. Delmedigo hoped to compromise to retain his patron and mentioned in his letter to Pico that he intended to help procure for Pico a qabbalistic commentary on the Torah by Menaḥem Recanati (d. 1290), a *s*ᵉ*pirot*ic Qabbalist.[106] Delmedigo complained about the road conditions but seems to have sent the Hebrew manuscript to Pico.[107] In any case, Flavius Mithridates completed a Latin translation of Recanati's *Commentary on the Pentateuch* sometime after May 10, 1486,[108] and this translation became Pico's most significant source for *Qabbalah*.[109] As much as Flavius's openness to hypostatic *s*ᵉ*pirot* facilitated his poaching of Pico's patronage, the dynamics of the competition for patronage affected what Delmedigo chose to communicate to Pico about *Qabbalah*.

Certain implications of Pico's interest in hypostatic *s*ᵉ*pirot* were a bridge too far for Delmedigo. For example, Pico's harmonization of astrology and a Christian interpretation of *Qabbalah* must have irked Delmedigo, a Jewish opponent of astrology. Pico, in his *900 Theses*, associated each *s*ᵉ*pirah* with a planet.[110] In the eighteenth thesis, he stated that "whoever joins astrology to Cabala will see that to sabbatize and rest becomes more appropriate after Christ on the Lord's day than on the day of the Sabbath."[111] While Saturn ruled Saturday, the Jewish Sabbath, the sun, associated with *tip̱ʾeret* (glory) and Jesus, ruled Sunday. Flavius Mithridates, though, was willing to accommodate that and other Christological interpretations of *Qabbalah* expressed, for instance, in the fifth thesis of Pico's 1486 "Cabalistic Conclusions Confirming the Christian Religion."[112] In fact, Flavius rendered into Latin a Jewish source for the integration of astrology into *Qabbalah*, Joseph b. Waqar's (fl. fourteenth century) *al-Maqāla al-jāmiʿa bayn al-falsafa wa-l-sharīʿa* (The Treatise Combining Philosophy and Religious Law). [113] Ibn Waqār saw philosophy and *Qabbalah* as parallel, though not perfectly reconcilable, systems[114] and likened the *s^epirot* to the planets.[115] Jewish backing for an astrological interpretation of *Qabbalah* was available to Pico.

After the break with Pico, Delmedigo landed on his feet. By 1487, he had returned to Padua, where the Venetian Girolamo Donato (1457–1511), the future bishop Antonio Pizzamano (d. 1512), and the future cardinal Domenico Grimani (d. 1523) became his new patrons.[116] Grimani was the dedicatee of five of Delmedigo's writings, including a translation of the proem to book 12 of Averroës's commentary on Aristotle's *Metaphysics*.[117] Grimani's personal physician and a correspondent of Delmedigo's, Abraham de Balmes (d. 1523), also rejected Neoplatonic interpretations of *Qabbalah* and instead interpreted *Qabbalah* in Aristotelian terms.[118] In a brief treatise entitled *Iggeret ha-ʿasiriyya* (Epistle of the ten), de Balmes described the ten *sᵉpirot* as principles (*shorᵉshim*).[119] In his interpretation, the *sᵉpirot* were not hypostases and God was not affected by human actions.[120] For de Balmes, the first principle was how all of the attributes with which we describe God must somehow be united in God.[121] The second was how God is known intellectually only through the effects that God produced as the first cause.[122] As evidenced by his correspondence with the Jewish physician de Balmes as well as his new, renowned Christian patrons, the loss of Pico's patronage simply brought Delmedigo more intellectually congenial contacts.

Delmedigo's Latin letter to Pico is evidence that the split between the merchant of knowledge and his contact is best explained by the interplay of an interpretive disagreement about the *sᵉpirot* with competition between Delmedigo and Flavius for patronage. The letter included a close reading of Averroës's *Meteorology*, as Delmedigo favored a *Qabbalah* that could be more closely harmonized with the Averroism in which he also instructed Pico. Indeed, after Pico wrote *900 Theses*, he found his way to rejecting astrology, an Averroist position that Delmedigo had always held.

Averroism and *Qabbalah* among the Merchants of Knowledge after Delmedigo

After Delmedigo's death, merchants of knowledge continued to explore the overlapping of *Qabbalah* and philosophy. The merchant of knowledge Saul Cohen Ashkenazi was a correspondent of Isaac Abravanel (d. 1508),[123] who was an influential scholar and community leader expelled from Spain.[124]

Saul addressed to Abravanel twelve questions and, in the ninth question, sought an interpretation of the prophet Ezekiel's renowned and tantalizing vision of God on a chariot. Esoteric speculation on Ezekiel's vision can be found in the Talmud, and the later scholars paid attention too. We read in the Bible, "Now as I beheld the living creatures, behold one wheel at the bottom hard by the living creatures, at the four faces thereof. The appearance of the wheels and their work was like unto the colour of a beryl; and they four had one likeness; and their appearance and their work was as it were a wheel within a wheel . . . upon the likeness of the throne was a likeness as the appearance of a man upon it above" (Ezekiel 1:15–16, 26). Abravanel responded that when Ezekiel described visions of otherworldly chariot wheels and God, Ezekiel spoke literally. But when he discussed matters (e.g., the firmament, the throne, and what is beneath it) that were amenable to demonstration, he spoke figuratively.[125] The determination of whether God, for Averroists, or the first effect (Heb. *ʿalul*), for Neoplatonists, was responsible for the daily motion impinged on the answer to the question of whether the throne should be understood as a metaphor for the first effect or for God as first cause.[126] Abravanel concluded that, according to Aristotle and Shem Ṭoḇ b. Falaquera's (d. 1290) commentary on the *Guide*, the first cause was the first mover of the orb(s).[127] Hence, the throne was a metaphor for God. Isaac Abravanel's son Judah (a.k.a. Leone Ebreo, d. after 1523), whom Isaac lauded as the choice philosopher in Italy, was a Neoplatonist, unlike his father.[128] Judah determined that because Aristotle "was not informed by the 'ancient theologians,' as his teacher Plato had been, he was limited and could not attain true *unio*."[129] Judah's Platonism facilitated a level of professional success that even his Aristotelian father acknowledged.

There was a significant exchange of another text of *Qabbalah* in 1539. The merchant of knowledge Moses Galeano/Mūsā Jālīnūs purchased from his student, the merchant of knowledge Abraham Algazi, a manuscript of qabbalistic commentary on the Torah, which was inspired by Menaḥem Recanati's scriptural commentary. This manuscript, which is now Vatican MS Ebr. 201, was sold to the Fuggers's agents shortly thereafter.[130] Thus, an interest in *Qabbalah* connected Galeano/Jālīnūs to another merchant of knowledge and to the Fuggers. In the statement of sale from Algazi to Galeano/Jālīnūs, the latter's name is accompanied by the qabbalistic abbreviation נר"ן, which stood for parts of the human soul (*nep̱esh*, *ruaḥ*, *nᵉshamah*). Although Galeano/Jālīnūs did not author a text in the genre of *Qabbalah*, it is difficult to imagine why he would have purchased the manuscript were he not intrigued by *Qabbalah.*

Passages in *Puzzles*, which Galeano/Jālīnūs completed in 1537, just before the acquisition of the qabbalistic Torah commentary, yield glimmerings of his interest in the esoteric. Galeano/Jālīnūs's term for the esoteric was *sod*, which also means "secret."[131] For example, Galeano/Jālīnūs mentioned the miracle of the talismanic brass snake, described in Numbers 21:4–9, that cured Israelites who had been bitten by fiery snakes sent by God: "Another religious example touches on the secret of the brass snake. The scholars of talismans said that in their books and in the books of the Jews there are many stories that a great many of them do not understand. They intend the likes of the [stories of the] cherubim[132] and the brass snake that was to cause the forces to descend. It came to pass that if one was bitten but looked to the snake of brass, he lived."[133] Galeano/Jālīnūs did not venture a naturalistic explanation for the effect of the brass snake, implying instead that the snake functioned as an amulet, the operation of which he turned to investigating. Kumaṭiano had considered and rejected the possibility that the efficacy of the copper snake was due to astral forces entering it,[134] forces that Schwartz has described as astral magic and noted as a concern of Qabbalists.[135] Kumaṭiano concluded that amulet construction was a craft but not a discipline of knowledge. Galeano/Jālīnūs's reason for exploring topics of *Qabbalah* in *Puzzles* was his desire to demystify.

Indeed, elsewhere in *Puzzles* we find specific examples of events perceived as mysterious or secret (*sodi*) but only when the proximate causes were overlooked: "Another religious and mysterious (*sodi*) example including [the error of taking] that which is not the proximate cause and ascribing to it the status of the proximate cause is how the books of the prophecies attribute actions to God, may He be exalted, without mentioning the proximate cause such as the cedars of Lebanon that were planted. But the tablets were God's work."[136] In Galeano/Jālīnūs's analysis, though the Psalmist wrote that God planted the cedars of Lebanon (Ps 104:16), God was not the proximate cause of the trees being planted, though God was the proximate cause of the revelation of the Ten Commandments. Hence, the perception of mystery stemmed from incomplete understanding. The planter might have been a person or the natural order in which the seeds fell off the cone. Likewise, in Numbers 11:34, God caused flocks of quail to drop from the sky at a place where the Israelites were complaining and lusting for food other than manna. Galeano/Jālīnūs was concerned that readers of the passage would be confused about the true cause for why the quail dropped out of the sky. By "confusing causes," he meant the process in which one exchanged (*t^e^murah*) the perceived cause (God) for the true cause, here the natural reason why the quail fell. He remarked:

> Galen said in his commentary on the *Epidemics*[137] that when we wish that newborns drop a loose stool, we give the wet nurse to drink or feed a goat or sheep laxative medicine like ecballium (*qiṭṭāʾ* [i.e., *qiththāʾ*] *al-ḥimār*) because the power of that same medicine remains in her [the nurse's] milk. This is potentially explained with individuals because there are people who passed loose stools when they drank the milk of goats that grazed in the fields of scammony or types of danewort and serpent melon. Many people who increasingly ate their fill of quail had contractions, which are a tugging of the muscle, or death due to Hellebore or deadly grasses. [138]

Confusion was sometimes resolved with discursive reason. Galeano/Jālīnūs debunked the apparent mystery of how the Israelites died from excessive consumption of quail: the quail, before being eaten, may have ingested substances that were toxic to them and to humans. That accounted for how they fell from the sky and why the Israelites seemed to be wiped out in a plague. Conversely, he cautioned that one also must not search in vain for the hidden causes of true miracles, such as the revelation of the tablets, namely the Ten Commandments.[139] The ability to discern when a naturalist explanation was appropriate was critical to avoiding the error of confusing causes.

The term *t^e^murah* (substitution) was the foundation for a marginal comment in the qabbalistic Torah commentary that Galeano/Jālīnūs acquired from his student Abraham Algazi. *À propos* the interpretation of Leviticus 1:2 ("Speak unto the children of Israel, and say unto them: When any man of you bringeth an offering unto the LORD, ye shall bring your offering of the cattle, even of the herd or of the flock"), the annotator noted that the soul of the sacrificial animal was substituted for the soul of the one offering the sacrifice.[140] The desire of Recanati and the annotator to render explicit the principles of ritual practice paralleled Galeano/Jālīnūs's goal of demystification in *Puzzles*, the lone surviving manuscript of which his student Algazi copied. Thus, Algazi may have produced the Byzantine-hand marginalia in Vatican MS Ebr. 201. Many, but not all, of the marginal insertions came from Recanati's Torah commentary. The thrust of the annotations in a qabbalistic scriptural commentary that Galeano/Jālīnūs owned echoed Galeano/Jālīnūs's attention to the complexities of *t^{e}murah* in *Puzzles*.

Other marginalia yield more information about the *Qabbalah* of which Galeano/Jālīnūs and Algazi were aware. At the beginning of the exegesis of Deuteronomy 3:23–7:11, the annotator commented that whenever one found the words "The Lord God" (*Adonai Elohim*), "the attribute of mercy (*middat raḥamim*) is clothed (*mitlabbeshet*) with the attribute of justice (*din*)."[141] The *s^{e}p̱irot* of mercy and justice were employed here as a hermeneutic tool. Likewise, the annotator also interpreted the symbolism of the afternoon, or *minḥah*, prayer in terms of the *s^{e}p̱irot*. Regarding Genesis 24:50 ("Then Laban and Bethuel answered: by the Lord"), the annotator remarked: "And Isaac went out walking in the field [at the eventide] (Genesis 24:63): He already interpreted the verse above through the secret of the *minḥah* [viz. afternoon] prayer. . . . At the time of the *minḥah*, Elijah was elevated in the flame, and Gabriel appeared to Daniel at the time of the evening *minḥah* (*minḥat ʿereḇ*) You should know because his name is derived from *g^{e}ḇurah*."[142] *G^{e}ḇurah* (strength) is the name of a *s^{e}p̱irah* and shares the same root as the name Gabriel. In another marginal comment, the annotator cited Saadia Gaon's (d. 942) statement in *Emunot w^{e}-deiʿot* (Beliefs and Opinions) that the Torah spoke in a general language (*lashon kollelet*). The annotator continued: "We could not know all of the hidden things of wisdom (*ḥoḵmah*) and the understanding (*binah*) that the Torah intends by wisdom (*ḥoḵmah*), even though he did not interpret it to mean that the words of Torah have an internal and external, revealed and hidden meaning; each of them is the word of God."[143] According to the annotator, the *s^{e}p̱irot* of wisdom (*ḥoḵmah*) and understanding (*binah*) functioned not necessarily as hypostases but as hermeneutic tools for these passages, leading one to a deeper understanding of the language of scripture.

In one case, with regard to Deuteronomy 10:14 ("Mark, the heavens to their uttermost reaches belong to the Lord your God, the earth and all that is on it!"), the annotator was dissatisfied with the interpretation, found in the main text, that the uttermost reaches of the heavens were nobler than those visible to us. The marginal insertion, which depended on Recanati's Torah commentary, continued:[144]

> This is the opinion of some of the scholars of *Qabbalah* and were it not for how I fear to dispute with them, I would interpret the opposite. A piece of evidence for my words is "Even the heavens to their utmost heights do not contain you" (First Kings 8:27) and in [the midrash collection] Tana Dᵉb̲ei Eliyahu of blessed memory.[145] And the utmost heights of the heavens: the heavens are the lower heavens, and the utmost heights are the upper [heavens]. Also, in the words of our rabbis of blessed memory: "The native is on the ground, and the stranger is in the utmost heights. The ground is the land of treasure (Malachi 3:12)."[146]

That is, the heavens (*ha-shamayim*) were the lower heavens, and the utmost heights were the upper heavens. Both were visible. Presuming Galeano/Jālīnūs agreed with the thrust of the marginalia in his copy of qabbalistic scriptural commentary, then he was open to others' reflections on the *sᵉp̲irot*. Since there was no indication from the contents of *Puzzles* that Galeano/Jālīnūs elevated *Qabbalah* above other disciplines, he likely concurred with the annotator's invocation of scripture and rabbinic texts against the views of some Qabbalists. From the 1460s to the 1540s, the merchants of knowledge and their contacts explored a multivocal *Qabbalah* that impinged upon and interacted with other disciplines but did not always supersede them. Amid this intellectual flexibility, there was ample scope for an interdisciplinary exchange of ideas and Hebrew and Latin manuscripts in a variety of fields among multiple merchants of knowledge and their Christian and Jewish contacts.

Qabbalah Leading Christians to Arabic, Islam, and Istanbul

The ripples of the waves of Christian engagement with *Qabbalah* reached Istanbul. Though *Qabbalah* directly implicated relationships between Jewish and Christian scholars, a preoccupation with *Qabbalah* whetted the appetite of Christian scholars for Arabic and Islamic texts.[147] For example, because Pico believed that *Qabbalah* reflected perennial truths, knowledge of those truths was not confined to scholars of *Qabbalah*. Consequently, Pico acknowledged Islamic sources for the ancient and timeless wisdom that he found in *Qabbalah*, as he recounted in the opening of his *Oration on the Dignity of Man*: "Most esteemed fathers, I have read in the ancient texts of the Arabians that when Abdallah the Saracen was questioned as to what on this world's stage, so to speak, seemed to him most worthy of wonder, he replied that there is nothing to be seen more wonderful than man."[148] Pico credited Arabs with understanding underlying truths, such as the foundational role of numbers.[149] He recognized how al-Kindī (d. 873/77) investigated the Platonic and Pythagorean mysteries that occupied so many ancient philosophers.[150] Pico also had this to say about his project of recovering the true wisdom: "Inasmuch as all wisdom has flowed from the barbarians to the Greeks and from the Greeks to us, what would have been the point of dealing only with the philosophy of the Latins—that is, of Albert, Thomas, Scotus, Giles, Francis, and Henry—while leaving the Greek and Arab philosophers aside? Thus, our authors have always seen fit, in matters of philosophy, to ground themselves in foreign discoveries and to perfect the doctrines of others."[151] The possibility of acquiring more of the *prisca theologia* motivated, at least partly, Pico's pursuit of Arabic learning.

The link between *Qabbalah* and Islam motivated a scholar outside the merchants of knowledge's web of Christian contacts to journey to Istanbul. A Jewish scholar, Moses Almuli, was an intermediary for Guillaume Postel (d. 1581), a French Orientalist, when Postel was in the Ottoman Empire in 1536 collecting books on astronomy and *Qabbalah*, among other fields.[152] The links between *Qabbalah* and cosmology intrigued Postel.[153] He paid attention to occult knowledge in Islamic culture as well and recorded an incident in which an Ottoman text on prognostication was used as a talisman.[154] In 1548, Postel began to translate the *Zohar*;[155] he eventually translated *Seiper Yeṣirah* and *Seiper ha-Bahir*.[156] On a 1549 trip East, Postel purchased a manuscript of *al-Tadhkira fī ʿilm al-hayʾa* (The Memoir on Astronomy) of Naṣīr al-Dīn al-Ṭūsī (d. 1274) and annotated it in Latin.[157] Paralleling what we have found, over and over, with the merchants of knowledge and their contacts, immersion in one field (*Qabbalah*) went along with proficiency in another (astronomy).

For Postel, *Qabbalah* was not a tool to convert Jews but a way to understand his own religious experiences.[158] His theory that everyone would accept the Gospel if they knew some ancient languages and the fact that Arabic was the most widely spoken of those ancient languages led Postel to Arabic.[159] He reported that he received books on mathematics and the interpretation of Aristotle from Sultan Süleiman the Magnificent (d. 1556) and that he procured other books for King Francis I of France in Istanbul.[160] In the same era, the cosmographer Sebastian Münster (d. 1552) aimed—through the study of Hebrew and the publication of Hebrew texts, often supplied with translations—to broaden the intellectual foundation of Christian *Qabbalah*.[161] In 1546, at Basel, Münster published an edition of *Qiṣṣur melʾeḵet ha-mispar* (The Abridgment of the Operation of Number) by Elijah Mizraḥi, a teacher of Galeano/Jālīnūs, with Latin translation.[162] The merchants of knowledge's and their contacts' interest in *Qabbalah* and its connections to philosophy and science not only brought Christians to polymathic merchants of knowledge, such as Delmedigo and Galeano/Jālīnūs, but also broadened Christians' reading and acquisition of texts to Islamic sources.

Conclusions

Three conclusions emerge from the foregoing discussion of the interdisciplinary role of *Qabbalah* in intellectual exchange among the merchants of knowledge and their contacts. First, losing an argument did not mean losing a place in the network of the merchants of knowledge or further opportunities for exchange with Christian scholars. As we saw, although Moses Cohen Ashkenazi lost the debate with Balbo over metempsychosis, Moses's son Saul Cohen Ashkenazi was Delmedigo's student and corresponded with Isaac Abravanel. Though the merchants of knowledge and their Jewish contacts tended to favor an Aristotelian interpretation of *Qabbalah*, while their Christian contacts preferred a Platonist interpretation, contact between the merchants of knowledge and Christians did not diminish. Though Pico eventually jettisoned Delmedigo for Flavius Mithridates, other elite Italians took Pico's place as students of Delmedigo, who remained intellectually relevant.

Second, we have seen that discussions of *Qabbalah* were as interdisciplinary and nuanced as the discussions of Averroist philosophy studied in chapter 4. For instance, though Balbo and Delmedigo disagreed about the doctrine of the unicity of the intellect, both scholars favored an Averroist cosmology. Without the epicycles and eccentrics of the Ptolemaic cosmos, God's control over the cosmos as first cause—for the Averroists Delmedigo and Balbo—was more evident and explicable. For Delmedigo, Averroism militated against hypostatic *s*ᵉ*p̱irot*. But for Balbo, Averroism was a plank in his argument for the qabbalistic doctrine of metempsychosis. Two scholars who agreed about Aristotelian philosophy reckoned its intersection with *Qabbalah* differently. Yet disagreement did not hinder exchange.

Third and most important, the intellectual connections between *Qabbalah* and other disciplines, above all Averroism, meant that exchanges in one field did not inhibit and often led to exchanges in another field. Though Pico preferred Flavius Mithridates to Delmedigo as a teacher of *Qabbalah*, Pico followed Delmedigo's rejection of astrology. Though many of the exchanges that I have described so far, including of *Qabbalah*, had practical applications, all of them also involved reflection, in one way or another, on the motions of the celestial orbs. In a sense, this is not surprising because scholars of all the disciplines treated in this book explored God's relationship to the cosmos. The orbs mediated that relationship. Theoretical astronomy, a field in which scholars proposed physical models to explain the motions of the celestial orbs, turned out to be the field in which exchange with momentous implications for the historiography of Renaissance astronomy took place.

SIX

THEORETICAL ASTRONOMY *between* RENAISSANCE ITALY *and the* OTTOMAN EMPIRE

In the world of the merchants of knowledge and their contacts, the planets did not orbit on their own, pulled by gravity. Rather, the orbs, sometimes called spheres, were physical bodies that moved the planets. The details of the physical structure of the heavens did not matter only for astronomy; rather, these details implicated many disciplines. In Aristotelian philosophy and in astrology, for example, the cosmos was a locus of meaning making. Interpreting the motions of the planets demystified much of human existence, from disease to political upheaval. In the preceding chapter, we discovered that a merchant of knowledge interested in *Qabbalah* held that the *s*ᵉ*pirot*, manifestations of divinity, were part of the cosmos. For Michael Balbo, studying the structure of the orbs that carried the planets was exploring the relationship of God to the cosmos that scholars of theoretical astronomy described. By "theoretical astronomy," I mean the field in which scholars theorized combinations of the physical yet translucent, uniformly revolving orbs that were thought to make up the heavens and carry the planets in their paths. Because planets' motions were complex, often more than one orb was required to account for the motions of a single planet. Modern astronomers call these paths orbits and account for them with universal gravitation. The orbs that were thought to carry closer planets, such as Mercury and Venus, were thought to be nested within the orbs that carried more distant planets, such as Jupiter and Saturn. Though astrology can be thought of as applied astronomy, we have already learned about how scholars' exploration of cosmology and the kinematics of the orbs mattered for other disciplines.

The challenge of theoretical astronomy, as far as the merchants of knowledge and their contacts were concerned, was determining whether the models that cohered with Averroism and *Qabbalah* were sufficiently precise for practical applications such as timekeeping and astrological forecasting. We have learned that Averroists theorized orbs that shared the same center, a feature that hindered the mathematical consonance of the theoretical models with observations. But we will see that the importance of Averroist philosophy to the merchants of knowledge and their contacts widened the exchange of theoretical astronomy. The exchange of theories that cohered with Averroism facilitated the exchange of non-Averroist, more mathematically precise theories that became significant in the history of Renaissance astronomy. The merchants of knowledge mediated a transregional conversation that was a part of the Mediterranean context for advances in astronomy.

A Shared Tradition of Theories of Homocentric Orbs

Many of the merchants of knowledge and their European contacts favored the Averroist interpretation of Aristotelian natural philosophy. As explained in chapter 4, Averroists mostly denied the existence of epicycles and eccentrics because both were orbs that did not share a center with the earth (see chapter 4, figure 4.4).[1] Rather, Averroists claimed that all orbs must share the same center, meaning the cosmos must be composed of homocentric orbs. Astronomers in the Ptolemaic tradition theorized epicycles and eccentrics, however, because models that included epicycles and eccentrics allowed a scholar to predict where the planets would be much more successfully. Given the application of astronomy to timekeeping and astrology—both of which required precision—it is surprising that so many scholars during the European Renaissance took homocentric theories seriously. But they did.

One reason why was because homocentric theories better mirrored some observations.[2] Specifically, Renaissance scholars believed that homocentric models were more effective than Ptolemaic theories (with epicycles and eccentrics) at accounting for observations of the sizes and distances of the planets. In premodern times, the sizes and distances of the planets were determined as follows. After the size of the earth was known through geodesy, scholars determined the distance of the moon from the earth by measuring the lunar parallax—a distortion in the apparent location of the moon resulting from the observer's location on the earth's surface rather than at the center of the earth. Once the moon's distance was known, the rest of the planetary distances could be calculated because the orbs were presumed to nest within each other. A signal weakness of homocentric models was that they entailed a fixed distance between each celestial body and the earth, which meant that it was impossible for homocentric models to explain observations of variations in the observed sizes of the planets. Yet astronomers in the Ptolemaic tradition, to predict planetary positions successfully, theorized epicycles and eccentrics with dimensions that entailed sizes and distances that diverged even more from observations! Such epicycles and eccentrics could not exist. Hence, for important Renaissance astronomers, homocentric astronomy was a philosophically coherent, and, relatively speaking, mathematically precise way of disagreeing with Ptolemy.[3]

To these Renaissance scholars, only homocentric orbs existed in reality and not just in theory. Regiomontanus (Johannes Müller von Königsberg; d. 1476) was the most talented European Renaissance astronomer before Nicholas Copernicus (d. 1543). Regiomontanus may not have believed that his teacher Georg Peuerbach (d. 1461) investigated real orbs in his *Theoricae novae planetarum* (New Theories of the Planets) precisely because Peuerbach accepted eccentrics and epicycles.[4] Scholars who favored homocentric astronomy aimed to do more than just demonstrate that epicycles and eccentrics were not absolutely necessary for predicting planetary longitude. In his *Homocentrica*, Girolamo Fracastoro (d. 1553)—an instructor of logic at Padua in the first decade of the sixteenth century[5]—investigated observations of comets, which, at the time, were believed *not* to be the result of the motions of the orbs.[6] Hence, theorists of homocentric orbs understood their task to be presenting systematic theories for all celestial phenomena, and not just arguing that planetary motions might be accounted for with homocentric orbs. Models composed of homocentric orbs resolved another contradiction of Ptolemaic astronomy brought to Regiomontanus's attention by Henry of Langenstein's (d. 1397) criticisms of the eccentric. That contradiction was Ptolemy's proposition that some orbs revolved uniformly, in place about a point, called the equant point, removed from the orb's center.[7] Put differently, a point on the circumference of the orb should not revolve uniformly from the perspective of an off-center observer.

Scholars who advanced homocentric theories did so because they believed that homocentric theories could account precisely for all observed celestial phenomena, especially the longitudes of the sun, moon, and planets.[8] Homocentric models, to their proponents, were not just an Averroist alternative that regrettably sacrificed mathematically accurate predictions.

An Andalusian Muslim scholar, Nūr al-Dīn al-Biṭrūjī (fl. 1200), was the first to attempt homocentric theories that might yield the mathematical precision of the Ptolemaic models in his *On the Principles of Astronomy* (Ar. *Kitāb fī al-hay'a*). That text was translated into Latin in 1217 and into Hebrew in 1259. Paralleling the greater popularity of Averroist philosophy among Christian and Jewish scholars than among Muslim scholars, Biṭrūjī had a greater impact on astronomy in the European Renaissance and in Jewish cultures than he did on astronomy in Islamic societies.[9] Regiomontanus took Biṭrūjī's work very seriously and aimed to improve on the homocen-

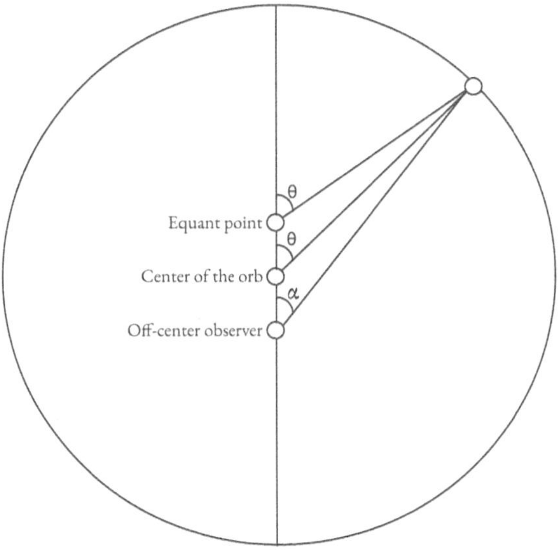

FIGURE 6.1. The equant point.

tric models presented in Biṭrūjī's *On the Principles.* Copernicus cited Biṭrūjī's *On the Principles,*[10] which was—like Copernicus's *Commentariolus*[11]—described as a *theorica planetarum* text in 1531.[12] *Theorica planetarum* was a Latin genre devoted to summary descriptions of the celestial orbs, initiated by the translation of Ibn al-Haytham's (d. ca. 1040) *On the Configuration of the World* (*Maqāla fī hay'at al-ʿālam*) into Latin.[13] The classification of Biṭrūjī's *On the Principles* with other astronomy texts, and not as philosophy, is evidence that Biṭrūjī's work was taken seriously as astronomy.

Still, Ptolemaic models enjoyed a superior mathematical precision, so by proposing innovative homocentric models for the sun and the moon in 1460, Regiomontanus aimed to close the gap between homocentric and Ptolemaic models.[14] To improve the mathematical precision of his homocentric theories without resorting to epicycles and eccentrics, Regiomontanus hypothesized a combination of uniformly rotating homocentric orbs that caused a point to oscillate back and forth.[15] This hypothesis was known as a reciprocating mechanism,[16] or a slider-crank mechanism, and functioned similarly to the rod on a locomotive wheel:[17] the locomotive wheels revolve, but the rod oscillates back and forth. Then, the orbs of the recip-

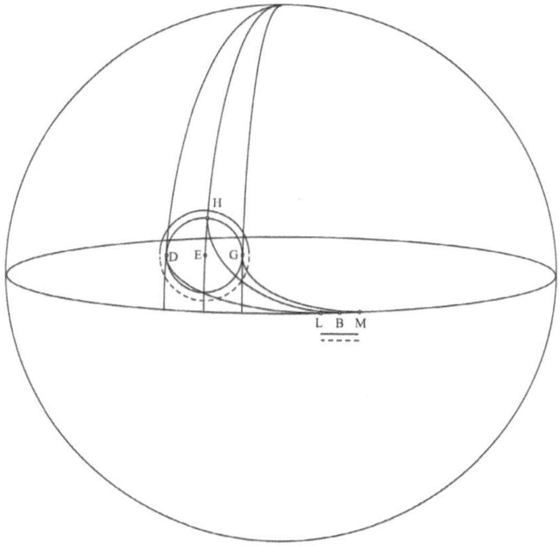

FIGURE 6.2. The reciprocation mechanism. The revolution of the pole of the outer, partially dashed orb from D to H to G moves a point from L to B to M on the inner orb, which also revolves uniformly.

rocating mechanism were combined with another uniformly rotating orb, resulting in a homocentric model that came much closer to accounting for the planets' observed positions than Biṭrūjī's solar and lunar models.

But there was a physical problem. How did the oscillating point stay on the arc without some sort of track?[18] A solution was available in astronomy from Jewish cultures. Around 1400, Joseph b. Joseph Ibn Naḥmias provided insightful solutions to that problem in *The Light of the World* (Ar. *Nūr al-ʿālam*), the most sophisticated text of homocentric astronomy to pass through the network of the merchants of knowledge. The text was composed in the Iberian Peninsula.[19] Ibn Naḥmias wrote in order to resolve the "true perplexity" described in book 2, chapter 24 of Maimonides's (d. 1204) *Guide for the Perplexed*. Maimonides was perplexed because if the cosmos was really composed of homocentric orbs, then one should be able to account precisely for observations with models of homocentric orbs. Maimonides did not believe that any previous theorist of homocentric astronomy had succeeded, as he named none. In *The Light of the World*, Ibn Naḥmias took up the challenge posed by Maimonides. The text survives both in its Judeo-Arabic original and in a Hebrew recension that may have been produced by Ibn Naḥmias himself.[20] Moving forward, when I attribute a statement to the author of the Hebrew recension, that statement is absent from the Judeo-Arabic original. Statements ascribed to Ibn Naḥmias are found in the Judeo-Arabic original and, in translation, in the Hebrew recension. Only the author of the Hebrew recension proposed theories that were free of the aforementioned shortcoming of the reciprocation mechanism.

Galeano/Jālīnūs knew the contents of *The Light of the World* and shared them with Ottoman Muslim scholars in a text that he composed in Arabic, in Arabic script. In that treatise, which is now bound in a codex in the Topkapı Library with Biṭrūjī's *On the Principles*, Galeano/Jālīnūs described the Ottoman court and reproduced long passages verbatim from *The Light of the World*.[21] Hence, there is every indication that he succeeded in his goal of informing Muslim scholars about *The Light of the World*. Galeano/Jālīnūs's Arabic text must have been completed by 1500 because Galeano/Jālīnūs's translation of the *Almanach perpetuum* into Arabic in 1505 evidenced a much better command of Arabic, indicating that he wrote on homocentric astronomy before his Arabic had improved.[22] In Giulio Bartolocci's (1613–87) *Bibliotheca magna rabbinica de scriptoribus*, a bio-bibliographical dictionary of Jewish literature, there was a report of *The Light of the World* being seen at

Padua.[23] Jewish astronomers knew of Regiomontanus: Vatican MS Ebr. 387, for example, contains astronomical tables from the 1460s, computed for the longitude of Padua, which were attributed to Regiomontanus.[24] Even if the attribution was spurious, it can be explained only by an awareness among Jews in Italy of Regiomontanus's achievements. Regiomontanus's desire to improve on Biṭrūjī led to the passage of *The Light of the World* to Padua.

Exchange of *The Light of the World* between Galeano/Jālīnūs and Christian scholars must have occurred when Galeano/Jālīnūs visited Venice between 1497 and 1503 and met with the printer Gershom Soncino.[25] Because Galeano/Jālīnūs referred to the trip in *Puzzles*, he must have completed *Puzzles* after the visit. Yet the chronology of his Arabic writings, discussed in the previous paragraph, means that he knew a lot about astronomy before traveling. The career of a printer like Soncino illustrates how commerce and scholarship intermingled. He printed numerous works in Hebrew; the modern *Soncino Ḥumash*, a volume containing the biblical selections read in synagogues throughout the liturgical year, has been published by a press named for the family. He also published extensively in Latin and Italian under the name Hieronymus Soncinus.[26] To procure texts to publish, Soncino was in sustained contact with Renaissance humanists,[27] and his printing activities contributed to the propagation of the curriculum of Renaissance humanism.[28] It was during this visit that Galeano/Jālīnūs acquired a copy of the *Almanach perpetuum*, which he translated for Muʾayyadzādah in 1505. Therefore, while Galeano/Jālīnūs was providing Christian scholars with texts from the Ottoman Empire, he was on the lookout for material that might interest potential Ottoman Muslim patrons.

Statements by Christian scholars reflect some interest in recent Arabic texts composed close to their time. Alessandro Achillini (d. 1512) was a professor of philosophy at Padua between 1506 and 1508 who also taught at Bologna before 1506 and after 1508.[29] Copernicus studied in Bologna from 1496 to 1500, overlapping with Achillini. Achillini was the unnamed target of Francesco Capuano's commentary on Sacrobosco's (d. 1256) *De sphaera* (On the Sphere), first published in Venice in 1498/99.[30] Incidentally, Capuano's commentary was known to Copernicus's circle.[31] Achillini's *De orbibus* (completed in 1498) was a text of homocentric astronomy and furnishes evidence that scholars in Italy recognized that the available homocentric theories needed to be improved.

In *De orbibus*, Achillini remarked that theories of homocentric orbs were less successful than other models at accounting for available observations—that is, saving the appearances. He specified that "therefore it is by the circular motions described through spirals, rather than through sphericals, from the true poles of imaginary spheres that the reasons for the appearances are restored."[32] The imagined spheres (*sphaerarum imaginariae*) that yielded accurate calculations were the eccentrics and epicycles of Ptolemaic astronomy. Achillini explained that Averroës called the spiral traced by a point on a revolving epicycle *lembab* or *leulebie*; both were derived from the Arabic *lawlab*.[33] In contrast, only homocentric orbs existed in external reality, though they did not account for observations or predict future positions.

Achillini averred that, with more work, Callippos's (d. ca. 300 BCE) and Eudoxos's (d. ca. 340 BCE) homocentric theories—which reportedly would save the appearances—might be recovered. However, to this day, their theories are known only indirectly through the reports of Aristotle and other ancient Greek scholars. An additional complication was that the Arabic reports of the theories of Eudoxos and Callippos were insufficient for a full explication of their theories. Achillini explained: "Some of what the first master [i.e., Aristotle] is to have said is not to be believed without a strong proof. But at hand is only a recitation of the opinion of Eudoxos and Callippos, themselves being silent, and following whom Averroës clearly rejected the epicycles and eccentrics that he recalled from his post-Aristotelian predecessors. But he [Averroës] truncated the appropriate way of saving the appearances as much as by his few words as by the Arabic."[34] In other words, as much was lost in transit and translation, Achillini did not view Averroës as having the last word in astronomy. In another case, Achillini criticized Averroës's Arabic translation of Aristotle's texts in other fields, for example noting that Averroës rendered the Greek *daemonia* in *Metaphysics* 1017b—translated in English as "divine"—as "idols" (*aṣnām*).[35] Achillini, given his pessimism about the accuracy of Arabic translations of Aristotle's works, left no indication that Eudoxos and Callippos's astronomy could be retrieved through Arabic translations of Aristotle's works. *The Light of the World*, though written in Arabic, was an original composition and, as such, not subject to the aforementioned criticisms. To an interested European scholar, it would have been a novel source for homocentric theories. Achillini, despite his thorough analysis, did not provide new homocentric models; there is no evidence that he read *The Light of the World*.

There is evidence that Christian scholars would have been motivated to learn about *The Light of the World* from Galeano/Jālīnūs when he was in Venice. Agostino Nifo (d. 1538), who studied at Padua with one of Pico della Mirandola's teachers, was more sanguine than Achillini about the value of Arabic sources. In *Expositions of the "Metaphysics,"* a super-commentary on Averroës's *Long Commentary*, composed in the 1530s, Nifo returned to the problems of the epicycle and the eccentric. He proposed, as Averroës[36] had, that Spanish mathematicians' models for the orb of the fixed stars—based on concentric orbs—be expanded to the models for all planets as solutions for the difficulties of the epicycle and eccentric. Nifo wrote:

> Thus, the mathematicians of Spain (*mathematici Hispaniae*) intend that the poles of the zodiac revolve about the poles of the entire world, describing a small circle, supposing consequently that these motions be granted for the zodiac and the entire world. Consequently, it is not unsuitable for such spheres. He [Averroës] adds that the Spanish mathematicians grant such a motion of the poles to the starry orb. They say that because the starry orb is moved about the poles of the whole, this is the ninth orb's motion in trepidation. From this the truth of the solution becomes clear, as is clear in the rest of the comment.[37]

Aspects of Biṭrūjī's homocentric models resembled those Andalusian models for the motion of the zodiac.[38] The merchant of knowledge Elijah Delmedigo also referred to these Spanish scholars in "Question on the Prime Mover" and listed Ibn al-Zarqāl (d. 1100), who theorized a new model for the motions of the zodiac, among them.[39] Even if Nifo did not meet Delmedigo, he could easily have read Delmedigo's printed works. Like Achillini, Nifo acknowledged that authors writing in Arabic (and *The Light of the World* was in Arabic) were a source for innovations in homocentric astronomy. With Delmedigo having paved the way, there was an audience at the University of Padua for anything Galeano/Jālīnūs shared about the improvements on Biṭrūjī's homocentric astronomy found in *The Light of the World*.

The Hebrew recension of *The Light of the World* contains three proposals for combinations of homocentric orbs that would make a point oscillate on a great circle arc. Those proposals did not suffer from the shortcomings of the reciprocating mechanism that Regiomontanus theorized. Regiomontanus failed to explain how the oscillating point would stay on track.[40] The first two proposals had some physical inconsistencies: the proposals en-

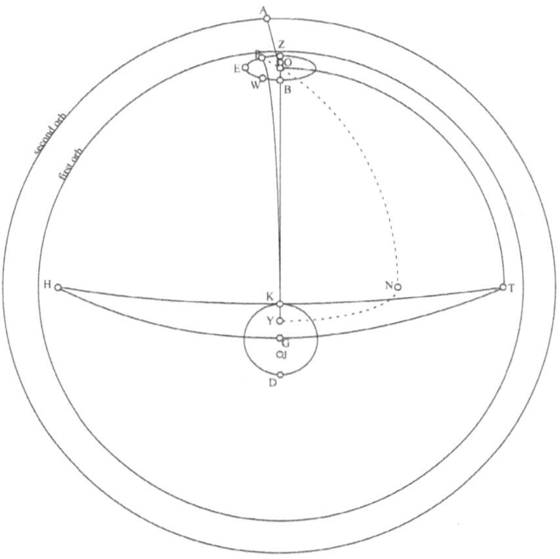

FIGURE 6.3. The Eudoxan couple.

tailed that entire orbs oscillated, rather than just a point. But the author of the Hebrew recension provided an elegant third option consisting of two homocentric orbs with the pole of the first inclined from the pole of the second.[41]

The two orbs were proposed to revolve the same amount but in opposite directions.[42] The theory functioned in the following way: first, the second orb revolved a certain number of degrees about pole A to move point K to point N. Second, the first orb revolved the same number of degrees in the opposite direction about pole Z to bring point K from point N down to point Y.[43] Though I have described the first and second motions sequentially, the motions occurred simultaneously so that point K moved on a great circle arc to point Y. As the author of the recension realized, the hypothesized solution did not work precisely, because the two equal motions in opposite directions were not about the same pole.[44] Rather than an oscillation precisely on a great circle arc, this configuration of homocentric orbs produced a narrow hippopede: a pinched figure-eight shape that nevertheless imparted variations from the mean motion.[45]

When these two orbs were combined with a third that revolved uniformly with the planet's mean motion, the author of the Hebrew recension improved markedly on the theories of Biṭrūjī and Regiomontanus. Scholars of homocentric astronomy at Padua would have been interested in such an innovation.[46] It makes sense that *The Light of the World* was seen at Padua. This hypothesis has been called a Eudoxan couple because it is identical with the most widely accepted reconstruction of Eudoxos's homocentric astronomy. Anyone frustrated with the shortcoming of the reciprocating mechanism would have found a solution in the Hebrew recension of *The Light of the World*.

In *The Light of the World* we also find a version of a famous innovation from late medieval Islamic astronomy: the Ṭūsī couple, first proposed by Naṣīr al-Dīn al-Ṭūsī (d. 1274), the director of the Marāgha Observatory.[47]

The Ṭūsī couple comprises two orbs, with one orb twice the size of the other. The circles in the figure indicate the circumference of the orbs. The edge of the circumference of the smaller orb or circle is flush with the circumference of the larger one. The smaller orb or circle revolves with twice the angular velocity as the larger one, and in the opposite direction. As a result, a point on the diameter of the larger orb oscillates back and forth. The Ṭūsī couple could be combined with a uniformly revolving orb to pro-

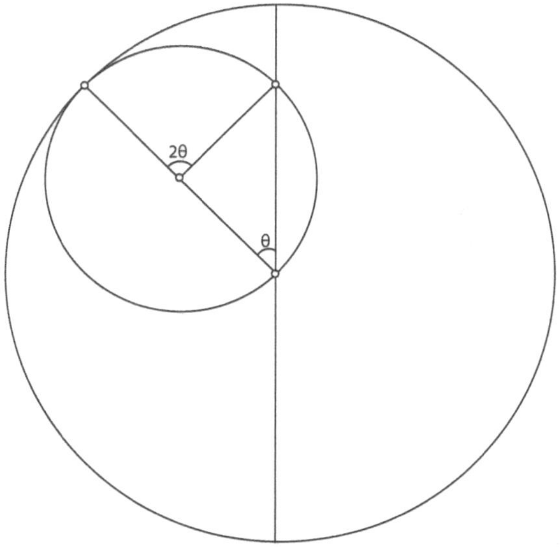

FIGURE 6.4. The Ṭūsī couple.

duce variations from uniform revolutions. There was another version of the couple in which the oscillation is on the surface of an orb and is produced through homocentric orbs.[48] In fact, Ṭūsī remarked that the Ṭūsī couple was a development of an Eudoxan couple. In other words, by placing a homocentric theory in the genealogy of the Ṭūsī couple, Ṭūsī suggested that his goals as a theoretical astronomer had something in common with the aims of scholars who favored homocentric theories. The author of the Hebrew recension of *The Light of the World* made the same connection, without mentioning Ṭūsī's name.[49] These theoretical connections were paralleled by exchange between the Islamic East and al-Andalus. Ibn Naḥmias did not claim to have developed the Ṭūsī couple found in the Judeo-Arabic original of *The Light of the World*.[50] Exchange is a more likely explanation for the presence of the Ṭūsī couple in *The Light of the World*, as there is evidence of mathematics and instruments passing from Marāgha to Andalusia before Ibn Naḥmias's lifetime.[51]

There is evidence that Galeano/Jālīnūs was a successful intermediary for the exchange of the Ṭūsī couple with scholars at the University of Padua in the late fifteenth and early sixteenth centuries. A version of the Ṭūsī couple appears in the homocentric astronomy of Giovanni Battista Amico (d. 1538), who wrote in the 1530s at Padua. Amico applied the Ṭūsī couple to solve the same problem faced by Regiomontanus, Ibn Naḥmias, and the author of the Hebrew recension of *The Light of the World*: producing an oscillation on a great circle arc on the surface of homocentric orbs.[52] With the exception of the addition of an orb to prevent the motions of the Ṭūsī couple from being passed to other orbs, Amico's version of the Ṭūsī couple is identical to one found in both versions of *The Light of the World*.[53] Amico read Hebrew.[54] Girolamo Fracastoro (d. 1538) was a Paduan scholar who innovated homocentric models, and paralleling Ṭūsī's remark about the Eudoxan couple, presented Amico's hypothesized version of the Ṭūsī couple within a chapter on Callippos and Eudoxos's astronomy.[55] Fracastoro did not explain why he did not incorporate the Ṭūsī couple into his own models.[56]

The applications of the Ṭūsī couple by Amico and Ibn Naḥmias sometimes differed. Amico used the Ṭūsī couple to account for variations in the sun's motion around the earth.[57] By contrast, Ibn Naḥmias introduced the Ṭūsī couple for a slightly different purpose: to eliminate a remaining displacement in latitude entailed by his solar theory but not observed. Nevertheless, Ibn Naḥmias and the author of the Hebrew recension of *The*

Light of the World, a text that Christian scholars saw,[58] proposed using the Ṭūsī couple alone to account for variations in the sun's velocity around the earth, just as Amico later did.[59] Amico also incorporated a Ṭūsī couple into the lunar model to account for additional variations in the motion of the moon around the earth.[60] Amico did not understand the Ṭūsī couple as well as Ibn Naḥmias and the author of the Hebrew recension did. Mathematical analysis reveals that while the Ṭūsī couple functions perfectly in the plane, it does not do so on the surface of an orb.[61] That is because the geometry of a plane surface is not the same as the geometry of a curved surface. Amico was not aware of this fact. Ibn Naḥmias, on his part, noted that a solar model with the Ṭūsī couple alone would not reproduce the asymmetries of the sun's observed motion.[62] While Ibn Naḥmias was aware of this fact, Amico was not. Amico's ignorance of this imprecision militates against the presumption that he developed a version of the Ṭūsī couple independently.[63] These strong similarities and slight differences between Amico and Ibn Naḥmias's astronomy can be explained more easily by the exchange just of diagrams or oral exchange of part of the content of *The Light of the World*; the whole text did not have to travel, though it could have.

The homocentric astronomy of Fracastoro came to the attention of Copernicus.[64] In a letter that Copernicus prefaced to *De revolutionibus*, his second, better-known text on theoretical astronomy, he mentioned that mathematical astronomers contested the foundations of their theories and surveyed the different approaches. Some used homocentric orbs, others used epicycles and eccentrics, and he alluded to some who used only eccentrics.[65] In that letter, Copernicus referred to but disagreed with Fracastoro's homocentric theories ostensibly because of their weak predictive accuracy.[66] When he wrote, Copernicus might have been preoccupied by the challenge that Fracastoro's *Homocentrica* posed to his own attempt to gain patronage and support for his distinct reform of astronomy.[67] Copernicus incorporated into the *Commentariolus* a version of the Ṭūsī couple found in, among other places, *The Light of the World* and Amico's *De motibus corporum coelestium* (On the Motions of the Celestial Bodies).[68] Thus, while Copernicus did not favor homocentric models for the moon and planets, he would have been open to some of the components of homocentric models. Whatever Copernicus's direct source was, homocentric astronomy texts such as *The Light of the World* contained highly sophisticated theories that were unknown to Eudoxos and Callippos.

Connections between Homocentric and Nonhomocentric Theories

Homocentric astronomy impelled Delmedigo, an Averroist merchant of knowledge, to think more broadly about theoretical astronomy. In his 1485[69] commentary on Averroës's *De substantia orbis*, Delmedigo made a tantalizing reference to attempts to reform Ptolemaic astronomy:

> All of them [i.e., Averroës, Ibn Ṭufayl, and Maimonides] said that the principles of physics upset what Ptolemy proposes regarding this [i.e., astronomy]. And it [physics] is true and, without a doubt, there are attached to Ptolemy's astronomy enormous gaps in his proofs, since he did not have completely satisfactory evidence for the epicycle and eccentric he proposed, as Averroës explained in many places. Even the doctrines of the modern astronomers and their like, who thought to save Ptolemy, necessitate that there be a heavenly body without any function so that there will not be proposed any void in a few of the models of the planets. This body nearly eliminates the difference, that is to say that a part of it is very thick and that a part is very thin and that it occurs with this that it moves in a way agreeing with the rest of the bodies that are with it until no void occurs nor interpenetration of [celestial] bodies as is known to whomever looks at their words. All of this is a worthless fancy.[70]

The quotation from Delmedigo's commentary began with references to recent scholars—Maimonides and Ibn Ṭufayl (d. 1185)—who, like Averroës, questioned the existence of epicycles and eccentrics. Next, Delmedigo referred to modern (i.e., post-Ptolemaic) astronomers who also sought to save Ptolemy by updating his theories to recuperate their mathematical precision and noted that these modern astronomers accepted the existence of the complementary bodies, bodies believed to fill the void between the partial and concentric orbs. At the end of the translated passage, Delmedigo rejected the existence of these bodies, indicating that he favored a homocentric astronomy because complementary bodies were necessary only if epicycles and eccentrics were theorized. Clearly, the would-be savers of Ptolemy whom Delmedigo invoked were not Biṭrūjī or Ibn Naḥmias, who favored a homocentric astronomy as Delmedigo did. It is possible that Delmedigo was thinking instead of Ibn al-Haytham or Jābir b. Aflaḥ (fl. twelfth century CE), critics of Ptolemy whom Averroës cited in *Compendium of the "Almagest."*[71] Given the young age at which Delmedigo left Crete for Italy, it is also possible that he learned of these attempts to save Ptolemy once he was in Italy. No matter the identity of the referent of Delmedigo's comment about modern astronomers, we have in Delmedigo a Jewish scholar who had Christian students, moved between Crete and the Veneto in the last two decades of the fifteenth century, and who informed his students of some attempts to reform Ptolemaic astronomy that incorporated epicycles and eccentrics. Delmedigo, a merchant of knowledge with a broad readership, wrote most about recent advances in homocentric astronomy but was aware of other paths forward in theoretical astronomy.

Who might have been those savers of Ptolemy to whom Delmedigo referred? In addition to the names mentioned above, there were a few European, pre-Copernican constructive critics of Ptolemy who theorized epicycles and eccentrics rather than homocentric orbs. One, Albert of Brudzewo (d. 1497), taught astronomy at the University of Cracow while Copernicus was a student there. Albert, in his commentary on Peuerbach's *Theoricae*, attempted to resolve a nagging contradiction of Ptolemaic astronomy.[72]

Albert called attention to the nonuniform motion of certain orbs Peuerbach had adopted from Ptolemy for the models for the planets. Those orbs that moved with nonuniform motions facilitated calculations but could not exist in the physical world because orbs had to revolve uniformly in place about their own centers, not another point (called the equant).[73] In addition, to explain observed variations in the moon's position, one of the moon's orbs in the Ptolemaic lunar model revolved in place uniformly about a point other than its center. Albert commented that tables calculated according to this point "frequently cannot be attributed or applied to the motions as they are in their nature or as they appear."[74] The motion of the moon's epicycle in the Ptolemaic system was uniform according to a moving point, and Ptolemy never proposed a mover for the apogee of the epicycle, the point according to which the epicycle's uniform motion was measured. To eliminate this difficulty, Albert hypothesized an additional, outer epicycle concentric with the lunar epicycle.[75] Hence, by incorporating a version of the Eudoxan couple into the lunar model, Albert accounted for the motion of the apogee of the epicycle by proposing orbs that revolved uniformly about their own centers. He borrowed a technique from homocentric astronomy to improve astronomy in the Ptolemaic tradition.

In *The Light of the World*, Ibn Naḥmias mentioned that he was aware of that same problem—the need to explain the motion of the apogee of the lunar epicycle.[76] Ibn Naḥmias classified the lack of an explanation as one of the repugnancies (*shināʿāt*) of Ptolemaic astronomy and acknowledged that it was unrelated to the philosophic shortcomings of the eccentric and epicycle.[77] Hence, among Christian and Jewish scholars, critiques of and improvements upon Ptolemaic astronomy appeared in both homocentric and nonhomocentric texts. Albert's proposal was evidence that homocentric modeling techniques, i.e., concentric epicycles, appealed to authors who accepted epicycles and eccentrics. Because a manuscript of the Hebrew recension of *The Light of the World* arrived in Europe, the overlapping concerns of Ibn Naḥmias and Albert of Brudzewo are significant. Disagreement over approaches to theoretical astronomy was not a barrier to exchange.

Other Critiques of Ptolemaic Astronomy among the Jewish Contacts of the Merchants of Knowledge

Besides the improvements and critiques of Ptolemaic astronomy found in texts of homocentric astronomy, Jewish scholars in the century before the merchants of knowledge working outside of homocentric astronomy noted faults of Ptolemaic astronomy. For example, Levi b. Gerson (d. 1344) knew that the equant point was problematic but found no alternative. A century later, Profayṭ Duran (fl. ca. 1415) displayed an extensive knowledge of the weaknesses of Ptolemy's astronomy in a commentary on Averroës's *Compendium of the "Almagest."* Duran called the equant a *sapeiq* (doubt) and remarked on other shortcomings of Ptolemaic astronomy not found in Averroës's *Compendium*, such as Ptolemy's failure to propose an orb to move the planets to the north and south of the ecliptic (the path of the sun), as observations indicated.[78] Duran rejected homocentric astronomy but was optimistic about humans' ability to find better mathematical models.[79] For Jewish scholars, as was the case for Christian scholars, homocentric astronomy was not the only path to better theories.

Among the merchants of knowledge and their Jewish contacts, some of the first discussions of theoretical astronomy appeared in Mordechai Kumaṭiano's 1480 commentary on Maimonides's *Guide of the Perplexed*, a text central to Romaniot intellectual life. Mirroring Maimonides, Kumaṭiano had greater confidence in practical astronomy than in humans' ability to find a mathematically precise physical model that would cohere with the homocentric orbs of Aristotelian philosophy. Kumaṭiano commented, "This remark belongs to me (*zeh hu haʿarah li*), that is to say that from the orb of the moon upward, the causes are not perceived as is appropriate."[80] In other words, the model (*tᵉkunah*)[81] that an astronomer proposed to account mathematically for observations was not necessarily the correct physical explanation of a planet's motions.[82] He agreed with Maimonides's statement at the end of the chapter that "regarding all that is in the heavens [i.e., above the orb of the moon], man grasps nothing but a small measure of what is mathematical; and you know what is in it."[83] To justify his skepticism, Maimonides questioned the correlation between a planet's distance from the uppermost orb and its speed in *Guide* II.19.[84]

Kumaṭiano concurred that the different velocities of the orbs were not caused solely by their different distances from the first cause because the velocities were not in precise proportion to the differences in distance.[85] Venus, the sun, and Mercury, for example, shared the same mean motion despite their different distances from the first cause.[86] Kumaṭiano found that Averroës's proposal that there were three causes for the celestial bodies' different motions—the motive power, the power opposing the motive power, and distance[87]—"was not an improvement" upon available explanations for how Venus, the sun, and Mercury disrupted the distance-velocity relationship.[88] Kumaṭiano, however, offered no alternative.

Kumaṭiano's student Caleb Afendopolo (d. 1525) carried on the critical study of Ptolemaic astronomy among Romaniot Jews. Afendopolo's study of Ptolemy's *Almagest* began with Jābir b. Aflaḥ's *Iṣlāḥ al-Majisṭī* (Correction of the *Almagest*; Heb. *Qiṣṣur al-Magisṭi*).[89] Marginalia in a 1482 manuscript of a Hebrew translation of Jābir's *Correction*, copied by Afendopolo, reflect the thoroughness with which Afendopolo and his students studied the text.[90] For example, one reader discovered that a certain passage was missing in the Arabic version.[91] Elsewhere Afendopolo realized that a figure was in the wrong place.[92] A flyleaf at the end of the MS contains three statements from students of Afendopolo, including one who hailed from Candia, stipulating that they were not permitted to teach Jābir's *Correction* without Afendopolo's permission, as long as Afendopolo lived, to anyone, either Rabbanite or Qaraite.[93] The commentary may have been a work in progress. Afendopolo mentioned that he authored a work on astronomy entitled *Gan ha-meleḵ* (The Garden of the King) to facilitate the study of the *Almagest*, but Afendopolo's text with that title is mostly poetry.[94] His personal library included a 270-

folio copy, now BnF hébreu 724, of part 1 of book 5 of *Wars of the Lord*, chapter 43 of which contains Levi b. Gerson's critique of Ptolemaic astronomy and the first twenty chapters of which contain Levi's innovative models for planetary longitudes.[95] The Qaraite Afendopolo purchased the manuscript from the Rabbanite Rabbi David b. Judah Messer Leon (d. ca. 1526 in Salonika),[96] an acquaintance of the merchant of knowledge Elijah Delmedigo, in 1497–98.[97] Afendopolo's quest for newer, mathematically precise models led him to forge connections with Rabbanites.

Elijah Mizraḥi (d. 1525/26)[98] was another student of Kumaṭiano, a teacher of Galeano/Jālīnūs, and a correspondent of the merchant of knowledge Elijah Capsali.[99] Mizraḥi's commentary on the *Almagest* contains the first detailed exploration by someone in the milieu of the merchants of knowledge of solutions to the problems of Ptolemaic astronomy.[100] In his commentary, Mizraḥi repeatedly referenced an Arabic version of the *Almagest*, as well as Ibn Rushd's *Compendium* and Jābir's *Correction*, suggesting that he had access to Ṭūsī's recension and/or to the commentaries thereon, as manuscripts of the original translations of the *Almagest* that have survived to the present are few.[101]

Mizraḥi's extensive comments on Ptolemy's final lunar model show that he grasped its physical inconsistencies. Mizraḥi argued that the opposite directions of the two motions of the orb (the eccentric deferent) that carried the epicycle posed a problem. As the framing of this critique of the Ptolemaic lunar model was not found in Ibn al-Haytham's *al-Shukūk ʿalā Baṭlamyūs* (Doubts about Ptolemy), it may have been Mizraḥi's own.[102] To elucidate, Mizraḥi imagined that the mover for each of these opposite motions was a line emerging from different points.

One of these lines emerged from the center of the ecliptic orb, the orb with the constellations, and terminated at the center of the epicycle.[103] The other line emerged from the center of the ecliptic and terminated at the center of the eccentric deferent. The first line moved the epicycle center to the east, and the second line moved the center of the eccentric to the west. Different lines caused different motions.

Mizraḥi wondered how a single body could move with two opposite motions in the same instant (*bᵉ-regaʿ eḥad*).[104] He invoked the example of an ant walking on a millstone that moved in the opposite direction of the ant to argue that there would have to be an instant between the opposite motions of the two lines that moved the epicycle center. Mizraḥi added that an orb

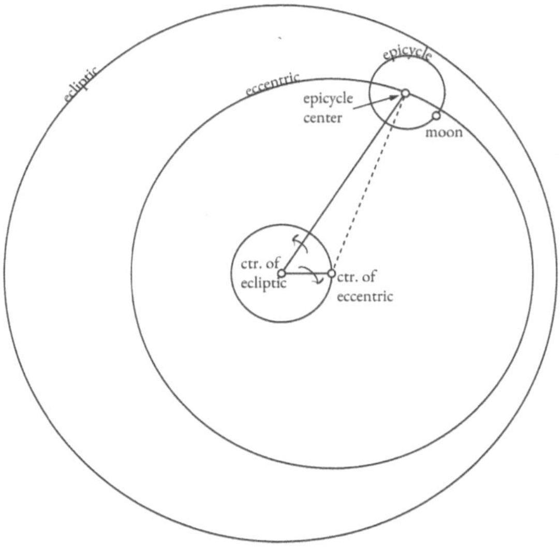

FIGURE 6.5. Mizraḥi's interpretation of Ptolemy's lunar model.

cannot be at rest, so it would be futile to propose that the opposite motions somehow occurred successively with a moment of rest between them. Mizraḥi also considered the more widely recognized complication of how the motion of the center of the moon's epicycle was uniform about the earth at the center of the ecliptic, even though the earth was not at the center of the orb that moved the lunar epicycle.[105] Put differently, the epicycle moved closer to and farther from the earth, the center of the epicycle's uniform motion.[106] Mizraḥi had a thorough, if idiosyncratic grasp of some problems of the Ptolemaic lunar model.

Mizraḥi attempted to resolve these doubts. On the matter of whether one orb could move with two opposing motions without there being an intervening moment of rest, Mizraḥi investigated the material substances of the orbs.[107] He remarked that there would have to be a moment of rest if the surface of the inclined orb that moves the orb carrying the center of the epicycle was a body composed of something hard, such as stone or iron, for the inner orb to be moved in a direction opposite to that of the motion imparted by the inclined orb.[108] But, he noted, the surface of the inclined orb was made of a different substance, something airy (*awiri*) and spongy (*s*ᵉ*pogi*).[109] Thus the inclined orb moved only the epicycle, and not the orbs that were theorized to move in the opposite direction. Were all the orbs made of the same substance, such as iron, there would had to have been a hollow space between them for them to move in opposite directions. Debates about whether the orbs were spongy (i.e., soft) or solid were found in Latin texts, suggesting that Mizraḥi was influenced by Hebrew scholasticism.[110] Another way to resolve the problem of a single orb moving with two motions would be to eliminate one of the two motions.[111] Mizraḥi did not say so explicitly, but doing so would entail phenomena that were never observed.

Mizraḥi took up the other doubt: that the motion of the eccentric deferent was uniform with respect to the center of the world and not with respect to the center of the eccentric deferent (G).

Because the center of the deferent did not move from its position, line AB would have to impart a repulsion (*d*ᵉ*ḥiyyah*) so that the center of the epicycle moved uniformly through line AB about the earth while varying its distance from the earth.[112] Though the observed size of the moon varied throughout the lunar month, there was no way to explain the changes in the length of line AB. Hence, Mizraḥi was unable to propose combinations

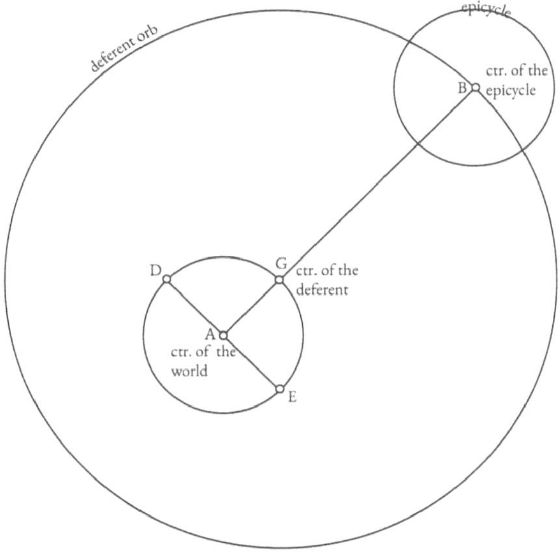

FIGURE 6.6. Mizraḥi's lunar model.

of uniformly revolving orbs that could explain the observed changes in the moon's distance from the earth.

Mizraḥi was Galeano/Jālīnūs's teacher, meaning that Galeano/Jālīnūs would have known of Mizraḥi's quest for solutions to the problems of Ptolemaic astronomy as well as of the challenges that faced scholars, such as Mizraḥi, who attempted solutions. Astronomers from late medieval Islamic societies successfully resolved the problems that tantalized Mizraḥi, so Galeano/Jālīnūs was well prepared to recognize the solutions when he encountered them at the Ottoman court.

The Availability of Islamic Theoretical Astronomy to the Merchants of Knowledge

We have already seen how the merchants of knowledge absorbed Islamic texts on practical astronomy. They procured texts of theoretical astronomy too. During the reign of Mehmed the Conqueror, the merchant of knowledge Moses b. Elijah Galeano, probably Galeano/Jālīnūs's uncle,[113] produced a Hebrew version of a popular Islamic introduction to theoretical astronomy. Muḥammad b. ʿUmar al-Jaghmīnī's *al-Mulakhkhaṣ fī al-hay'a al-basīṭa* (The Summary of Plain Astronomy), composed in 1205–6 CE,[114] was a widely commented-upon summary astronomy text that was a common starting point for the study of astronomy in the Ottoman Empire.[115] The oldest copy of Jaghmīnī's *Mulakhkhaṣ* in a Turkish library dates to 1246–47 CE,[116] and there are three hundred copies of Qāḍī Zādah's (d. after 1440) commentary on the *Mulakhkhaṣ* in libraries in Istanbul alone.[117] ʿAbd al-Raḥmān Mu'ayyadzādah (d. 1516), who patronized Galeano/Jālīnūs's Arabic version of the *Almanach perpetuum*, studied Qāḍī Zādah's commentary on the *Mulakhkhaṣ* in Shiraz with Jalāl al-Dīn al-Dawānī (d. 1502).[118] Moses b. Elijah Galeano brought this classic Arabic text to readers of Hebrew.

The *Mulakhkhaṣ* belonged to a genre of astronomical literature that the Ottoman taxonomer of the sciences Ṭāsh Kubrī Zādah (d. 1561) categorized as *hay'a basīṭa* (plain astronomy). Authors of astronomy texts in the genre of *hay'a basīṭa* described the models of orbs that move the planets but omitted the lengthy mathematical demonstrations of how the models were derived from observations. The contents of Latin *theorica* texts resembled those of *hay'a basīṭa* texts. Ṭāsh Kubrī Zādah began his teaching career in 1525, at the tail end of the network's intellectual activities, so he was well positioned to reflect upon the study of theoretical astronomy in the Ottoman Empire in the late fifteenth and early sixteenth centuries.[119] Ṭāsh Kubrī Zādah classified Ṭūsī's *Tadhkira* (Memoir), which contains the final version of the Ṭūsī couple, as an example of a text on the summary (*mukhtaṣar*) level. Ṭāsh Kubrī Zādah also named texts at the intermediate (*mutawassiṭa*) level, such as Mu'ayyad al-Dīn al-ʿUrḍī's (d. 1266) *Kitāb al-Hay'a* (The Astronomical Work; also "On the Principles of Astronomy").[120] *Nihāyat al-idrāk fī dirāyat al-aflāk* (The Highest Attainment in Comprehending the Orbs) by Ṭūsī's student Quṭb al-Dīn al-Shīrāzī (d. 1311) was a text at the fully elaborated (*mabsūṭa*) level.[121] *Hay'a basīṭa* texts were plain (*basīṭa*) due to the absence of proofs; the contents could nevertheless be complex.

The title of the Hebrew version of the *Mulakhkhaṣ*, produced by Moses b. Elijah Galeano,[122] is *Seiper Meᵉzuqqaq* (A Refined Book).[123] Moses b. Elijah Galeano had composed at least part of *Seiper Mᵉzuqqaq* by 1459, meaning that the book predated Mizraḥi's commentary on the *Almagest*.[124] Moses b. Elijah Galeano mentioned the assistance of an Ottoman Muslim intermediary, Mevlānā Aḥmeṭ, a judge in the Macedonian city of Veria, who was the father of Maḥmūd Çelebī.[125] Mevlānā Aḥmeṭ also translated a text from Greek into Turkish.[126] The inconsistent transcriptions of numerous technical terms in *Seiper Mᵉzuqqaq* indicate that Mevlānā Aḥmeṭ most likely explained the technical terms of the *Mulakhkhaṣ* to Moses b. Elijah Galeano in person.[127]

As a merchant of knowledge, Moses b. Elijah Galeano prized translation. In *Seiper Mᵉzuqqaq*, the theme of translation as a way to recover lost knowledge emerged once again. Galeano argued that Jews' knowledge of astronomy had been lost and that the only way to regain it was through translation:

> Moses Ben Elijah the Greek said: This book I translated from the mouth of a Muslim scholar named Mevlānā Aḥmeṭ. After God caused me to know all this, I undertook to translate it, due to its great importance, from the mouths of scholars, but not from the mouth of books.[128] Even though I heard, in truth, that this book had been translated into our language before by someone who preceded me, the need has come to me to do it because that translation is not to be found with us. Had I desisted from translating it, I would have caused it to be lost, since forgetfulness is commonplace. I intended [it] for my own benefit and for whomever de-

> sired. It is possible, as well, that with the differences between the translations there might be made known things that are not observed in one translation, these being from among the things that would behoove students to possess; they are prepared deeply and broadly. If that [earlier] translation be clearer and more encompassing, we still have not lost a thing. So let it be [lit. "let us think that"] as if we have done nothing at all. From God we solicit help.[129]

This comment has a context: consider the 1393 translation, from Latin into Hebrew, of Arnaldo da Villanova's (d. 1313) *Aspects in Judgment*. The translator, Solomon b. Abraham Avigdor, lamented the loss of Hebrew books in the Diaspora and observed that the books that the Jews still possessed were mixed with refuse.[130] Both translators used similar literary tropes to naturalize foreign knowledge in Jewish cultures.

With the *Purified Book*, Moses b. Elijah Galeano introduced Jewish readers to a text that was the first step in the study of theoretical astronomy at Samarqand and Istanbul.[131] In it, he blended abstract geometrical and physical approaches to astronomy, and, like Mizraḥi, theorized lines that would cause epicycles to revolve. Moses b. Elijah Galeano concluded *Seiper Mᵉzuqqaq* by counseling the interested reader to consult the *Almagest* and other longer books on astronomy for treatment of more complex questions. To a reader of Hebrew, next steps might have been Mizraḥi's *Almagest* commentary, Levi b. Gerson's *Wars of the Lord*, or *The Light of the World*. Somehow, Galeano/Jālīnūs's appetite was whetted to learn more.

Theoretical Astronomy in Galeano/Jālīnūs's *Puzzles of Wisdom*

From his time at the Ottoman court, Galeano/Jālīnūs became familiar with the most sophisticated late medieval Islamic astronomy. In a key passage of *Puzzles of Wisdom*, a passage first studied by Tzvi Langermann in 2007, Galeano/Jālīnūs described the main challenge faced by scholars of theoretical astronomy: combining orbs that individually revolved uniformly and, when combined, produced the observed nonuniform motions of the planets, moon, and sun. Then, he outlined four approaches to modeling the motions:

> This may occur (a) from the compounding of motions of uniform direction, models, and centers, as we maintain, along with "the man [whose theory] shook [the world]" and *The Light of the World* of R. Joseph b. Yaᶜish; or (b) from varying centers, models, and directions of the motions, according to the principles of Ptolemy; or (c) from varying centers, directions of motion [and] models on the epicycle alone, according to the astronomy of Ibn al-Shāṭir; or (d) from varying the models, directions, and centers, but by means of eccentrics alone, as in the new astronomy of Gersonides,[132] of blessed memory.[133]

Galeano/Jālīnūs has listed the four dominant approaches to representing the nonuniform motions of the planets with combinations of uniformly rotating orbs. Three of the four approaches cited were innovative post-Ptolemaic approaches.

Although no Muslim scholar of the period cataloged theoretical approaches as Galeano/Jālīnūs did, other Jewish authors did. For instance, Isaac b. Samuel Abū al-Khayr, a Jewish author in Padua, surveyed theoretical astronomy in his 1497 commentary on Farghānī's *Elements of Astronomy*. Abū al-Khayr concluded that "the human intellect had not yet arrived at the knowledge of the true situation of the stars in their orbs in the heavens."[134] Eliyahu al-Faji, who composed *Mik̲tab̲ Eliyahu* (Elijah's Letter), articulated the same question about astronomy that Galeano/Jālīnūs did: What were the ways in which one could represent nonuniform motions with uniformly rotating orbs?[135] Al-Faji was probably from the Ottoman Empire, as his responsa (Heb. *teshub̲ot*; answers to questions of Jewish law) appeared in a collection of responsa from Asia Minor and Palestine.[136] The three approaches[137] al-Faji suggested were those proposed by Ptolemy, Biṭrūjī (referenced as *ha-marᶜish*; the one who shook the world),[138] and Gersonides (Levi b. Gerson; 1288–1344). These were also three of the four that Galeano presented in *Puzzles of Wisdom*.

Al-Faji commented that astronomers were not able to prove which of the three approaches corresponded with external reality.[139] Neither computational proofs nor instruments (*keilim*) could distinguish decisively between competing demonstrations, both of which account for observations.[140] While al-Faji did not cite Ibn al-Shāṭir's approach, al-Faji did refer to an attempt to account for celestial motions with epicycles alone, which Galeano/Jālīnūs justifiably perceived as the hallmark of Ibn al-Shāṭir's approach.[141]

Most important, Galeano/Jālīnūs was the first person we know of with knowledge and understanding of Ibn al-Shāṭir's theoretical astronomy who was also in the Veneto at the same time as Copernicus.[142] While parts of *Puzzles* were written after Galeano/Jālīnūs's ca. 1500 trip to Venice, other evidence in the text shows that he, like his uncle, had been at the Ottoman court before he traveled. The timing of his work on practical astronomy and his composition on homocentric astronomy dating the composition of *Puzzles of Wisdom* to circa 1500 mean that Galeano/Jālīnūs certainly knew of *The Light of the World*, and most likely of Ibn al-Shāṭir's astronomy, before journeying.[143] Copernicus's theoretical astronomy provides circumstantial evidence that one or more Christian scholars learned of the theories of Ibn al-Shāṭir from Galeano/Jālīnūs.

The Light of the World contains a version of another innovation of Islamic astronomers—namely Ibn al-Shāṭir's lunar theory—that also appeared in Copernicus's work. Noel Swerdlow observed that Ibn Naḥmias's most advanced lunar theory[144] was "in essence a spherical form of Ibn ash-Shāṭir's double-epicycle."[145]

While Ibn al-Shāṭir's astronomy was mathematically precise in a way that Ibn Naḥmias's was not, scholars partial to homocentric astronomy would see the improved precision of *The Light of the World* as a step in the right direction.

The highly sophisticated astronomy of Ibn al-Shāṭir, which built on astronomy in the Marāgha tradition and which may have influenced Ibn Naḥmias, was available to Ottoman astronomers. İzgi and Fazlıoğlu identified the Marāgha and Samarqand observatories as the two most important institutional sources for fifteenth-century Ottoman astronomy.[146] Yet Ibn al-Shāṭir's *Nihāyat al-Sūl fī taṣḥīḥ al-uṣūl* (The Ultimate Quest in the Rectification of the Hypotheses/Principles), which contains his influential non-Ptolemaic models, was from Mamlūk Damascus, not the Ilkhanid (Marāgha) and Timurid (Samarqand) milieux. Still, an Istanbul manuscript of Ibn al-Shāṭir's *Nihāyat al-sūl*, in the Kadizade Mehmed Effendi (d. 1635) collection of the Süleymaniye Library, bears an ownership statement from 879 AH (1473–74) from Warsanīn, a district of Samarqand.[147] Given the close connection between the scholars from the Samarqand *madrasa* and early Ottoman intellectual life, if *Nihāyat al-sūl* was available to astronomers in Samarqand, it would have been known to Ottoman scholars. Galeano/Jālīnūs was aware of recent science from the Mamlūk lands because he translated a treatise

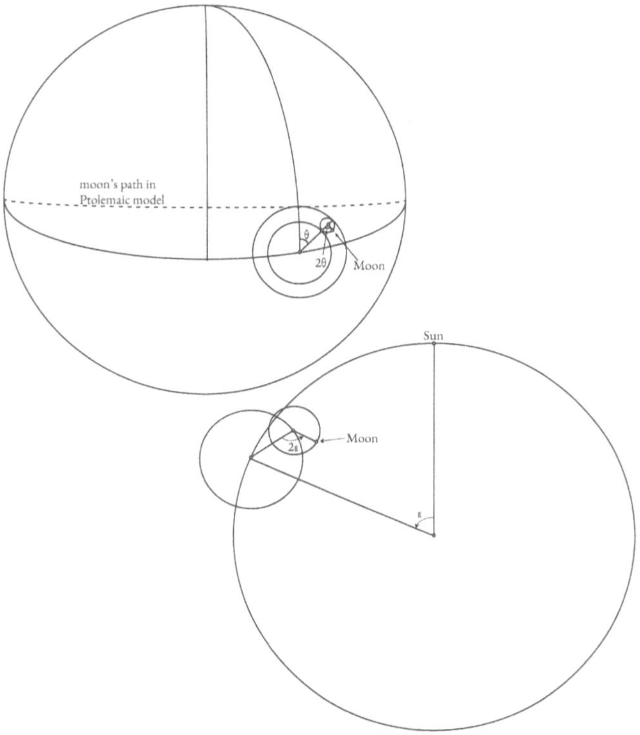

FIGURE 6.7. Ibn Naḥmias's lunar model (top) compared with Ibn al-Shāṭir's lunar model (bottom).

on the quadrant by the Mamlūk astronomer Sibṭ al-Māridīnī (d. ca. 1495).

Despite his acquaintance with Ibn al-Shāṭir's theories, homocentric orbs were Galeano/Jālīnūs's preferred ("as we maintain") solution to the problem of representing the planets' nonuniform motions with combinations of uniformly rotating orbs. In the passage I quoted from *Puzzles*, he mentioned two sources of homocentric models. The first was "the man who shook the world" (Heb. *ha-marʿish*), i.e., Biṭrūjī, the most famous exponent of astronomical models composed of homocentric orbs. The second was *The Light of the World*, in which Ibn Naḥmias improved on the predictive accuracy of Biṭrūjī's astronomy, as discussed above. Though Galeano/Jālīnūs mistakenly attributed *The Light of the World* to R. Joseph Ibn Yaʿish, he was nonetheless referring to Ibn Naḥmias's text.

After all, the fourth chapter of Galeano/Jālīnūs's Arabic composition on homocentric astronomy, entitled *Dhikr baʿḍ al-maḥallāt* ([*sic*] *muḥālāt*; An Account of Some of the Impossibilities), reproduced the text of *The Light of the World* verbatim.[148] The unique manuscript of *An Account* was in the sultan's personal library, the III Ahmet Collection of the Topkapı Library. The presence of the manuscript in the III Ahmet Collection suggests Galeano/Jālīnūs's connection to scholars who could access the most advanced texts of Islamic astronomy, e.g., Ibn al-Shāṭir's *Nihāyat al-sūl*. Elsewhere in *Puzzles*, Galeano/Jālīnūs endorsed the theories of *The Light of the World* along with Biṭrūjī's as the only ones faithful to the Aristotelian dictum of the uniform motions of the celestial bodies:

> The foundation that astronomers made as a starting point for their science is that the motion of a single celestial body does not speed up or slow down on its own. . . . According to the hypotheses of Biṭrūjī and the author of *The Light of the World*, however, which is the truth, [it is] through the compounding of motions of fully uniform movers, that is to say through [their] distance from the earth, their [external] form, and their motion. From the uniform and harmonious there may emerge a motion that is not uniform and not harmonious. That is, in equal time periods it does not traverse equal arcs from the ecliptic, one time speeding up and one time slowing down.[149]

Galeano/Jālīnūs was confident that a homocentric astronomy could save the appearances. In that respect, he agreed with astronomers such as Regiomontanus that homocentric theories could potentially be mathematically precise improvements on Ptolemy's models.

Given his appraisal of Ibn Naḥmias and Biṭrūjī's theories, it is not surprising that Galeano/Jālīnūs argued forcefully that homocentric models corresponded to reality more than the Ptolemaic models. In the third chamber of *Puzzles* in which he explored how people attributed certain effects to the wrong causes (*sibbot*), Galeano/Jālīnūs reiterated that while Ptolemy's models served as a basis for calculation, they were physically flawed.[150] The variations of the planets' diameters predicted in the Ptolemaic models were not observed. He wrote that these variations were not caused by "the variation of their locations (*aniyyuteihem*) on their own, but rather by the instruments."[151] Atmospheric phenomena, such as moisture in the air, were other possible reasons for the variations in the planets' observed sizes.[152] Elsewhere in *Puzzles*, Galeano/Jālīnūs drew a parallel from judicial astrology (*mishpaṭim*) to argue against the existence of epicycles and eccentrics: "When they say in judicial astrology that whenever there is such and such a conjunction, such and such occurs, one should not understand the opposite, that when there is that effect, there is that cause because that effect is particular to that cause."[153] Thus, Galeano/Jālīnūs concluded, errant observations of changes in the planets' sizes did not mean that the distances of the planets from the earth actually varied. If the distances of the planets from the earth did not actually vary, homocentric models would have a better chance of saving the appearances, i.e., accounting for available observations.

Earlier in this chapter, we learned that Maimonides characterized the then-intractable conflict between mathematical and physical approaches to astronomy as "the true perplexity" in the *Guide of the Perplexed*. To Maimonides, both approaches corresponded to reality, but in different ways. By contrast, Galeano/Jālīnūs pinpointed some instances in which mathematics did not describe the physical world perfectly because the modes of investigation of mathematics and physics differed.[154] One case in which Galeano/Jālīnūs distinguished mathematical and physical approaches to reality[155] involved the parallels postulate from Euclid's *Elements*, which reads: "That, if a straight line falling on two straight lines makes the interior angles on the same side less than two right angles, the two straight lines, if produced indefinitely, meet on that side on which are the angles less than the two right angles."[156] Reflection on the parallels postulate was an opportunity to explore whether the mathematical and physical approaches to studying nature were always equally valid.

Maimonides famously raised the same question in *Guide* I.73, noting that Apollonios showed in the *Conics* that two lines that are not parallel may also never meet.[157] On that basis, Maimonides concluded that "it has been demonstrated that something that the imagination cannot imagine or apprehend and that is impossible from its point of view, can exist."[158] Galeano/Jālīnūs remarked that mathematical proofs of the parallels postulate entailed one imagining that the lines approached each other, which meant that the lines must move. But, he averred, the investigation of motion belonged to the physicist because motion depended on a mover. Galeano/Jālīnūs wrote that the lines approached each other only in one's mind, and "when you investigate going from the intellected (*muskal*) to the actual or the opposite, you err just as John Philoponus[159] doubted the refutation of the creation of actual motion from infinite halving."[160] Galeano/Jālīnūs contended that the privileging of mathematical precision should be discounted a bit because mathematics did not perfectly model nature. Since the motion of the orbs was a topic in physics, not mathematics, homocentric theories were preferable because they cohered with Aristotelian physics even when they were mathematically imprecise.

Galeano/Jālīnūs's critiques of the confusion of mathematical and physical approaches to studying motion were due in large part to his familiarity with European science and philosophy.[161] The evidence was that Galeano/Jālīnūs made perhaps the only reference to the Oxford Calculators in Hebrew literature.[162] The Oxford Calculators were fourteenth-century thinkers affiliated with Oxford University who investigated topics such as motion mathematically. Scholars close to Galeano/Jālīnūs's orbit disagreed. One such scholar was Pietro Pomponazzi (d. 1525), who taught at Bologna and Padua during Galeano/Jālīnūs's lifetime and who dedicated a book to Domenico Grimani, a patron of Elijah Delmedigo. Pomponazzi reproved the Oxford Calculators for moving from one discipline to another in a single demonstration, a practice known as metabasis.

The *Conics* was the locus for Galeano/Jālīnūs's remarks on metabasis. The relevance of the *Conics* to discussions of metabasis became known to European scholars only at the beginning of the 1500s when they accessed the *Conics* through Greek manuscripts from the Byzantine Empire.[163] Galeano/Jālīnūs may have been relying on the Arabic version of the *Conics* for three main reasons: because his comments reflect his Ottoman context, because scholars at the Samarqand Observatory studied the *Conics* as well,[164] and because there is no trace of a Hebrew version of the *Conics*.[165] While Galeano/Jālīnūs learned of metabasis from European scholars, he expanded upon what he knew with a text available only to readers of Greek and Arabic.

With a comment that could easily apply to Galeano/Jālīnūs's hybrid intellectual milieu, Shefer-Mossensohn noted, "Ottoman scientific activity occurred in a multilayered, eclectic and practical manner. . . . Ottoman culture based itself on a rich past and an even richer present."[166] Discussions in Latin texts overlapped with intellectual life in Istanbul. Seemingly discordant approaches—e.g., homocentric and mathematical astronomy—combined to form for European scholars the context for Galeano/Jālīnūs's mediation of the theories of Ibn al-Shāṭir and the Ṭūsī couple as well as homocentric astronomy. We do not know to whom Galeano/Jālīnūs spoke, other than Gershom Soncino, when he was in Venice, but the circumstantial evidence from two genres of theoretical astronomy strongly suggests that he spoke to a Christian scholar or scholars while there and shared a great deal.

Ragep and Feldhay observed that "history in general and the intellectual history of the fifteenth and sixteenth centuries in particular show that the production of knowledge tends to be enhanced wherever and whenever circulation of knowledge across boundaries of languages, disciplines, and cultures occurs."[167] In this chapter, we have found that two approaches to theoretical astronomy, homocentric and Ptolemaic, met at times. The results were fascinating: scholars in both traditions recognized the homocentric antecedents of the Ṭūsī couple, and Ibn al-Shāṭir's lunar model was adapted to homocentric orbs in *The Light of the World*. Merchants of knowledge amplified this intellectual crossover through transregional exchange. Readers of Elijah Delmedigo's Latin works became aware of more recent contributions to homocentric astronomy made by scholars writing in Arabic. And there is evidence that the contents of *The Light of the World* reached Muslim and Christian scholars of both homocentric and Ptolemaic astronomy.

Consideration of Transregional Exchange in Earlier Scholarship on Copernicus

This exchange between Galeano/Jālīnūs and a European scholar somehow connected to Copernicus is a significant instance of exchange that the merchants of knowledge and their contacts did not record. Whether Copernicus, the most renowned and widely studied Renaissance astronomer, benefited from direct or indirect[168] exchange with scholars from Islamic societies has been a long running debate in the history of science. His fame is due to his theory that the sun is static while the earth is in motion with the other planets. The *Commentariolus* (Little Commentary) composed between 1508 and 1514 was the first text in which Copernicus presented heliostatic theories, i.e., those with a stationary sun, for planetary motions.[169]

For several decades, historians have recognized striking parallels in Copernicus's *Commentariolus* with the theories of astronomers from Islamic societies, specifically the theories of Ibn al-Shāṭir (d. 1375).[170] As noted above, Ibn al-Shāṭir built on the theories of astronomers from the late thirteenth and early fourteenth centuries associated with the observatory at Marāgha, in Northwest Iran. The theories of Copernicus and Ibn al-Shāṭir differed in terms of the dimensions of some partial orbs and regarding which body was motionless—Copernicus's were heliostatic while Ibn al-Shāṭir's were geostatic, i.e., with a stationary earth. But because these differences are outshone by the similarities, in 1984, Swerdlow and Neugebauer remarked of Copernicus that the "question therefore is not whether, but when, where, and in what form he learned of Marāgha theory."[171] The merchants of knowledge are the best answer to the question posed by Swerdlow and Neugebauer.

Certain recent authors on Copernicus have not rooted the genesis of Copernicus's heliostatic arrangement of the planets in matters of theoretical astronomy, the area in which there was most likely exchange with scholars connected to the Ottoman Empire. In *The Copernican Question*, Robert Westman contended that the controversy over astrological forecasting touched off by Pico's *Disputationes* led to Copernicus's reconsideration of the order of the planets.[172] Though the order of planets changed in a heliostatic system, Westman did not explain how the new order of the planets affected astrological forecasting. Putting the earth in motion should not affect the distance of the planets from the earth, and the planets' distances from the earth—not the order of the planets—are what matter for Pico's critique of astrology.[173] Yet we have seen that astrology, as an impetus for the practice of astronomy and as a locus of exchange, was relevant to the exchange of astronomy.

André Goddu, in *Copernicus and the Aristotelian Tradition*, argued that the impetus for Copernicus's proposal of the heliostatic arrangement lay in his Aristotelian (but not Averroist) philosophy. False hypotheses about the orbs ought to yield false results, but in astronomy the data can cohere with multiple hypotheses, including false ones. Goddu highlighted how Aristotelian logic affected how Copernicus evaluated competing hypotheses.[174] Goddu emphasized the significance of Copernicus's education in Poland, downplayed his subsequent time in Italy, and concluded that the elements of his theoretical astronomy that had no precedent in Europe and resembled Islamic astronomy owed to independent discovery rather than intellectual exchange.[175] By contrast, Matjaž Vesel, in *Copernicus: Platonist Astronomer-Philosopher*, held that Platonism motivated Copernicus's search for a more proportioned universe.[176] After all, in a heliostatic cosmos, the period of each planet is proportional to the distance of the planet from the sun.

Presenting Copernicus's heliostatic turn as the result of a crisis of judicial astrology, as Westman argued, or as due to Copernicus's foundation in Aristotelianism, as Goddu proposed, or as a ramification of Platonism, as Vesel contended, may be premature, since no author was able to argue, over the course of several hundred pages, that there must be a single reason why Copernicus proposed a heliostatic arrangement of the planets. There must have been multiple reasons, one of which was exchange. Goddu, in a 2018 article, modified his position on the role of scholarly exchange: "It seems likely that Maragha and other Islamic models or, at least, their ideas were known in western Europe in the fifteenth and sixteenth centuries, in which case we would be talking about 'idea diffusion' rather than 'blueprint copying' or 'independent development.'"[177] Vesel also acknowledged that Copernicus needed Islamic models to move to a heliostatic arrangement.[178] While authors who have focused on the European context of Copernicus's heliostatic turn may have turned the spotlight away from intellectual exchange, they do not deny the possibility.[179] There was transregional exchange in astrology and philosophy, fields that Westman and Vesel highlighted.

Other researchers have presumed that discourses within theoretical astronomy were the impetus for the heliostatic turn. They have reconstructed the details of how Copernicus transformed available geostatic theories into heliostatic ones. Reconstructions divide into two paths over whether Copernicus experimented with replacing eccentrics with additional epicycles (Ragep,[180] Wilson,[181] and Birkenmaijer[182]), or with transforming Ptolemy's epicycles into eccentrics (Swerdlow).[183] Even Birkenmaijer, who wrote before research on Marāgha astronomy commenced, allowed that Copernicus may have depended on predecessors.[184] The other authors mentioned in this paragraph presumed scholarly exchange with Islamic societies as a foundation for the heliostatic turn: Copernicus somehow learned of late medieval Islamic theories and transformed them into heliostatic models. Earlier researchers had identified Byzantine Christian scholars as a possible route for exchange of the Ṭūsī couple,[185] but a merchant of knowledge knew of a text containing versions of the Ṭūsī couple and a version of Ibn al-Shāṭir's lunar model, as well as of Ibn al-Shāṭir's models in general. Byzantine Christian scholars, who are beyond the scope of this study, are relevant, but they were not the only necessary intermediaries.

In contrast, Viktor Blåsjö has denied the necessity of any connection between Marāgha astronomers and Copernicus.[186] According to Blåsjö, Copernicus's heliostatic models were "natural" developments of Ptolemy's. However, those who have rejected any role for intellectual exchange have failed to account for not only all the details of Copernicus's achievements in their historical context but also the rest of European Renaissance astronomy through solely a European context. [187] Denying Galeano/Jālīnūs's role as an intermediary would mean attributing the parallels between Amico's astronomy and *The Light of the World* to independent discovery. One would also have to argue that readers of Delmedigo's printed Latin works would have been uninterested in the contents of recent texts in Arabic on homocentric astronomy. Above all, one would have to maintain that the exchange mediated by the merchants of knowledge in other fields did not occur in theoretical astronomy.

Conclusions

There is no question that the merchant of knowledge Galeano/Jālīnūs was in the Veneto around 1500 and that he knew the theories of Ibn Naḥmias, Ṭūsī (though not by name), and Ibn al-Shāṭir. The only specific individual whom we are sure Galeano/Jālīnūs spoke with when he was in the Veneto was Gershom Soncino. But the other exchanges detailed in earlier chapters, such as Galeano/Jālīnūs's translation of the *Almanach perpetuum* into Arabic, as well as the hunger of scholars at Padua for better theories of homocentric astronomy, make an overwhelming case for Galeano/Jālīnūs's exchange with Christian scholars when he was in the Veneto. *The Light of the World* certainly arrived in Padua by the mid-1600s and was probably known earlier given the parallels between the theories of Ibn Naḥmias and Amico, in addition to Nifo's acknowledgment of the value of Arabic sources.

It is possible that Amico and Copernicus were unaware of the sources of their information. It is also possible that Amico and Copernicus had a difficult time grappling with their own influences. In his 1973 classic, *The Anxiety of Influence*, Harold Bloom wrote, to characterize the central point of his argument, "Poetic Influence—when it involves two strong, authentic poets—always proceeds by a misreading of the prior poet, an act of creative correction that is actually and necessarily a misinterpretation."[188] Galeano/Jālīnūs's contacts in the Veneto may have had their own anxieties about the relationship of their theoretical astronomy to that coming from the Ottoman Empire. Given all that Galeano/Jālīnūs and the other merchants of knowledge did share with their contacts, and given the absence of an explanation of how Copernicus independently invented several late medieval Islamic astronomical models, we must conclude that Galeano/Jālīnūs's activities as an intermediary included the exchange of late medieval Islamic astronomy with someone.

Though Galeano/Jālīnūs's mediation of Islamic theoretical astronomy to the Veneto may seem, in retrospect, to be the culmination of this book's narrative, the importance of the heliostatic arrangement is more evident with our chronological remove from its historical context. Despite the relevance of the exchange of theoretical astronomy to the history of science, the actual explanation for the passage of these theories to the West lies in the motives of the merchants of knowledge and their contacts, not the telos of modernity. From the vantage point of the merchants of knowledge, there was no reason to stop exchanging. The final chapter of this book covers exchanges in logic, medicine, and mechanics, many of which occurred after Galeano/Jālīnūs's trip to the Veneto. Logic was the scaffolding for Galeano/Jālīnūs's intellectual outlook and, like medicine, was an area in which Galeano/Jālīnūs brought ideas from Latin into Hebrew. Galeano/Jālīnūs's mastery of mechanical devices, like his command of languages, was critical to his role as a merchant of knowledge.

SEVEN

TRICKS *of the* TRADE *in* MECHANICS *and* MEDICINE

The exchange of theoretical astronomy facilitated by Galeano/Jālīnūs depended on connections and interdisciplinary conversations initiated by earlier merchants of knowledge and their contacts such as Delmedigo and Kumaṭiano. Yet, the exchange of theoretical astronomy was not the telos of earlier exchanges, as will be demonstrated by studying those in mechanics (*taḥbulot*; also machinations or mechanical devices) and medicine involving Galeano/Jālīnūs that began before his trip to the Veneto and continued long after it. Medicine was Galeano/Jālīnūs's profession, and problems of mechanics, like the practice of medicine, presented prime opportunities for accruing social capital at the Ottoman court—that is to say, building relationships with patrons. Both medicine and mechanics involved transregional goods like technology and materia medica and treated transregional challenges such as war and disease. While we may be tempted to think of Galeano/Jālīnūs's mediation of theoretical astronomy as the culmination of his career, given the connection between astronomy from Islamic societies and European Renaissance astronomy, Galeano/Jālīnūs did not cease to be a merchant of knowledge after his return to Istanbul from the Veneto around 1503. Indeed, the following analysis will show, above all, how exchange and social prominence were intertwined, thus revealing that social considerations motivated exchange as much as the debates within disciplines.

Galeano/Jālīnūs's interest in medicine and mechanics foregrounds his unrelenting quest for prestige. In the passages in *Puzzles* about mechanics, Galeano/Jālīnūs consistently demystified matters to the reader, and in so doing he staked an implicit claim to the social and professional status he evidently hoped would come with superior knowledge. Occasionally, he stated baldly how he weaponized his knowledge to undermine his rivals, who were often physicians, before potential patrons. To do so, he sometimes mined Latin texts on mechanics for information that others lacked. In medicine too, Galeano/Jālīnūs translated knowledge from other cultures, knowledge that he was the first to introduce. After his return from the Veneto, he composed "Treatise on the Natures of Medicines and Their Use" in Ottoman Turkish, in which he included information from Latin texts that was otherwise unavailable in Islamic languages.[1] From Galeano/Jālīnūs's perspective, the information he imported from the Veneto into Istanbul was just as important as the astronomy he brought to the Veneto from Istanbul. The contents of "On the Natures" must have impressed Ottoman elites, as he dedicated the text to Ahī Çelebī, who was Sultan Bayezit II's chief physician and, for a time, the chief physician of Sultan Selim (r. 1512–20). Galeano/Jālīnūs succeeded in locating a prominent customer for his intellectual inventory.

Logic was Galeano/Jālīnūs's precision instrument in his pursuit of status. Just as Renaissance-era physicians studied a curriculum structured according to logic,[2] logic was the organizing principle of *Puzzles* and the tool he used to uncloak the workings of technology.[3] Galeano/Jālīnūs commented in *Puzzles* about how religious disputations with Christians led him to the study of logic. Through logical analysis, he showed Christians' arguments to be just another form of machination. Christians' conclusions did not follow from the evidence. His presentation of the connections and distinctions between disciplines through logic was a feature of Renaissance learning.

The Machinations of Mechanical Devices

Mechanical devices, a topic about which there were no original compositions in Hebrew in Galeano/Jālīnūs's lifetime, were his obsession.[4] The Hebrew term he deployed, *taḥbulot*, connoted the ruses and trickery he perceived in the operation of such devices.[5] Certain military technologies were less awe inspiring once their operation was demystified. Ignorance of the true physical causes of devices led one to be deceived. At times, harmless deceptions were a chance for Galeano/Jālīnūs to flex his intellectual muscles. For instance, he wrote that sometimes, at a burial, a wick might be soaked in turpentine (*ṭrimenṭina*) or kerosine (*neipeṭ*) so that an observer from afar might imagine that a divine fire descended during the burial.[6] More serious deceptions pervaded other fields. Galeano/Jālīnūs criticized machinating surgeons (*mitḥabbᵉlei ha-garāḥim*) who put something strange or something alive inside the cups used for cupping.[7] These surgeons would then exclaim to the patient in feigned surprise at what they managed to extract through the practice. Earlier analyses of medical fraud exist—such as those found in part of the tenth-century *Kāmil al-ṣināʿa* (The Complete Book of the Art) by al-Majūsī—but Galeano/Jālīnūs distinguished himself by spotting interdisciplinary cases that demanded debunking.[8]

He categorized instances of deception through the logical fallacies into which spectators were tempted. For example, he explained how the subterfuge of invisible handwriting depended on people confusing what existed in potentiality for what existed in actuality. That is, a piece of white paper seemingly without writing in fact might have writing upon it in potentiality. Galeano/Jālīnūs described the ruse in this passage from *Puzzles*: "They write on the paper previously with water of gall or onion juice or the water of sal ammoniac [from Pers. *nawshādir*]. Afterward, when they want to find what is written, they moisten or pass across the paper water of vitriol (*wiṭriʾol*), and that which is written emerges. It is written with onion juice at first, then they bring the paper close to fire. With the heat of the fire, the place of the writing becomes black with the onion juice, and likewise with sal ammoniac."[9] From uncovering the invisible ink, Galeano/Jālīnūs elicited a broader lesson about avoiding confusion: "Thus it is with other potential effects which are naturally actualized over time. Then if it is settled that no [new] effect be created from the actualization, it will be thought that that effect is realized and is created without a new cause. There are many like these. Pay attention."[10] Many seemingly wondrous feats were, in fact, deceptions. Galeano/Jālīnūs cast the penetrating light of his erudition on other scholars' attempts to impress.

In *Puzzles*, Galeano/Jālīnūs also described devices that confused the unsuspecting by inducing them to take the accidental reason for the essential. In other words, if someone died from a heart attack right after opening a book, the contents of the book and the act of reading are only the accidental causes of death. One device he mentioned to illustrate this fallacy was a basin containing a bearded man's head and a note that read "do not soak my beard."[11] If the basin filled until water reached the beard, it would empty. It induced the viewer to mistake the accidental for the essential: the basin emptied once the water level forced it past a bend in a siphon that then drained the water. But since this physical cause was hidden from view, observers erred and attributed the emptying of the basin to the sign on the head.[12] Galeano/Jālīnūs's aim was to uncover what was unseen by most.

He was as attuned to his social context as he was to the technical details of the devices he demystified and well understood that demystifying the machinations of others brought him patronage. The role of patronage and gifts in scientific production during the careers of the merchants of knowledge is well known.[13] The patronage system endured in the next century, for instance at the Accademia dei Lincei in the Papal States in the early seventeenth century. There is an evocative parallel between the word *taʿalumot* (puzzles) in the Hebrew title of *Puzzles of Wisdom*, with its overtones of things hidden (Heb. *neʿelam*), and the justification of the lynx as the symbol of the Accademia dei Lincei, as the lynx also viewed what was hidden inside.[14] Paula Findlen explained that, within the Accademia dei Lincei, patronage accentuated social hierarchies.[15] In this context, Galeano/Jālīnūs's hard-to-come-by knowledge was a gift to a potential patron of a higher socioeconomic status. In the vignettes that Galeano/Jālīnūs recounted from the Ottoman court, the public elucidation of a device's operation had a theatrical effect. In Islamic societies as in Europe, patronage was an exigency that shaped scholars' many moves.

In the economy of performance and patronage, more exotic devices circulated more widely due to their potential to spark curiosity.[16] For example, Galeano/Jālīnūs reported that the master of mechanical devices (*baʿal ha-taḥbulot*) at the sultan's court could recount a long list of names without writing anything down. Galeano/Jālīnūs hoped to uncover how those present were fooled by what he believed must be a ruse. The complexity of the technology used for this particular machination remained beyond Galeano/Jālīnūs's mastery until he traveled to Italy and met with the printer Gershom Soncino around 1500. In Soncino's company in Venice, Galeano/Jālīnūs observed the same feats being performed.[17] That experience helped Galeano/Jālīnūs understand the mechanics behind the ruse. He claimed that the device would work for many languages, enable the blind to write, and that someone who could not write would be able to master the device after five days of work.[18] This device numbered among those that Galeano/Jālīnūs classified as depending on change and transformation (*t*^{e}*murah*) in sensibilia as language was transformed from something heard to something felt.[19]

Because the illustrations of this device found in the manuscript of *Puzzles* were labeled with Hebrew letters, and because Galeano/Jālīnūs's discussion focused on Hebrew, the details of my provisional description of the functioning of the device focus on how it would function with Hebrew words. The Hebrew alphabet contains twenty-two letters, three of which function both as vowels and consonants. Other vowels are indicated by marks above and below the consonants. Each Hebrew vowel was assigned a numerical value based on the dominant vowel in the Turkish numbers. *Bir* (one), given the emphasis on the *b* when pronounced, meant emphasis (Heb. *diggush*). Emphasis either doubled consonants or made a soft consonant (e.g., ḵ or p̱) hard (e.g., *k* or *p*). Galeano/Jālīnūs also wrote that *bir* represented a sixth Hebrew vowel sound, the *sh*^{e}*wa* (schwa: a light, almost silent vowel). *Iki* (two) represented a *ḥiriq*, an "i" sound. *Üç* (three) represented the *shuruq*, an "u" sound. *Dört* (four) represented the *ḥolem*, an "o" sound. *Beş* (five) represented the *ṣeirei*, an "ei" sound. *Altı* (six) represented a *pataḥ*, an "a" sound. He incorporated a chart indicating correspondences between the vowels of Hebrew, Arabic, and Latin and Greek (which he believed had the same vowels), even if the vowels were written differently in each alphabet.

The chart reflected his view of the underlying similarity of the languages in his ambit. While he described how the device functioned for the Hebrew alphabet, recall that Galeano/Jālīnūs sometimes observed this device being demonstrated by non-Jews who would not have been writing in Hebrew. Thus, he had reason to believe that the device enabled one to write in other languages as well.

The device itself included a disc, perhaps of wood, divided into twelve stalks, segments or spokes of which corresponded with the signs of the zodiac (see figures 7.1 and 7.2).[20]

In the first segment, or house, were the letters *ʾ/q*; *b/r* were in the second, *g/sh* in the third, *d/t* in the fourth, *h/n* in the fifth, *w/s* in the sixth, *z/ʿ* in the seventh, *ḥ/p* in the eighth, *ṭ/ṣ* in the ninth, *h/k* in the tenth, *l/m*

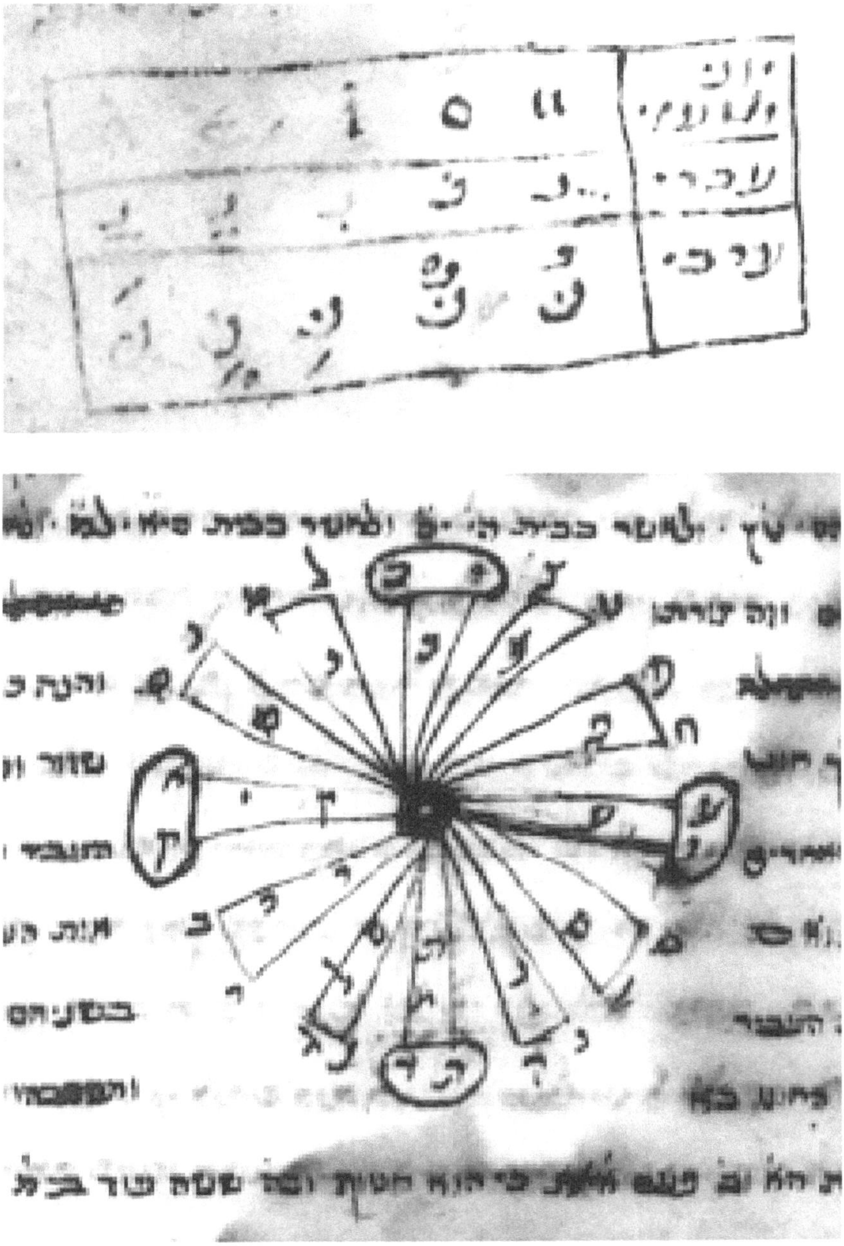

FIGURE 7.1. Figures in the manuscript of *Taʿalumot ḥoḵmah* (*Puzzles of Wisdom*), Cambridge University Library MS Add 511, 1, 10a. Reprinted with permission.

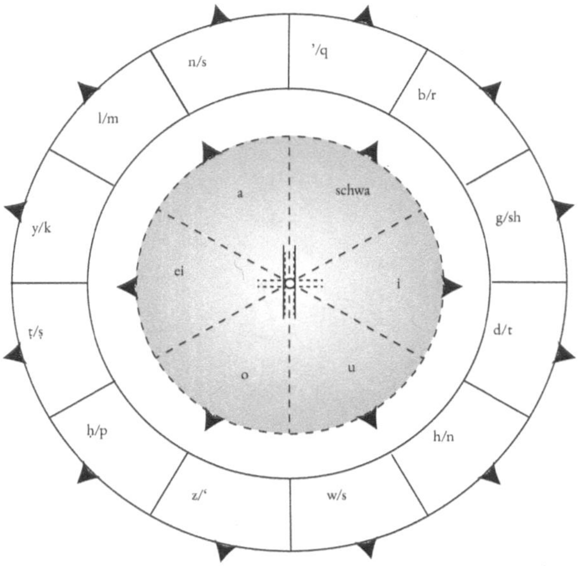

FIGURE 7.2. The transcription device.

in the eleventh, and *n/s* in the twelfth. One way in which the device may have been used to record words follows.[21] There were probably pegs to mark each letter and a crack between each spoke. Galeano/Jālīnūs did not explain how one knew without looking which spoke corresponded with each letter. To record a word, one used a thread or wire that was attached to the middle of the disc. Galeano/Jālīnūs called the letter that was higher (i.e., *b*) in the alphabet the "units" letter and the letter that was lower (i.e., *r*) the "tens" or "hundreds" letter based on their alphanumeric value. To mark the letter of the dyad in each segment or spoke that was higher in the alphabet, one threaded the wire between the pegs. Or, one peg could be closer to the center. For the letter that was lower in the alphabet, one threaded the wire around both pegs. One would then wrap the wire around the peg or pegs the number of times that corresponded to the numeric value of the vowel that went with the letter. To finish writing, one passed the thread between the spokes.

Galeano/Jālīnūs also described either a second device or another way to use the device that enabled the user to write in a variety of languages.[22] The second device, or the second use of the first device (figure 7.2), would have facilitated the feats of memorization that drew Galeano/Jālīnūs's attention to the device in the first place. He described an outer wheel resembling the one mentioned above, with twelve stalks or spokes. Each corresponded to the houses of the zodiac. I propose that this wheel was fixed over a base to which paper was attached. On the underside of each stalk, except one, were two letters of the Hebrew alphabet. To select the letter, one revolved the wheel. To select the letter closer to the center, one kept the wheel centered; to select the other letter, one shifted the outer wheel. Because Galeano/Jālīnūs asserted that a blind person could read and write with this device and because ink was not mentioned, one must have used the wheel to impress or puncture the paper with the selected letter.

As the outer wheel revolved, an inner wheel with six spokes, either flush or elevated above the outer wheel, revolved with it. That circle contained the six vowel markings mentioned above. Due to the revolution of the outer wheel, the starting point of the revolution of the inner wheel coincided with the selected consonant. In the manuscript, Galeano/Jālīnūs drew geared wheels, suggesting that the revolution of one wheel affected the revolution of the other. From there, one revolved the inner wheel to find and impress the correct vowel. There must have been a way for the position of the outer wheel not to have been affected by the revolution of the inner wheel. To accommodate larger alphabets, one needed to add stalks or spokes and presumably set aside the parallel with the zodiacal signs.

Galeano/Jālīnūs described another small wheel with six spokes that might have been beneath the wheel responsible for indicating the vowels. By revolving, aligning, and decentering this wheel, one specified emphasis, weakness, hyphenation (*mappeiq*), penultimate stress, ultimate stress, and the lack of any of these morphological features unique to classical Hebrew. This level of morphological detail was unnecessary for the aforementioned feats of transcription that he witnessed, which is another reason why I believe Galeano/Jālīnūs was describing either two devices or one device with two functions. Because some of the morphological details do not matter for languages other than Hebrew (and some only for classical Hebrew), and because Galeano/Jālīnūs claimed that one could use the device with Indian script,[23] he could not possibly have witnessed every application for the device that he claimed to. Given the importance of *Qabbalah* for the merchants of knowledge and their Christian contacts, Galeano/Jālīnūs's composition on geomancy, the interest of the Ottoman sultans in lettrism (*ʿilm al-ḥurūf*), and the division of the outer wheel to correspond numerically to the cosmos, Galeano/Jālīnūs must have been concerned with demonstrating that all languages shared an underlying mathematical reality.

Recall that with the first device he classified the Hebrew letters as belonging to the categories of units, tens, and hundreds. He drew parallels between the letters and the zodiacal signs.[24] He found a correspondence between the nine orbs and the nine uses of the wheels for vowels and accents. When Galeano/Jālīnūs asserted without further explanation that each letter served 2,624 uses, the claim reflected his belief that all languages shared an underlying mathematical and cosmic unity.

Because languages were the currency of the merchants of knowledge and their contacts, Galeano/Jālīnūs believed that he would accumulate cultural capital by unshrouding this mechanical reflection of the unity of languages. Christian, Jewish, and Muslim devices could all be made and understood according to the same logical principles, even or especially when the underlying principles were occult.[25] In this respect, Galeano/Jālīnūs's exceptional argument for the unity of languages and for the use of logic to unify disciplines and to crosscut cultures has a context. Hava Tirosh-Rothschild likened the Jewish idea of *ḥakam koleil* (universal sage) to the Renaissance ideal of universal man (*homo universalis*). She believed that this comparison was exemplified by Rabbi David ben Judah Messer Leon, an Italian émigré to Istanbul in the 1490s who mastered the Renaissance humanist curriculum and who was an acquaintance of Elijah Delmedigo and Caleb Afendopolo.[26] *Puzzles* was unprecedented, but Galeano/Jālīnūs was not the first Jewish scholar to bring Italian Renaissance learning to Romaniot Jews.

Military Technology

The transcription devices were transregional, and so was war. The second Ottoman-Venetian War raged between 1499 and 1503, around when Galeano/Jālīnūs visited Venice. As military stratagems were of more than academic interest, he advertised his value by analyzing military technology. He devoted a paragraph to explaining sailors' use of a compass (*busula*), a piece of equipment that was widespread and nonlethal.[27]

Galeano/Jālīnūs also delved into a more exotic weapon: the flambeau (Turk. *yelmumi*; lit. "wind candle"). The Turks, he wrote, made candles that could not be extinguished by the wind; indeed, by dipping the wicks in sulfur, they created candles that would reignite if extinguished; this feature gave the appearance of the wind causing the burning.[28] The same technology resurfaced in a longer anecdote about the fallacy of confusing the accidental (the wind) and the essential (the wick dipped in sulfur) causes. In the vignette, Galeano/Jālīnūs recounted events on the banks of the Danube, perhaps during Mehmed II's attempt to conquer Wallachia in 1462.[29] The Ottoman army was confronted with a fortified city on the other (lit. Hungarian) side of the river. The Ottomans affixed to pigeons, cats, and dogs "gradually burning threads like those of saltpeter."[30] At the end of each thread was a package of tar and resin. When the wind blew at night, the animals fled to the attics of the buildings in the fortified city, and, as the burning tar and resin was inextinguishable, even when the wind blew, many buildings were incinerated. Galeano/Jālīnūs quoted Psalms 106:18: "A fire blazed among their party, a flame that consumed the wicked." The Hungarians' errors in reasoning sealed their fate. Galeano/Jālīnūs, by sharing this proprietary technology, offered to protect the reader.

In other words, the ability to reason precisely distinguished the victors from the vanquished. The message emerged once more in the following narrative from *Puzzles*, in which Galeano/Jālīnūs recounted the capture of the city of Rhodes by Cassius from the Greeks in 42 BCE. The Romans first besieged the city but could not capture it.[31] Then, the Romans learned that a group of shepherds was going to return the Greeks' flocks to the city at a certain time. Thus, the soldiers disguised themselves in sheepskins, entered the city walking on all fours with the sheep at the appointed time, and conquered it. According to Galeano/Jālīnūs, the use of disguise induced the people of Rhodes to succumb to the fallacy of substitution, i.e., not recognizing that the sheep were actually soldiers. He added that, in his own time, the people of Minorca were misled by the Muslims when Khayr al-Dīn Barbarossa's (d. 1546) landing party (Heb. *ha-Rugiyyot*) arrived disguised as Christians and the people of Minorca were deceived by the trick. Though this last explanation appears in a marginal addition, probably in Galeano/Jālīnūs's own hand, the reference to Barbarossa is in the main text.[32] This vignette has implications for the chronology of the composition of at least this part of *Puzzles*. If *ha-Rugiyyot* indeed referred to Barbarossa, then Galeano/Jālīnūs must have been chronicling the Ottoman sack of Mahón (on Minorca) in 1535, two years before he completed *Puzzles*.[33] Candia was a safe place to divulge Ottoman secrets and a setting where he would find an eager audience.

There was another case in which Galeano/Jālīnūs's desire to share Ottoman stratagems would be best satisfied after his time at the Ottoman court. The Ottomans may have awed the enemy by appearing to destroy ships through prayer. In another passage that is probably also in Galeano/Jālīnūs's handwriting, he argued that military technology was the true reason why that prayer seemed to be answered. He observed:

> One might say, "I pray," and the rigging of the enemy's ships collapses. It is thought that his prayer is the cause. But he might machinate by having a sailor pass a wire under and over it [viz. the rigging] hidden inside the sea, and by means of the same [hidden] wire, from the place he was sitting, pass a cable wrapped around it that resembles an iron wire which has teeth like an irritant (*m^e^gareh*). By means of the motion of the irritating wire and its rubbing, through the motion of the sea, or by its being drawn from opening to opening, the rigging of the ship at sea is cut. One night is sufficient.[34]

Galeano/Jālīnūs then compared this misplaced confidence in the power of prayer to the mistaken presumption that tithing to Sultan Bayezit II brought good fortune. Presumably, Galeano/Jālīnūs would have made that statement only if he were outside of Ottoman lands, perhaps on Candia, where he arrived between 1523 and 1525. Galeano/Jālīnūs's interventions in the debate shed light on his biography. Because he modified the manuscript of *Puzzles* after his student, the merchant of knowledge Abraham Algazi, copied it in Candia in 1539, Galeano/Jālīnūs must have remained on Crete through the 1530s.[35] With military technology, timing mattered as much as knowledge.

When demystifying technology, Galeano/Jālīnūs foregrounded his linguistic skills. His desire to disempower the machinators impelled him to seek out Latin texts that had never been translated into Hebrew or an Islamic language. And his sharing of a paraphrase of a Latin text was unbeatable advertising. In the following passage from *Puzzles*, Galeano/Jālīnūs referred to Pietro d'Abano's (d. 1316) commentary on the *Problemata*, *Expositio problematum Aristotelis*, a popular text during the Renaissance.[36] The *Problemata* was a text attributed to Aristotle that offered solutions to a variety of problems, mostly in medicine. The *Problemata* circulated in two printed versions, one of which was accompanied by Pietro's *Expositio.*[37] From the commentary on the *Problemata*, Galeano/Jālīnūs gleaned ideas about how to disprove the machinators. He wrote,

> The machinators say, "I am writing names on my palm. Then I will grasp in it molten lead or some other hot thing and it will not burn me." They attributed, mistakenly, the reason to the names written in the palm or to spoken things that he says and swears. The truth is like what Pietro d'Abano says in the commentary on the *Problemata*: Anoint the hand beforehand well with juice of *mercurialis* or the juice of sweet clover (*meliloṭ*) and the juice of *laglagut* which is purslane (from Ar. *buqlat al-ḥamqāʾ*). I know that, if you cover the palm beforehand also with a thick coating of moist spittle of sebesten (from Ar. *luʿāb sabistān*) or with arterial tragacanth (from Ar. *kathīrāʾ sharawiya*), they also tend to cause this, that is to say the absence of being burned for a short period of time.[38]

This passage from *Puzzles* paralleled d'Abano's commentary on the *Problemata.*[39] Imported knowledge uncloaked the deception. Mauro Zonta has drawn our attention to Hebrew translations of Latin Aristotelian philosophy texts carried out in the late fifteenth century, and we learned in chapter 4 that the merchant of knowledge Elijah Delmedigo composed Averroist texts in Latin.[40] In this context, Galeano/Jālīnūs built on the foundation of Hebrew scholasticism to trade in knowledge that others did not stock; as a result, patrons would be attracted by his skills and hustle.

These material incentives aside, Galeano/Jālīnūs's immersion in Latin texts was profound: he drew on da Vinci's use of *polo* to mean axle and used the Hebrew *sadan* (usually "pole") to mean "axle."[41] This nonstandard usage reflected Galeano/Jālīnūs's unique position at the junction of Ottoman, Romaniot, and Renaissance European cultures. In this instance, his word choice was not motivated by the need to self-promote. Rather, his linguistic competencies facilitated access to texts that solved the puzzles of military technology. To make a connection with the previous section, the transcription device(s) were a synecdoche for the centrality of language to uncloaking secrets of machinations. Differences between languages were bridged by their shared mathematical foundation. Thus, language was the currency that brought the keys to the logical puzzles of machinations. Galeano/Jālīnūs possessed riches that could not be taxed on Candia.

Religious Machinations

Galeano/Jālīnūs's logical analysis of military technology, particularly his finding that the ships did not collapse due to prayer, motivated his conclusion that machinations never depended on metaphysical insights. Religion, however, was a matter of metaphysical insight, but sometimes what was attributed to the metaphysical arose, in fact, from physical causes. The confusion of the accidental with the essential led people to overlook physical causes. Galeano/Jālīnūs referenced the non-Israelite soothsayer Balʿam (Numbers 22–24), "who was given to think that from his blessing or curse good or bad things would result, but he was not the reason." [42] Rather, according to Galeano/Jālīnūs, Balʿam's blessings and curses were effective because of his understanding of astral forces. Such demystifications of non-Jews' supernatural claims would have appealed to Jewish readers.

Scholastic logic enabled Jewish scholars to engage more effectively in disputations with Christians.[43] In an example of the confusion of the accidental with the essential, Galeano/Jālīnūs argued that machinators and Christians shared misconceptions about causes. He described how Christians made a crucifix from blood stone[44] and informed the reader that the redness was due only to polishing. They placed the crucifix of blood stone "on the stomach of a woman who was leaking a great deal of menstrual blood. It [viz. the bleeding] stopped, and they thought that the crucifix was the cause."[45] He added that crucifixes made of peony (from Ar. *fāwāniyā*), also known as *ʿūd al-ṣalīb* (from Ar., lit. "the branch of the crucified one"), were used for epilepsy, and greenstone (from Ar. *ḥajar al-yarqān*) for jaundice. Again, the healing force was in the substance, not its cruciform shape. Logic and science helped him debunk claims about the role of Christian faith in healing.

I provide below a translation of the vignette to illustrate how Galeano/Jālīnūs was confident in the application of amulets to heal to emphasize how the effective cause is the metal, not a higher power. His source is his grandfather, R. Elijah Galeano, who reported that it was

> well-known among the Christians, to put a gold coin from Constantinople on the skull of the leper, at the hairline, under the skin and after cleansing. It will help him or cure him, and they think that it is the power of Jesus that cured him of his leprosy. My grandfather says that

> he experimented with other gold coins and that it was no less effective. For this [healing property] belongs to gold qua gold, which brings the temperament of the brain into equilibrium and prevents overheating and the corruption of the humors there.
>
> It is possible that the first [to employ this] did not specify a gold coin from Constantinople on account of a power that the Constantinople coin acquires from the crucified one, but rather because of its convex shape, which makes it easier to affix it to the top of the skull, at the point of the juncture, which is called the commissure,[46] which is a good and efficacious place to intervene in the brain.[47] There are many such [things] among the general populace.[48]

Neither Galeano balked at the application of a gold compress to the head as a therapy for leprosy. Rather, in their estimation, the error Christians made was presuming that the gold had to be in the form of a gold coin from Constantinople and that such a coin cured because Jesus acted through it. Galeano/Jālīnūs wanted to inform his Hebrew readers that Christians' errors were sometimes subtler than they might have imagined so that Jews, through the application of logic, could salvage practical knowledge of amulets.

There were other cases in which Galeano/Jālīnūs argued that the religious practices of non-Jews functioned by machination. For example, the oath-swearers (*ba^calei ha-hishshab̲^cot*) advised taking a stick dipped in the blood of a billy goat and writing something on it to keep fleas away from humans. But when fleas gathered on the stick, away from humans' skin, Galeano/Jālīnūs explained that it was because of the blood, not because of whatever was written, as the masters of oaths claimed. One would be mistaken in attributing the fleas' fleeing to the inscriptions; the true cause was the smell of the blood in which the inscriptions were written. In support of his claim, Galeano/Jālīnūs cited Maimonides's explanation that a miracle is something for which the cause was misapprehended. In *Guide* III.37, Maimonides argued that the real possibility that one might "believe that accidental matters are essential causes" motivated the Talmudic ban on profiting from idolatry. Galeano/Jālīnūs quoted from the *Guide*: "God forbade deriving any benefit from idol worship so that the masses will not err by thinking that from the day that it [the idol] came in their house there was success," even if one brought the idol home intending to destroy it.[49]

Idolatry was, in Galeano/Jālīnūs's view, but a case of the misapprehension of causes and effects. Logic neutralized the claims of other religions.

The logical fallacies that were the foundation for the metaphysical machinations of non-Jews were also present in their scriptural interpretations. The longest example of faulty religious reasoning in *Puzzles*—indeed, the longest account of any error in reasoning given in the whole text—was the Christological interpretation of Isaiah 7:14: "Therefore the Lord Himself shall give you a sign: behold, the young woman [*ʿalmah*; LXX: ἡ παρθένος (virgin); NRSV: "young woman"] shall conceive, and bear a son, and shall call his name Immanuel." According to Galeano/Jālīnūs, Christian interpreters of this verse erred by ignoring the categories of time (*matai*) and quality (*eik̲*), meaning that the author of Isaiah omitted information in two critical categories. Galeano/Jālīnūs commented that a) the verse from Isaiah did not foretell the son's death and b) that it did not specify that the woman was a virgin. Thus, when Christians deployed this verse in polemics against Jews, they excluded other possible interpretations.[50]

In another instance of Galeano/Jālīnūs rebutting Christian interpretations of the Hebrew Bible, he adduced Deuteronomy 31:29 to answer charges that Moses forecast the eclipse of Judaism by Christianity. The verse reads: "For I know that after my death ye will in any wise deal corruptly, and turn aside from the way which I have commanded you; and evil will befall you in the end of days; because ye will do that which is evil in the sight of the Lord, to provoke Him through the work of your hands." Galeano/Jālīnūs provided a competing interpretation and argued that the Israelites' failings after Moses's death were caused not by Moses's words but by the Israelites' bad choices. The possibility that scripture contained predictions of the future did not necessitate that everything found in scripture foretold the future. Galeano/Jālīnūs explained: "The upshot is that whatever the prophet says, then the prophecy is true without necessitating that its entire story be true. That is, it is true that such and such will be created, but not that the created is true [just] because the story of its being created is true. For in the example, the prophet said that a new religion will arise. Behold, it is not necessary that the established religion be true on account of how his word and the prophecy of its rising are true."[51] From Galeano/Jālīnūs's perspective, it would be deceptive to conclude that a correct forecast of the rise of Christianity entailed the truth of Christianity.

Jews were not immune to such errors. Galeano/Jālīnūs brought up the incident at the Waters of Meribah (the waters of quarreling) where the Israelites did not understand, though they should have, that the cause of the water bursting forth from the rock was God, not Moses hitting the rock.[52] Galeano/Jālīnūs identified additional cases in which Jewish readers mistook something other than the cause for the cause. He reported:

> It is said in the books of the Muslims that whoever says that the religion (*dat*) of the Christians is better (*ṭob̲ me*) than that of the Jews is a heretic (*kop̲eir*). Behold, the hearer errs if one thinks from this statement that the reason for this is that the religion (*dat*) of the Jews is more esteemed. This would be the reason only if he had said more good [*yoteir ṭob̲*; also: "better"], for the meaning affirmed with it is goodness. But with this it is heresy. It would have been a valid judgment (*din*) if he had said the opposite, that is, that the religion (*dat*) of the Jews is more good [*yoteir ṭob̲ah*; also: "better"] than that of the Christians, which is heretical.[53]

Jews should not be distracted, he argued, from the reality of the Islamic perspective: both Judaism and Christianity were equally inferior to Islam. Attention to the precise wording of statements about Judaism in Islamic texts mattered.

I conclude this section by presenting a passage from *Puzzles* that illustrates how Galeano/Jālīnūs and his readers' concerns about logic and religion were a pretext for him to highlight his rivals' errors. Here, he once more criticized idolators for misunderstanding the category of quality (*eik̲*), namely how things happened. Changes always occurred in nature for natural reasons, he argued; direct requests in the form of prayers did not matter. Not only could he highlight rivals' errors, but he could also promote his linguistic skills. Galeano/Jālīnūs exploited his rivals' confusion as an opportunity to showcase his knowledge of words in Greek and Romance:

> Another religious example from the category (*sug*) of "how" is the waters of *marah* (lit. "bitter"; cf. Ex. 15:23–25) that were bitter "and the LORD showed him a tree, and he cast it into the waters, and the waters were made sweet" (Ex. 15:25). Indeed, the water became sweet through the combination of the tree and the water. I have seen with well water that was a little salty that when they cast into it a resinous tree[54] that is called *dadi*[55] in Greek and *ṭeiʾa*[56] in Romance, it returned to being sweet.
>
> It appears to me that the secret of the ban on breeding, crossbreed-

> ing, and botanical grafts pertains to how the practitioners of idolatry were publicizing them and machinating with them before the masses by relating those strange actions to strange worship.[57]

The false assertion, Galeano/Jālīnūs wrote, was that the changing taste of the waters was due to idol worship and not to the properties of the tree. From Galeano/Jālīnūs's perspective, there was no supernatural reason why the talismans worked. He had pointed out that same misunderstanding of talismans elsewhere in *Puzzles*, and without the Greek and Romance words. Though the anecdote about resinous wood was not necessary to solve the puzzle, languages were the source of his wisdom.

Medicine

By uncloaking others' machinations in *Puzzles*, Galeano/Jālīnūs was selling transregional secrets such as the transcription device(s) and military technology. He functioned similarly as a merchant of knowledge in the field of medicine, as both disease and materia medica were transregional.[58] He marketed his medical skills to figures who, most likely, did not read *Puzzles* and his other Hebrew texts. We have already seen in chapter 5 that Galeano/Jālīnūs translated the Latin canons of the *Almanach perpetuum* into Arabic in 1505, after his return from the Veneto. After carrying Islamic knowledge west, he conveyed Latin knowledge east to the Ottoman Empire in a medical text that he composed.

A few years after the translation of the canons of the *Almanach perpetuum*, Galeano/Jālīnūs authored the Ottoman Turkish medical text "Treatise on the Natures of Medicines," dedicated to Ahī Çelebī (Muḥammad b. Kamāl al-Tabrīzī), Sultan Bayezit II's (r. 1481–1512) chief physician from 1507 to 1512 and, later, Sultan Selim's (r. 1512–20) chief physician. The treatise was probably written between 1507 and 1512 or late in the next decade because, in the dedication, Galeano/Jālīnūs referred to Ahī Çelebī as "chief of the physicians" (*raʾīs al-ḥukamāʾ*). As such, "Treatise on the Natures" was also an early example of scientific literature in Turkish, and, in it, Galeano/Jālīnūs explored how to determine the strength and characteristics of a compound medicine.[59]

On the first page of the treatise, Galeano/Jālīnūs stated that he would draw on "the words of Islamic, Frankish, Greek, and Jewish physicians."[60]

Though Islamic sources predominated, Galeano/Jālīnūs also referenced the works of Arnaldo da Villanova (1234–1310) and Bernard de Gordon (fl. 1270–1330), whose works were unavailable in Islamic languages. Jews' medical practice in Europe had long been informed by Latin medical texts.[61] Some of Arnaldo and Bernard's texts had been translated into Hebrew, but the Latin sources that I identify for Jālīnūs's citation of Arnaldo (i.e., Arnaldo's *Speculum medicinae* or the *Antidotarium*) had not been.[62] Also, the statements that Galeano/Jālīnūs attributed to Arnaldo are not found in either Bernard or Arnaldo's treatises entitled *De gradibus*. But Joshua Lorqi (d. ca. 1419) did include this information in *Gerem ha-maʿalot* (The Cause of the Degrees)—a treatise on materia medica composed originally in Arabic.[63] The wording of "Treatise on the Natures" in which Galeano/Jālīnūs cited Arnaldo and Bernard followed Lorqi's Hebrew closely. Even if the Latin information present in "Treatise on the Natures" was mediated by the Hebrew translation of Lorqi's text, Galeano/Jālīnūs's summary of Arnaldo's ideas was an example of him bringing the fruits of Hebrew scholasticism—in this case Lorqi's text—to Muslims.

Ahī Çelebī was attuned to what European physicians were up to. He authored a text in Ottoman Turkish entitled *Risāla fī al-ṭibb* (Epistle on Medicine). At one point in the text, Ahī Çelebī described a surgical instrument made out of a pipe that could absorb water and penetrate the bladder. He explained that, in Turkish, the instrument was called *ṣūʾiçeq* (perhaps "water drinker"?). Then, he remarked that surgeons coming from Europe (*Firengistān'dan gelir cerrahlar*) used the same instrument in this place, that is, on the bladder.[64] Somehow, Ahī Çelebī's curiosity had been piqued by the contents of Latin medical texts.

Ahī Çelebī was not the only elite Ottoman scholar Galeano/Jālīnūs cultivated with medical knowledge. Galeano/Jālīnūs described a medical experience from his time in Venice in an Arabic-script version of Maimonides's (d. 1204) "Treatise on Poison and Antidotes" (Ar. *al-Maqāla al-Fāḍiliyya fī al-ṭibb*). Galeano/Jālīnūs dedicated the text to Şāh Çelebī (d. 1550), a scholar during the reigns of Bayezit II, Selim, and Süleiman.[65] In the text, Galeano/Jālīnūs intervened several times and provided additional information about poisons and antidotes. He commenced the final intervention in Arabic and concluded in Turkish. He warned of an herb that resembled parsley but was as fatal as 100 poisons. Galeano reported that when he was in Europe, he saw (*ben Firengistān'da gördüm*) someone ingest it and die, for he took no precautions.[66] He counseled that older antidotes were ineffective, hence his advice to abstain (*parhīz etmek*). The correct identification of transregional medicines and poisons depended on intermediaries.

In that respect, Galeano/Jālīnūs's tendency in *Puzzles* to refer to foreign pharmacological terms found in Latin texts that were unavailable in Hebrew and Islamic languages in order to show up his competitors made sense. In the third chamber of *Puzzles*, Galeano/Jālīnūs mentioned an herb, known in Latin as *verbena*, from a text he called *Pandiqṭa*. The *Pandiqṭa* was Matthaeus Silvaticus's (d. 1342) pharmacopoeia entitled *Pandectarum medicinae*.[67] The *Pandectarum* was known to other Jewish authors but had not been translated into Hebrew or Islamic languages.[68] In the *Pandectarum*, *verbena* was classified as *gerebotanum*, a Latinization of its Greek name, ιερόβοτάνη.

Galeano/Jālīnūs recounted that "the physician takes and grasps in his hand the herb called *berbena* in Latin and *eirovotano* in Greek when he goes to visit the patient, and the patient does not know that he is holding something in his hand. He asks the patient when he enters, 'How are you?' If he answers 'Well,' he lives, and if he answers 'Badly,' he dies."[69] Galeano/Jālīnūs explained that the foreknowledge was not due to the physician's wisdom, again excluding metaphysical factors, but rather to a property of the herb. That was why the herb, in Greek, was called ιερόβοτάνη (lit. "holy plant," transcribed in Hebrew as יירווטנו). Galeano/Jālīnūs did not elucidate how ingesting the herb yielded an outcome that accorded with the patient's words, but his ability to find unfamiliar and original knowledge made him a most effective merchant of knowledge.

Galeano/Jālīnūs was not the only figure at the Ottoman court who curried favor by bringing European medical knowledge to the attention of powerful figures. Ilyās b. Ibrāhīm al-Yahūdī (d. after 1512), known as ʿAbd al-Salām al-Muhtadī after his conversion to Islam, was present at both Bayezit II's and Sultan Selim's court. He ascended to the office of *defterdar*, minister of finance, and dedicated a treatise on the plague to Sultan Selim. The treatise was entitled *Kitāb majannat al-ṭāʿūn wa-l-wabāʾ* (The Book of the Shield from Plague and Pestilence).[70] Like Galeano/Jālīnūs's "On the Natures," *The Book of the Shield* is an example of how Jewish physicians gained expertise from working with Christians and then wielded that expertise to their professional advantage at the sultan's court. Barkai noted that "though he wrote his treatise in Arabic for Muslim readers, Ilyās praised and recommended the use of medicaments whose composition he learned from Christian physicians."[71]

We learn from the treatise how the plague motivated transregional intellectual exchange. Because Muslim scholars considered the plague to be a matter of social and religious responsibility, Ilyās's ideas, derived from Christians, reflected his willingness to learn from non-Muslims. While he accepted the view common among Muslim physicians that the plague was a result of God's will, Ilyās counseled evacuation to escape the putrid air that was thought to be the cause of the plague.[72] Christian and Muslim responses to the plague were more nuanced and less divided than previously believed, but some Christians in the sixteenth and seventeenth centuries depicted the Ottomans as foolishly fatalist.[73] Still, Varlık identified the beginning of the sixteenth century as a turning point at which Ottoman legal scholars began to sanction flight from the plague.[74] Clearly, the plague was a pressing transregional concern. Muslims' received attitudes were reconsidered with input from outside.

The life and death consequences of many illnesses bred cut-throat competition. Physicians were vigilant about the threat that their competitors posed to their own livelihood, and accusations of trickery abounded. Galeano/Jālīnūs accused competing physicians of machinating. In his narrative that accompanied one such accusation, one physician, a refugee from Spain named Samuel Abulafia, feigned a disease in order to confound a competing Jewish court physician. That other physician, Isaac, presuming the false symptoms were real, forecast that Samuel would die. Samuel's equally feigned recovery led to Isaac's replacement with Samuel. Subsequently, Samuel died from bloody diarrhea, ironically one of the symptoms he had performed as part of the ruse.[75]

According to *Puzzles*, even the sultan was not above the chicanery that festered at the court. Galeano/Jālīnūs reported that the sultan asked that a court physician be shown the urine of a healthy young man but be told that the urine was the king's. That same urine, if it had come from an old man, would be evidence of a fever, and so the physician told the sultan as much. As the young man was not ill, the king mocked the attendants for praising this physician. Then, some attendants told Galeano/Jālīnūs what happened as a way to slander the physician (*higidu li qᵉṣat sᵉrisim la-lashon ha-raᶜ mei-ha-ropeiʾ*). Galeano/Jālīnūs naïvely (*bᵉ-tomi*) responded that they had wronged the king (*ḥamsu ʿal ha-melek̲*) because the physician actually made the correct diagnosis on the basis of the available information. Ultimately, the attendants praised Galeano/Jālīnūs for helping another physician, and the king conceded as much (*eik̲ ropeiʾ ʿozeir le-ropeiʾ be-ṣedeq we-gam ha-melek̲ hodah*).[76] Note that Galeano/Jālīnūs described the attendants' gossip as an attack against the king even though he acknowledged the king's role in the deception. In pursuit of his goal of exposing the logical fallacies that led people to succumb to machinations, Galeano/Jālīnūs nevertheless went to great lengths not to appear to the reader to be showing up the sultan at court. Failing in that regard would have cost Galeano/Jālīnūs professional connections.

After Galeano/Jālīnūs's lifetime, Jewish physicians continued to compete for the sultan's favor.[77] For example, Moses Hamon dedicated a treatise on dentistry to Selim's son Sultan Süleiman (r. 1520–66). Moses was the son of Joseph Hamon, an emigrant from Granada. Joseph Hamon had been Selim's personal physician and accompanied him on military campaigns.[78] Galeano/Jālīnūs's teacher Elijah Mizraḥi (d. 1526) mentioned Joseph Hamon in a legal responsum.[79] His success as Selim's physician brought accusations of poisoning Bayezit II, Selim's father. Successful physicians were attractive targets.

In this atmosphere of intellectual one-upmanship, acquiring prestige was a central motivation for sharing knowledge. In "Treatise on the Natures," Galeano/Jālīnūs intervened in an Islamic debate about materia medica that kicked off with the appropriation of Galen's pharmacology in Kindī's (d. 873) *On Degrees*.[80] Galen postulated that drugs were hot, cold, dry, and moist to varying degrees of intensity, as well as in a state of equilibrium. The various degrees of hot, cold, dry, and wet in simple medicines determined how hot, cold, dry, wet, or equilibrate a compound medicine was. Kindī, clarifying what Galen had written, arrived at a couple of conclusions about the relationship of something in equilibrium to something in the first degree. First, Kindī determined the relationships between the degrees mathematically, not through observation or experimentation. Second, he concluded that the relationship between the degrees was one of doubling (2, 4, 8, 16) and not integers (1, 2, 3, 4).[81] To Kindī, "something hot in the fourth degree is 8 times as hot as something hot in the first degree and 16 times as hot as an equilibrated compound."[82] The relationship between these ratios and experience was not settled.

Hence, Kindī's conclusions became the starting point for subsequent discussions among physicians in Islamic societies about the computus of simple and compound medicines. Because there was little empirical evidence, i.e., that a simple medicine in the fourth degree was eight times as strong as a simple medicine in the first degree, for the ratio of doubling, Averroës (d. 1198; Ar. Ibn Rushd) took a different tack. He noted that, theoretically, one should be able to reduce a fever by administering a hot substance that was, however, not as hot as the temperature of the patient.[83] In the *Kulliyyāt* (Lat. *Colliget*), a text of general medicine, he argued that the more appropriate ratio between degrees was that of integers.[84] His most developed views on the subject appeared in his treatise on theriac, a compounded antidote, where he forsook mathematical analysis in favor of qualitative assessments.[85] Thus there was scope for scholars inclined to mathematical approaches.

Following the translation of Arabic medical texts into Latin, scholars writing in Latin joined the debate. Arnaldo da Villanova produced a Latin text based on Kindī's *De gradibus*. Yet Arnaldo's work was not derivative. Arnaldo's theory of degrees was novel and, as such, sufficiently independent of the Arabic authorities respected at Montpellier where Arnaldo worked.[86] On the question of the relationship between the strengths of the degrees, Galeano/Jālīnūs reported that Arnaldo wrote that one dram of a medicine hot (or cold, etc.) in the fourth degree was equivalent to one and a half drams of a medicine hot in the third degree, and to two drams of a medicine hot in the second degree, and to three drams of a medicine hot in the first degree. Galeano/Jālīnūs's report concurred with the contents of Arnaldo's Latin text.[87] Bernard de Gordon, in his treatise *De gradibus*, which he completed in 1303 and which was translated into Hebrew, defended Arnaldo's methods and presented them in a more accessible manner.[88] Bernard outlined an experimental foundation to support the mathematical theory, showing that scholars in the Latin West answered questions posed by their predecessors who wrote in Arabic.

In "Treatise on the Natures," Galeano/Jālīnūs picked up the thread of the discussion from Arnaldo and Bernard. Although Galeano/Jālīnūs thought that Arnaldo's theory was worth repeating, he did not agree completely with Arnaldo. Galeano/Jālīnūs pointed out, citing Averroës, that two drams of sandalwood that were cold in the second degree balanced out twenty drams of honey that were hot in the second degree.[89] In this sense, different simple medicines were of different intensities, even if of the same degree.

Still, Galeano/Jālīnūs acknowledged that while the work of Avenzoar père (twelfth century) on materia medica was also valuable, Arnaldo's theory was simpler.[90] Thus, Galeano/Jālīnūs strongly implied that Latin knowledge was necessary along with Islamic knowledge, justifying his role as a merchant of knowledge. Jewish intermediaries were the source for key updates of Averroës's pharmacological computus.

In *Puzzles*, Galeano Jālīnūs summarized the debate over the relationship between degrees of simple medicines:

> If, thus, what the physicians opined, saying, "This is from such and such a degree and this is from such and such," I say that pepper is hot in the second degree when a small quantity of it is eaten, and clove is hot in the fourth when much is taken. Not everything from a single degree is of a single measure as Isaac Israeli in his epistle *The Cause of the Degrees* (*Gerem ha-maʿalot*)[91] and Arnaldo da Villanova [wrote]. This is because, as we said, we find their measures different because the effect of euphorbium and pepper on the body is not equal. A lesser amount of euphorbium burns more than twice its measure of pepper. The response to this is in our treatise on the degree of the compound [medicine].[92]

Earlier in *Puzzles*, Galeano/Jālīnūs complained that not only were some physicians unfamiliar with the degrees of simple medicines, but they did not know that some medicines from one degree exerted the effect of that degree while their measures varied.[93] Not only would the provenance of the sources of "On the Natures" have impressed Ahī Çelebī, but the information found in Latin medical texts was the basis for a productive intervention in debates over pharmacological computus against physicians whom Galeano/Jālīnūs deemed less worthy. As was the case with Galeano/Jālīnūs's analysis of military technology, he brought foreign knowledge that had immediate practical applications.

Medical Astrology and the Causes of Disease

Besides the debate over identifying the correct ratios between the degrees of simple and compound medicines, pharmacological computus drew Galeano/Jālīnūs into an interdisciplinary debate over the identification of precise causes. In "Treatise on the Natures," he criticized scholars who asserted the essential truth of medical theories just because cures based on those theories were always effective. The recognized computational advantages of Kindī's pharmacological computus, he argued, were an insufficient basis upon which to conclude that these ratios existed in the same way that ratios between musical notes existed. Galeano/Jālīnūs wrote that ingesting a simple medicine in the fourth degree could be fatal, but simple medicines in lower degrees were not.[94] Ratios were descriptions, not causes. Physicians who relied on these ratios were similar to astronomers who asserted the external reality of the philosophically objectionable hypotheses of the eccentric and epicycle, which were, to his mind, but an instrument for calculating planetary positions. Galeano/Jālīnūs believed that those who alleged the external reality of the epicycle and eccentric erroneously inferred the middle term of a syllogism. He asserted, remarkably, that even Ptolemy held the epicycle and eccentric to be impossible, perhaps on the basis of the uncertainties recognized in *Almagest* IX.2.[95] Likewise, the ratios of simple medicines were, at most, approximate tools for calculation.

The way physicians misconstrued causes of disease was also paralleled by natural philosophers' imprecisions. As an example, Galeano/Jālīnūs mentioned Aristotelians' view that there must be rest between two opposite motions. He conceded that rest was required between two natural motions, or between a natural motion and a compelled motion, but not between two compelled motions. That is, an ant could ascend and descend a mountain without ever stopping crawling.[96] According to Galeano/Jālīnūs, physicians were similarly confused when it came to understanding critical days. Critical days were the days, counted from the beginning of the disease, on which diseases tended to reach critical points. Hippocrates was the first scholar to mention the critical days, and Galen suggested a connection to the phases of the moon. Galeano/Jālīnūs remarked:

> Thus, if the doctor says that an odd (*nip̱rad*) number of seeds be taken because the nature of the form and the prime mover are indivisible (*nip̱rad*), as Muḥammad says[97] that God is indivisible (*nip̱rad*) and loves that which is indivisible (*ha-nip̱radim*), then this is not [a matter] for the physician. Thus, why is it said that the critical days (*buḥranim*) are in pairs, if odd things are better? This is because it is by accident, for it is not due to odd things inasmuch as they are odd. For if it were due to how the occurrence of the aspects of the moon were truly the reason, then they would be the odd numbered days.[98]

For Galeano/Jālīnūs, critical days were a tool for marking the course of a disease, not a cause of the stages of a disease.

Galeano/Jālīnūs's skepticism of the relationship of the critical days to astral causes had a context in European medicine. Galen's *De diebus criticis* was studied in the first year[99] at the University of Bologna, but the curriculum for theoretical medicine in 1405 did not include a single text on astrology.[100] Scholars other than Galeano/Jālīnūs also detected the logical contradictions between the science of astrology and the way it was applied to medicine.[101] Pietro d'Abano went further and identified problems with the doctrine of critical days, both with the broader foundations and with the details. He commented, "Assignment of the cause of critical days is almost always in error, because the subject leads the *medicus* away from his own art, since it has more to do with astrology than natural philosophy."[102] According to d'Abano, the complexity of lunar motions meant that the astral forces could never be sufficiently understood. Even in his (copious) writings on astrology, such as the *Conciliator*, d'Abano acknowledged the difficulty of the correct determination of a patient's horoscope.[103] Galeano/Jālīnūs cautioned that in medicine, as in other disciplines, causation should not be carelessly inferred from elegant mathematical descriptions.

Yet astrology and medicine were both important disciplines at Renaissance courts and the Ottoman court.[104] Thus, apart from denying astral causes for the critical days, in *Puzzles*, Galeano/Jālīnūs took a nuanced position on astrological prognostication of the course of disease. Galeano/Jālīnūs was skeptical that the death of the native, i.e., the person whose chart was the subject of the forecast, could be consistently and confidently predicted:

> Behold; the judicial astrologers (*shopṭim*) and the physicians say that one born in the eighth month will not live. But Aristotle wrote that there are places where those born in the eighth [month] live. It is apparent that this is by accident (*b*ᵉ*-miqreh*) because there are places with hot climates where Saturn, which rules that month, does not affect them with its cold. Thus, another planet does away with the harm of Saturn ruling there. Therefore, we do not say that the judgment of the astronomers is false, just that their premise is truly decisive [only] if there is no other reason impeding or preventing or improving or harming.[105]

Abraham Ibn Ezra's statement in *The Book of Judgments* that the upper planets made no impression on the seasons served as additional evidence for Galeano/Jālīnūs's denial that astral causes were decisive.[106]

He found more support in Latin texts. In *Puzzles*, he classified a statement by Bernard de Gordon in his *Seiper ha-gᵉbulim* (On Prognosis) about the celestial causes of seasonal temperatures as an example of the fallacy of confusing the accidental cause with the essential cause.[107] Galeano/Jālīnūs criticized Bernard for stating that the summer was hot and dry because the sun was overhead *and* because of the effect of the sun's position in Sirius in Leo in heating and drying the air. As counterevidence, Galeano/Jālīnūs cited the *Introduction to the Phenomena* of Geminos, where Geminos declared that the rise of fixed stars had no effect on seasonal temperatures.[108] Hence, he argued, astrological forecasts should not foreclose possible treatments of illnesses. The role of astral causes should not be heedlessly overstated.

Given what Arnaldo and Bernard had to say, it is no surprise that Galeano/Jālīnūs's nuanced position on medical astrology was shared by his contemporaries at the University of Padua, likely the largest medical faculty in Europe with a body of medical students nearing 100 in the fifteenth century.[109] Medical astrology was most developed at Padua, and Pietro d'Abano acknowledged that astrology was relevant to medicine.[110] Girolamo Fracastoro, a scholar of homocentric astronomy and also an instructor of logic at the University of Padua during Galeano/Jālīnūs's time in the Veneto, weighed in on the relevance of astrology to medicine in *De causis criticorum dierum libellus* (Treatise on the Causes of Critical Days). He accepted the theory of critical days but denied that the cause was astral.[111] In his *Syphilis, sive morbus gallicus* (Syphilis, or the French Disease), Fracastoro acknowledged the relevance of astral causes, particularly for the appearance of the disease, but favored a material theory for syphilis transmission.[112] Astrological forecasts, he claimed, were the basis neither for therapies nor preventive measures.

Likewise, Galeano/Jālīnūs, in his criticisms, favored medical experience over astrological theory. These criticisms were reflected by changes in medical instruction occurring at the University of Padua in the mid-sixteenth century.[113] The emerging sense that judicial astrology could not be a basis for medical decision-making meant that physicians had to turn either to logic or to empirical experience as a basis for their diagnoses and therapies.[114] Hence, the organization of *Puzzles of Wisdom* around errors of reasoning rather than around disciplines, along with an acknowledgment of the relevance of astrology to medicine, suggests that Galeano/Jālīnūs was affected by the disciplinary situation of medical instruction at Padua.[115]

The spread of syphilis and the plague created a space for exchange between scholars grounded more in Latin and those grounded more in Middle Eastern languages.[116] Physicians of Islamic societies knew that syphilis and the plague extended beyond their own realm, as they called syphilis the Frankish Chancre (*al-ḥubb al-ifranjī*).[117] Fracastoro confirmed the potential for the disease to transit regions but commented that he would not send the divine solution to Europe's woes to "Turkestan and Ammon's realm"—Turks and Arabs did not merit the *lignum sanctum*, a treatment for syphilis.[118] Challenging medical problems impelled physicians to seek new sources and to be open-minded. For instance, Galeano/Jālīnūs's contemporary Ilyās b. Ibrāhīm, in his treatise on the bubonic plague (*Majannat al-ṭāʿūn wa-l-wabāʾ*), reinforced the etiological flexibility evinced by Galeano/Jālīnūs. In the treatise, God was at the top of the hierarchy of causes, with astral factors just beneath.[119] These two classes of effects were transmitted through the air, which fluctuated with geography. Human natures were at the bottom of the chain of causes. Open-mindedness about the causes of communicable diseases was a notable feature of medical thought at the time of the merchants of knowledge. Nükhet Varlık observed that "as long as there was a hierarchy of causal explanations, one could always find a way of explaining plagues; if one causal factor failed to explain it, another would do it. . . . Having the flexibility of using multiple systems of etiology, treatise writers could establish connections between seemingly incongruent notions."[120] Flexible etiology suited intellectual exchange about transregional diseases with particular local manifestations.

Conclusions

Though Galeano/Jālīnūs was a critical intermediary for exchange in several disciplines, medicine and mechanics showcased his performance as a merchant of knowledge. Much of his activity as an intermediary in these fields, which were central to his professional identity, took place after his trip to Venice. He was occupied with bringing technology from Europe as well as information from Latin texts to Jews and Muslims. Technology, e.g., the transcription device(s), diseases, and materia medica, had tangibly transregional effects, while theoretical astronomy and Averroist philosophy did not. Because medicine and mechanics were the fields in which scholars depended on transregional goods (e.g., devices, diseases, and materia medica), Galeano/Jālīnūs was most interested in debunking and demystifying in these areas. Wars and disease had a pervasive impact that, say, the doctrine of the unicity of the intellect lacked. In earlier chapters we saw how the merchants of knowledge trafficked in knowledge; here, Galeano/Jālīnūs traded in secrets.

And he never stopped trading. In 1529, after leaving the Ottoman Empire for Candia, Galeano/Jālīnūs translated a letter addressed to the lordship (*signoria*) of Candia written in Turkish in Turkish (i.e., Arabic) script into either Italian or Latin.[121] The linguistic skills that he had honed in intellectual exchange remained useful in nonscholarly contexts. His career as an intellectual intermediary enhanced his value as a go-between at the ducal court.

CONCLUSION

During my undergraduate years, I never went anywhere special just to buy a cup of coffee or tea. Only the cognoscenti seemed to frequent cafés. Ten years later, when I began my first tenure-track job, the United States had changed. I was comforted to find Starbucks coffee was proudly brewed in the dining halls and to know that there was a campus café with inviting furniture upholstered in soothing fabrics. Though I have left the devil's brew behind, portions of this book took shape in Portland, Maine's aptly named Arabica. How did academia ever function without widespread coffee culture? In hindsight, the explosion of coffee chains and their ubiquity on or near college campuses may seem inevitable and natural, just like Copernicus's heliostatic arrangement of the planets or the shared Muslim, Jewish, and Christian arguments against astrology. But neither were predetermined. In *Merchants of Knowledge*, I have provided precise explanations, based on historical contingencies and without resort to teleological thinking, for the exchange critical for intellectual life in the Eastern Mediterranean between 1450 and 1550.

Intermediaries were the key. Miri Shefer-Mossensohn has observed that the eclectic character of Ottoman science, including the contents of Latin and Greek texts, was due in part to how Greek and Jewish physicians at the Ottoman court strove to advance their own careers with little concern for the prerogatives of Ottoman elites.[1] The merchants of knowledge conformed to that model. Their ability to ferret out opportunities and texts, always to accrue social capital, accounted for their effectiveness as scholarly intermediaries. The merchant of knowledge Galeano/Jālīnūs was always on a quest for social capital; he acquired information that his contacts lacked, whether in Arabic, such as Fārābī's commentary on *De interpretatione* and Rāzī's *Manṭiq al-Mulakhkhaṣ*, or in Latin, such as Arnaldo da Villanova's theories of pharmacological computus and the *Almanach perpetuum*. As an indication of Galeano/Jālīnūs's immersion in Renaissance learning, which facilitated the acquisition of Latin sources, consider the unprecedented organization of his most fascinating text, *Puzzles of Wisdom*. His decision to structure *Puzzles* around the logical fallacies that transected scholarly disciplines was influenced by medical education in Renaissance Italy.

Similar dynamics of the quest for social capital accounted for the merchants of knowledge's activities as intermediaries for other Jewish scholars. A case in point is the Hebrew version of Elijah Delmedigo's commentary on *De substantia orbis*, which he produced after he composed the Latin original. In the Hebrew translation, he flaunted the potential of his knowledge of Christian thought to shield Jews from Christians' errors. And while Galeano/Jālīnūs translated into Arabic and composed in Ottoman Turkish, he translated also into Hebrew, the language in which he composed *Puzzles of Wisdom*. *Puzzles* was an exceptional text, though accessible only to readers of Hebrew. There we learned of his desire to outshine other Jewish scholars at the Ottoman court. The merchants of knowledge fostered their own vigorous intellectual culture that was nourished by texts in Arabic and Latin, the respective scholarly languages of their Muslim and Christian contacts.

Excavating transregional exchange in multiple disciplines has increased our appreciation for the depth of intellectual life in the Ottoman Empire during the reign of Sultan Bayezit II. *Puzzles of Wisdom* is a trove of information on the Ottoman court, in a format influenced by the European Renaissance, written in a language unknown to the author's most privileged contacts. Even more, it is evidence of how, most significantly, the achievements of Ottoman scholars were inextricable from the transregional connections that the merchants of knowledge maintained. As was the case during the reign of Bayezit II's father, Sultan Mehmed the Conqueror, exchange was essential to Ottoman intellectual life. Aspects of Galeano/Jālīnūs's exceptional work as an intermediary could be gleaned from earlier scholarship on him, but in *Merchants of Knowledge*, I have shown in detail how his work as an intermediary responded to the desires expressed by Ottoman Muslim scholars. By highlighting the substantive, ongoing exchanges that Galeano/Jālīnūs facilitated, the dichotomy between East and West that is foundational for modern Western exceptionalism is hard to maintain. The merchants of knowledge recognized and bridged differences between regions. The Eastern Mediterranean was a mediating space, and the merchants of knowledge filled it.

The connections engendered by the merchants of knowledge have become as worthy of study as the entities they connect. To borrow Bruno Latour's language, the merchants of knowledge were not only intermediaries but mediators.[2] Without understanding their own intellectual production, in which material from Hebrew, Arabic, Latin, and Turkish texts was hybridized in varying proportions, we would not fully grasp how the exchange occurred. If we thought of the merchants of knowledge as pieces of yarn linking pushpins that represent locales or Christian and Muslim scholars who were better known than their Jewish teachers and contacts, then, in *Merchants of Knowledge*, I have put the yarn under a microscope. Viewed that way, the yarn's multidimensionality has become hard to ignore. Though the segments of yarn might be of similar lengths and of a single color, each segment represented connections that were forged through multiple steps in consistently distinct ways.

The level of specialization necessary for survival in modern academia occludes the extent to which scholars in earlier times excelled in multiple fields, as well as the extent to which scholars pursued in one field the questions that had been posed in another. My own range has been tested by the details of the interwoven, multidisciplinary competencies of the merchants of knowledge and their contacts as well as by the complexities of the records of the transregional commerce that paved the way for their careers. Despite these shortcomings, I hope to have shed light on at least some of the connections between *Qabbalah*, astrology, Aristotelian philosophy, medicine, astronomy, and mechanics. These intersections of disciplines deepen our understanding of the dynamics of intellectual exchange. In so many cases, the merchants of knowledge and their contacts achieved in ways that would be inconceivable without exchange.

It is obvious that the linguistic abilities of the merchants of knowledge enabled them to traverse and puncture geographic and disciplinary boundaries and create their own distinctive culture. Hebrew scholasticism was the immediate context of Elijah Delmedigo's career and also paved the way for Galeano/Jālīnūs. Though he composed in Turkish for his Ottoman patrons, he recognized and capitalized upon their interest in Latin texts. In the commercial and political realms, Galeano/Jālīnūs and other members of merchant-scholar families reaped the social and economic rewards of being able to translate into a European language. Knowledge traveled out in the open through a shared intellectual space created by commercial connections.

In the preface I wrote that my goal was to depict a more complex Renaissance. I have aimed to narrate aspects of the Renaissance that suit a West that is aware of how much it is changing. It might come as a surprise that the Renaissance was more heterogeneous than we thought, but the merchants of knowledge knew what multiple cultures could do for each other. Communities with multiple cultural competencies produced the Renaissance. Today's West is full of recent and not-so-recent immigrants, transnational individuals who, in addition to preserving links with their previous homes, are creating new ways of existing in new lands. The merchants of knowledge are their precedent. And the Christian and Muslim contacts of the merchants of knowledge were self-aware enough to learn from them.

The merchants of knowledge belonged to families that were heavily taxed, and they lacked the citizenship status of their Muslim and Christian contacts. Despite their brilliance and productivity, university teaching positions were not in the cards. Rather, they flourished on the margins. I defined the network as scholars who were connected to Crete, which was a colony of the Venetian Republic during their era. The merchants of knowledge were not allowed to dwell permanently in the metropole, Venice. Despite their peripheral political status, European Renaissance and Ottoman intellectual history in the late fifteenth and early sixteenth centuries cannot be fully understood apart from the activities of intermediaries like the merchants of knowledge. Edward Said's theory of contrapuntal reading explains both the intellectual interchange between the merchants of knowledge, their contacts, and their readers, as well as why developments in Europe and the Ottoman Empire have heretofore received more attention. Crete was not an Ottoman colony, but it was on the periphery of the

Ottoman Empire. Future researchers may answer the "imperative to excavate the material inequalities of texts as events in the world and the asymmetry of cultural transactions."[3] As researchers better understand how much the more powerful learn from the less powerful, we may find ways to assess in greater depth the effect of power imbalances on transregional exchange in this period.

ACKNOWLEDGMENTS

At the completion of a book about intellectual exchange, it is a pleasure to be able to thank my teachers. I was fortunate to grow up at Temple Israel in Boston, where, in seventh grade, Rabbi Bernard Mehlman used fat chalk to introduce my classmates and me to the documentary hypothesis, the significance of Arabic as a Jewish language, and the importance of Abraham Ibn Ezra. During my high school years, at Hebrew College, Ms. Gabi Mezger held unwavering expectations and modeled a level of intensity and dedication to language study and pedagogy that I fully appreciated only in graduate school.

When I was an undergraduate and A. M. student in the Department of the History of Science at Harvard University, my adviser, the late Abdelhamid I. Sabra, never failed to be patient and encouraging even amid the presence of graduate students who were far more competent. If I had to reprise graduate school, I would hit rewind and return to study again with George Saliba at Columbia University. In addition to bestowing upon me unparalleled specialized training, George remains a paragon of a passionate commitment to academia. His own influential research on transregional exchange is one starting point for this book. From Professor Raymond Scheindlin, at the Jewish Theological Seminary, I gained a profound appreciation of the interdisciplinarity of medieval Jewish thought and of the interplay between medieval Arabic and Hebrew literature. Without exception, my mentors have set a high bar.

So have my colleagues. This book began as my contribution to a working group called Before Copernicus, which met between 2006 and 2009 under the auspices of the Max Planck Institute for the History of Science. Conversations with and presentations from Nancy Bisaha, Christopher Celenza, Raz Chen-Morris, İhsan Fazlıoğlu, Rivka Feldhay, Maria Mavroudi, Jamil and Sally Ragep, Michael Shank, and the late Edith Sylla revealed the depth of intellectual exchange in ways that were new to me. I learned more from the disagreements than from the points of agreement. I am grateful to the codirectors of the project for including me when I was an early-career scholar. I was fortunate to have the opportunity to begin to work with Rivka Feldhay and to deepen my relationship and friendship with Jamil Ragep. His and Sally's guidance and unstinting support since the late 1980s have been a gift. Their intellectual and professional generosity during the past twenty years have been immeasurable.

For scholars of science in the Jewish communities of premodern Islamic societies, the oeuvre of Y. Tzvi Langermann is, metaphorically, one-stop shopping for striking findings, erudition, and clarity. For two decades, Tzvi has graciously shared his own work before publication and, on several occasions, gently corrected me. He has blazed a tantalizing path for younger scholars. For three decades, Glen Cooper has been selflessly saving me from myself. As this book was coming together, he read and commented profusely on the entire manuscript and vastly improved my exposition.

Colleagues in many subfields made me feel as if this project mattered even as they pursued their own fascinating research. In addition to those I've acknowledged elsewhere, much appreciation to Asad Ahmed, Peter Barker, İ. Evrim Binbaş, Sonja Brentjes, Jonathan Decter, Dallas Denery, Khaled El-Rouayheb, Nahyan Fancy, Paula Findlen, Frank Griffel, Bruce Hall, Richard Kremer, Clark Lombardi, Hannah Marcus, Sajjad Nikfahm-Khubravan, Judith Pfeiffer, Lawrence Principe, Francesca Rochberg, Josefina Rodríguez-Arribas, Arielle Saiber, Dana Sajdi, Fateme Savadi, Sabine Schmidtke, Tunç Şen, Miri Shefer, Nathan Sidoli, Justin Stearns, Birgit Tautz, Hasan Umut, and Robert Wisnovsky. Though I cannot imagine enjoying a more illustrious intellectual milieu, any errors in the book are my responsibility alone.

Institutions matter. I began drafting this book while a fellow at the Stanford Humanities Center in 2012–13. Director Aron Rodrigue and his staff created a magical setting that led to the most productive year of my career. The bulk of the work on this book occurred during 2018–19 at the National Humanities Center. There, Director Robert Newman and his staff engineered an environment that maximized collaboration and collegiality. I have been blessed with these opportunities. In the second half of 2019, a fellowship from the Guggenheim Foundation funded an additional semester of sabbatical leave. Without the vision of Editor-in-Chief Kate Wahl at Stanford University Press, along with series editors Nükhet Varlık and Ali Yaycıoğlu, this book might still be a work in progress. Their capacious definition of an Ottoman world will enrich Ottoman studies. Thane Hale, Gigi Mark, and Adriana Smith got the book past the finish line.

I have always felt valued at Bowdoin College. More important, I am constantly motivated to be my best by my colleagues in the Department of Religion (Todd Berzon, Elizabeth Pritchard, and Claire Robison) and in the Middle Eastern and North African Studies Program (Nasser Abourahme, Oyman Başaran, Meryem Belkaïd, Barbara Elias, and Batool Khattab). They, in addition to my friends and colleagues in other departments and programs, make it impossible to give anything less. My colleagues at Bowdoin College's Hawthorne-Longfellow Library assisted me with my research for this book at every turn. Bowdoin supports its faculty with a strong sabbatical program and generously funded the services of Bridgette Werner at Throughline Editing. Bridgette's perseverance and penetrating feedback has left me with the book I wanted to write and with a book I could never have written on my own.

My parents (the late Alan Morrison and Peggy Morrison) and brother (Jeremy Morrison) model the life of the mind. Amid the competition, we have, I believe, learned that the arc of an intellectually engaged life is long with plenty of setbacks. They have been in my corner. Completing this book was challenging, but enduring the COVID-19 pandemic was more trying. Through it all, my wife (Dana Gold), daughter (Aziza Gold Morrison), and I did everything we could for each other. I look forward to the next chapter with them.

NOTES

Preface

1. North, *Cosmos*, 206.

2. For a study in which the author notes parallels between fifteenth-century Ottoman philosophical culture and Renaissance European intellectual life, see Balıkçıoğlu, *Verifying the Truth*.

3. Bisaha, "European Cross-Cultural Contexts," 39.

Introduction

1. For examples of the use of "transmission." see, e.g., Ragep and Ragep, *Tradition, Transmission, Transformation* and Wallis and Wisnovsky, *Medieval Textual Cultures*.

2. Marino, "Economic Encounters," 282–83.

3. Appuhn, "Tools for the Development," 270–75.

4. Marino, "Economic Encounters," 281.

5. Howard, "Cultural Transfer," 152–53.

6. Rozen (*A History of the Jewish Community*, 38) called Bayezit II's welcoming of the Jews "a myth" unconfirmed by any other source. See also Shmuelevitz, "Capsali as a Source," on Capsali's subjective position. Shmuelevitz did not mention Capsali's account of Bayezit II.

7. Hacker, "The Rise of Ottoman Jewry," 79–81.

8. See Jacoby, "David Mavrogonato from Candia" (Heb.), 391 on how the Venetian Republic attempted to intervene with the Sublime Porte through Mehmed the Conqueror's personal physician, Yacup Pasha. See İhsanoğlu, "Some Remarks on Ottoman Science," 64–65 on Yacup's conversion to Islam.

9. Jacoby, "Production et commerce," 253.

10. Norton, "Blurring the Boundaries," 10. See also Necipoğlu, "Visual Cosmopolitanism."

11. See Galante, *Türkler ve Yahudiler*, 22. Galante wrote that a synagogue in Istanbul came to be known as the Sürgün Synagogue.

12. Lauer, *Colonial Justice*, 25. The future Cardinal Basilios Bessarion (d. 1472) moved to Italy from Constantinople and also spent time on Crete. On Bessarion's time on Crete, see Nicol, *The Last Centuries of Byzantium*, 353. See also Holton, "The Cretan Renaissance," 7 on how Bessarion started a Uniate school on Crete soon after the fall of Constantinople. See also Holton, 3 on other Byzantine scholars who fled to Crete after 1453: Michael Apostolis (d. 1480); Apostolis's son Arsenios (d. 1535), who was an editor and copyist; and Ianos Laskaris (d. 1534).

13. Socrates Scholasticus, *The Ecclesiastical History of Socrates*, 378–79. See also Marcus, "Crete," 289–91.

14. Canard and Mantran, "Iḳrīṭis̲h̲."

15. Paudice, *Between Several Worlds*, 32. See also Lauer, *Colonial Justice*, 85–87 and her citations of David Jacoby's scholarship on the role of Jews in economic life. For an example from sericulture, see Molà, *The Silk Industry*, 65.

16. Corazzol, "Gli ebrei," 56.

17. See Lauer, "Cretan Jews," 129–30 for information on the movement of Sephardic Jews to Crete before 1391.

18. Jacoby, "Jewish Physicians and Surgeons on Crete" (Heb.), 436.

19. Marcus, "The Composition," 64–65.

20. ASV, NDC 17, folder 2, 75b. Lyas Thoroz, a Tatar, appeared before the Venetian notary in 1503 to record his sale of *abuffalo* skins, which he sold for two ducats apiece. See Poliak and Slutsky, "Crimea," 298–301 for information about how the Genoese conquered some of the Tatar lands in the fifteenth century. The Ottomans took over the Genoese Tatar possessions in 1475.

21. On the presence of Jews from Arab lands on Candia see Marcus, "The Composition" (Heb.), 67. See Marcus, 71 for information about how Italian Jews sometimes came to Crete because they were expelled from other parts of Italy due to the Inquisition.

22. Lauer, *Colonial Justice*, 30.

23. Malkiel, "The Ghetto Republic," 120–21.

24. Carpi, *L'Individuo*, 226. Brief biographical entries of Candiote Jews living in Padua from 1402 to 1508 are found on 225–33.

25. Holton, "The Cretan Renaissance," 7. Over one thousand Cretans went to Padua for education between 1500 and 1700.

26. Barzilay, *Yoseph Shlomo Delmedigo*, 35. In note 2, one reads that there were Jewish students studying at the University of Padua in 1501 under assumed names.

27. Carpi, *L'Individuo*, 209–11, 230.

28. Capsali, *Seider Eliyahu Zuṭa*, 2:254.

29. Carpi, *L'Individuo*, 194.

30. These studies include Goldberg, *Trade and Institutions*. For a later period, see Trivellato, *The Familiarity of Strangers* for how reputation and trust carried across linguistic, cultural, and religious borders. Sarah Abrevaya Stein, in *Plumes: Ostrich Feathers, Jews, and a Lost World of Global Commerce* (New Haven: Yale University Press, 2010), studies a more recent period and situated Jewish merchants within global trading and cultural networks. Avner Greif, in *Institutions and the Path to the Modern Economy: Lessons from Medi-*

eval Trade (Cambridge: Cambridge University Press, 2006), investigates the place of trust and reputation in long-distance trade.

31. Tirosh-Rothschild, *Between Worlds.*

32. García-Arenal, *A Man of Three Worlds.*

33. Pfeifer, *Empire of Salons*, 17.

34. Binbaş, *Intellectual Networks in Timurid Iran*, 77.

35. Pfeifer, *Empire of Salons*, 168–69.

36. Pfeifer, 130–31.

37. See, e.g., Brentjes, "Cross-Cultural Exchanges," 411–14, and Stearns, "Review of *Cross-Cultural Scientific Exchanges*," 760–63.

38. Hasse, *Success*, xii.

39. Saliba, *Islamic Science*, 193–232.

40. Hasse, *Success*, xv.

41. See Cassuto, *I manoscritti*, and Lehmann, *Eine Geschichte der alten Fuggerbibliotheken*, 1:93. For the list of manuscripts that Elijah Capsali was known to have owned personally, see Benayahu, *Rabbi Elijah Capsali* (Heb.), 148–52.

42. For the identification of the three agents, see Richler, *Hebrew Manuscripts*, 71. A third agent is named as Battista. See ASV, ADC 34bis/35, folder 18, 82a for a reference to Azalino, one of the agents. For more references to him, see ASV, NDC 35, folder 4, 53a and ASV, NDC 123, folder 5, 213b–214a. For a reference to Azalino in October 1548, see ASV, ADC 34bis/35, folder 20, 47b. For a reference to Thoma Sacellani, see ASV, NDC 123, folder 3, 82a–b.

43. Steimann, "The Story," 28.

44. See Paudice, *Between Several Worlds*, 33 for the suggestion that the agents were working for Ulrich Fugger III (1526–84). See also Cassuto, *I manoscritti*, 35 for the suggestion that the Jews on Candia sold the books to a bibliophile who, at a later date, sold the books to the Fuggers. See Otto Hartig, *Die Gründung*, 320 on how, in 1566, the Fuggers needed the help of a Jew for cataloging the Hebrew manuscripts. See also Cassuto, *I manoscritti*, 38–40 for the argument that the numbering of the acquired manuscripts, which involved numerals in Hebrew, was evidence of a Jewish assistant.

45. Forin, "A Padova col Caiado." On the Fuggers and Venice, see Strieder, *Jacob Fugger*, 103.

46. Costil, "Le Mécénat," 27.

47. Eric Dursteler, by researching Venetian merchants' activities in Istanbul a century later, arrived at a similar conclusion: commerce depended on translation, both written and oral. See Dursteler, *Venetians.*

48. ASV, DDC 33bis, folder 6a, 286a.

49. ASV, DDC 34, folder 13, 61b–62a.

50. On David Mavrogonato's role as a translator for Greek priests, see Jacoby, "David Mavrogonato," 389. See Paudice, *Between Several Worlds*, 223 for how David and Joseph Mavrogonato appear in a list of taxpayers from 1542 in Candia.

51. Setton, *The Papacy*, 2:296. See also Jacoby, "Un agent juif," on how Mavrogonato unveiled an anti-Venetian Greek plot. On David Mavrogonato's death date, see Manoussacas, "Le recueil," 363, and Jacoby, "David Mavrogonato," 394.

52. Dursteler, *Venetians.*

53. Cf. Trivellato, *Familiarity of Strangers*, 156. "Even a global diaspora such as the one formed by Western Sephardic merchants could not count on the presence of coreligionists in every corner of the world."

54. Paudice, *Between Several Worlds*, 30. On the tax burden in general, see Lauer, *Colonial Justice*, 90–92, 168–69. See also Lauer, 176 for how physicians who were paid by the government of Crete were exempt from taxes. See Corazzol, "Gli ebrei," 42 on how David Mavrogonato and his descendants had to pay only a single *iperpera* (a Byzantine unit of currency), while the Jews were taxed annually.

55. Lauer, *Colonial Justice*, 86.

56. Segre, "Juifs à Venise," 71, and Noiret, *Documents inédits*, 425n1. See also Lauer, *Colonial Justice*, 93.

57. Lauer, 5.

58. When he was introduced in the document, we read, "dicti Mauri."

59. ASV, DDC 33bis, folder 5, 251a.

60. ASV, DDC 33bis, folder 1, 77b–78a.

61. ASV, DDC 33bis, folder 6a, 264a. For more on divorce in the Cretan Jewish community, see Lauer, *Colonial Justice*, 126–32.

62. ASV, DDC, 33, folder 1a, 37b–38b.

63. For more on how the archival sources referred to Turks, see Corazzol, "Gli ebrei," 49–51.

64. Dursteler, *Venetians in Constantinople*, 117–18. See also 119 for a discussion of Ottoman Muslims who converted to Christianity. On Greek Ottoman subjects on Crete, see Noiret, *Documents inédits*, 182.

65. ASV, DDC 26bis, folder 11, 9a–b. In order to pay for the expenses of storing the calfskins, pledges (*plezarie*) had to be collected from Jews. On medieval Jews trafficking in furs, see Toch, "Economic Activities," 377. On Jews engaged in the fur trade in Eastern Europe in the early modern period, see Trivellato, "Jews and the Early Modern Economy," 147.

66. ASV, NDC 17, folder 2, 25a. "I am Manoli the Turk, son of Xo, an inhabitant of Candia as well." See Sebastien, "Turkish Prosopography," 1:291 on how Ottoman ambassadors did not all have Muslim names.

67. ASV, NDC 17, folder 2, 212a.

68. ASV, DDC 32, folder 98, 109b.

69. ASV, DDC, 34bis, folder 23, 14a.

70. ASV, DDC 34, folder 12, 407a. Though the document is smudged, Mehmed may have been complaining of corsairs. On Ottoman corsairs, see Sebastien, "Turkish Prosopography," 1: 230–33, 236–37, 265–80, 281–89. On the risk of pirates for merchants in the Mediterranean, see Noiret, *Documents inédits*, 261, 289, 365, 386, 390, 437, 455, 505. In June 1541, the same Mehmed appeared again before the chancellery in Candia. See ASV, DDC 34, folder 13, 209a.

71. ASV, NDC 17, folder 1, 199b.

72. On the sale, see Richler, *Hebrew Manuscripts*, ix.

Chapter 1

1. See Paudice, *Between Several Worlds*, 33, and Richler, *Hebrew Manuscripts*, 31, and passim for references to the statements of sale from Candiote Jews being in "Fugger's cipher."

2. On the sale, see Richler, *Hebrew Manuscripts*, ix.

3. On the considerable prices that the agents paid for the manuscripts, see Steimann, "The Story," 36.

4. Cook, *Matters of Exchange*, 28.

5. Lehmann, *Eine Geschichte*, 1:61. On Ulrich Fugger's connection to Venice, see Lehmann, 1:91.

6. Costil, "Le Mécénat," 34.

7. Costil, 166.

8. Stam, *International Dictionary*, 1:202. For the names of these copyists, see Hartig, *Die Gründung*, 254. On Johann Jakob Fugger's service, later in his life, as librarian for Albrecht of Bavaria, see Meadow, "Merchants and Marvels," 191.

9. Olszowy-Schlanger, "The Science of Language," 402.

10. Medan, "Levita, Elijah," 730–32.

11. Hartig, *Die Gründung*, 254. See Olszowy-Schlanger, "The Science of Language," 403 on how, by the 1520s, Levita was in Venice teaching Hebrew to Christian Hebraists.

12. Holton, "The Cretan Renaissance," 3.

13. Paudice, *Between Several Worlds*, 33–37. See also Cassuto, *I manoscritti*, 43–44.

14. For the statement of sale to Elqanah Capsali, see Albo, *Seiper ha-ʿIqqarim*, 310b.

15. See Marx, "Gershom (Hieronymus) Soncino's Wanderyears," 479–80 for how Soncino criticized the second printed edition produced by Daniel Bomberg in 1521.

16. The presence of manicules is evidence that the manuscripts were read. See Albo, *Seiper ha-ʿIqqarim*, 4b, where a manicule points to the beginning of the chapter in which Albo explained how God was divested of shortcomings. On 11a, a manicule points to the beginning of the chapter on reward and punishment. On 63a, a manicule points to a discussion of the possible contradictions arriving from God's knowledge of human actions. On 116a, the manicule points to a section alluding to how an individual might know contradictory information.

17. On Nathan Ha-Meʾati's Hebrew translation, which was corrected by Lorqi, see Bos, *Novel Medical and General Hebrew Terminology*, 3–4.

18. Hartig, *Die Gründung*, 255–56. See also Richler, *Hebrew Manuscripts*, 186 for an ownership statement from David Capsali dating from 1511 in Vatican MS Ebr. 249, 333b. On David's relationship to Elijah Capsali, see Richler, *Hebrew Manuscripts*, 186 for a statement in which David Capsali may have said that he was a son of Elqanah Capsali.

19. Hartig, *Die Gründung*, 256. Hartig notes how Elijah Delmedigo exchanged letters with Grimani, Grimani's physician Abraham De Balmes, and Vitalis Dactilomelos, a Jewish scholar who worked for Grimani and translated Averroës's middle commentary on the *Physics* from Hebrew into Latin (found now in Paris, BnF MS Lat. 6507).

20. Howard, "Cultural Transfer," 153.

21. Lauer, *Colonial Justice*, 45. See Arbel, "Notes," 125–27 for the suggestion that the Ashkenazic Delmedigos were not the same as the Delmedigos who came from Negro-

ponte. Arbel posited a 1359 *terminus ante quem* for the arrival of the first Delmedigo, who hailed from Negroponte, in Candia.

22. See Arbel, "Notes" on how one Judah Delmedigo arrived on Crete from Negroponte by 1359 and another person with the same name arrived at about the same time from Germany. See also Bercovy, "La famille Delmedigo," 13. See Corazzol, "Gli ebrei," 76 for the argument that there were three different Delmedigo families.

23. See Benayahu, *Rabbi Elijah Capsali of Crete* (Heb.), 97 on how the office of *condestabulo* could be occupied only for a nonconsecutive term of three or four years. For more information about the position of *condestabulo* and other offices in the Candiote Jewish community, see Papadia-Lala, "The Jews in Early Modern Venetian Crete," 144.

24. Carpi, *L'individuo*, 230–31.

25. Stahl, *Zecca*, 30–31.

26. Ruderman, "The Italian Renaissance," 385.

27. See Cassuto, "Gli ebrei," 290 on the end of the relationship between Pico and Delmedigo in 1486.

28. Cassuto, 282–83. Depending on the determination of the completion of *The Examination of Religion* (*Beḥinat ha-dat*), Delmedigo returned to Crete in 1490 or 1491. On the possibility that Delmedigo's return to Padua was due to a philosophical disagreement between him and Rabbi Judah Minz of the yeshivah at Padua, see Cassuto, 292.

29. Lauer, *Colonial Justice*, 192.

30. ASV, DDC 33, folder 1a, 427b–428a. In this document, we read that Moses sold a house for 650 ducats to Dominico Venerio.

31. ASV, NDC 17, folder 1, 211b.

32. ASV, DDC 34, folder 13, 254b. This document was recorded in the archives in Hebrew. Another Delmedigo also did business with a Casani. See ASV, DDC 32, folder 99, 182b. This record from October 1488 reads: "A visiting sailor Gastaldio reported by order of the lord on the 22nd day of the present month, having affirmed and surrendered into the hands of Lyachi Casani, son of Moses, the Jew, all of that for which he was obligated to him. By the reckoning of S. Ayti Capello, son of S. Aitus, he [Lyachi Casani] is bound to give to the agent of Isaac Delmedigo, the Jew, 20 iperpera."

33. Bercovy, "La famille Delmedigo," 20.

34. See Corazzol, "Gli ebrei," 43 and Bercovy, "La famille Delmedigo," 20 on how Meyuḥas Delmedigo was likely from the same branch of the family (or same Delmedigo family) as Menaḥem.

35. Richler, *Hebrew Manuscripts*, 145 (on MS 206) and 292 (on MS 343).

36. Littmann, "Relations between Egypt and Candia" (Heb.), 49.

37. Marcus, "The Composition" (Heb.), 68. Eliezer Delmedigo, a famous rabbi on Crete and yeshivah head, was the grandfather of Joseph Delmedigo (the Yashar, 1591–1655). See also Littmann, "Relations between Egypt and Candia," 53.

38. Carpi, "Ebrei Candioti," 232. Carpi explained that Manuel and Menaḥem were the same person.

39. See Vincent, "Money and Coinage," 288 on how one Cretan *perpero* equaled thirty-two *soldini*. See Stahl, *Zecca*, 61 on how a *soldino* was 0.55 grams of pure silver. Thus, a *perpero* was $16.20 at today's prices.

40. ASV, DDC 34, folder 13, 429v. Paudice reproduced and transcribed this document in *Between Several Worlds*, 214–26.

41. Lauer, *Colonial Justice*, 28.

42. Carpi, *Ebrei Candioti*, 230.

43. Grätz, *History of the Jewish People*, 1:67. See also Bowman, *The Jews of Byzantium*, 316–18, document 141. According to an account in *Seider Eliyahu Zuṭa*, Capsali did not know Turkish when he first met Sultan Mehmed II in 1454 or 1455. The implication was that Greek remained the lingua franca of many Romaniot Jews even after the conquest.

44. Paudice, *Between Several Worlds*, 51. See also Corazzol, "Gli ebrei," 26.

45. ASV, DDC 33, folder 1a, 37b–38b.

46. ASV, DDC 33, folder 1a, 37b.

47. ASV, DDC 33, folder 1a, 410a.

48. ASV, DDC 33, folder 1, 545a. See also Paudice, *Between Several Worlds*, 49–50.

49. I am reading *arisicum* (risk) because I have consulted ASV, DDC 33bis, folder 1, 139a. There, one finds the phrase (in Italian) *arisico del grippi* in a 1512 document.

50. Paudice, *Between Several Worlds*, 52.

51. Roth, "Capsali, Elijah," 455–56. On taxation, see Paudice, *Between Several Worlds*, 29–34. See ASV DDC 34, 13–181a for how Capsali, Mossani Delmedigo, and Cuda Todesco, the chamberlain of the Jewish community, came to the ducal court to negotiate the community's tax burden in May 1541.

52. Benayahu, *Rabbi Elijah Capsali*, 79–80.

53. Littmann, "Relations between Egypt and Candia," 51. The wine was called Malvasia di Candia.

54. Lauer, *Colonial Justice*, 227n184.

55. Lauer, 62. See also Lauer, 109, where Lauer notes that the Balbo family, like other elite Candiote Jewish families, frequently took their inheritance disputes to the ducal court.

56. ASV, DDC 26bis, folder 11, 56a–b.

57. See Gottlieb, "The Metempsychosis Controversy" (Heb.); Ogren, *Renaissance and Rebirth*; and Ravitzky, "The God of the Philosophers."

58. ASV, DDC 33bis, folder 6a, 256a. In this record from 1525 to 1526, Shabbetai was listed as the son of Michael. See also Elior, "Rabbi Yedidyah Rakh," 30. Elior found that Michael Balbo's father was named Shabbetai, meaning that Michael Balbo would most likely have named a son Shabbetai.

59. ASV, NDC 18, folder 4, 67a. In a notarial record from 1480, we read of a Solomon b. Michael Balbo. On another son, Isaac, see Horowitz, "Balbo, Michael ben Shabbetai Cohen," 84–85.

60. ASV, DDC 32, folder 99, 184a.

61. ASV, DDC 33bis, folder 6a, 226b.

62. He appeared in a tax census from 1505 (ASV, DDC 33, folder 1, 37b–38b) and 1547 (ASV, DDC 34bis/35, folder 20, 57a).

63. Corazzol, "Gli ebrei," 26.

64. Richler, *Hebrew Manuscripts*, 192.

65. David, "Moses Esrim ve-Arba," 556.

66. See ASV, DDC 26bis, which contains notarial records from 1453 to 1455, when Benetto Vittori was duke of Crete.

67. Cohen Ashkenazi, *Urim wᵉ-ṯummim*, 1b.

68. See also Capsali, *Seider Eliyahu zuṭa*, 2:205. Capsali lists Saul among those who died of the plague in 1523.

69. Anonymous, *Taqqanot Qandiya*, 1:66–67, 75, 81–82, 87 on the involvement of Saul b. Moses ha-Cohen Ashkenazi in Candiote Jewish affairs.

70. Hacker, "The Immigration" (Heb.), 150–51.

71. ASV, DDC 33, folder 1a, 381b–382b. No member of the family appeared in this 1505 tax census or in any previous tax census.

72. ASV, DDC 15bis, folder 6, 48a–b.

73. ASV, DDC 15bis, folder 6, 69b.

74. See Capsali, *Seider Eliyahu Zuṭa*, 2:205.

75. ASV, DDC 15bis, folder 6, 146b. For additional confirmation of the father-son relationship, see ASV, NDC 123, folder 3, 74b.

76. ASV, DDC 34, folder 13, 424a.

77. For more on Samuel Algazi, see Corazzol, "Gli ebrei," 222n951. On the contents of the manuscripts that Samuel Algazi owned, see Richler, *Hebrew Manuscripts*, 44, 177. On Samuel Algazi being a student of Elijah Capsali, see Paudice, *Between Several Worlds*, 53.

78. See Vatican MS Ebr. 201, 240b for a 1539 statement of sale from Abraham Algazi to Galeano/Jālīnūs, naming him as a rabbi, physician, and his (Algazi's) teacher.

79. Jālīnūs/Galeano , *Kitāb al-Zīj*, 1b–2a.

80. Morrison, "An Astronomical Treatise," 386.

81. Langermann, "A Compendium," 287n4. On the relationship between the two, see Langermann, "From My Notebooks," 376. There, Elijah Galeano was named as Galeano/Jālīnūs's grandfather, making Moses b. Elijah Galeano the uncle of Galeano/Jālīnūs. On Moses b. Elijah Galeano, see also Morrison, "The Role of Oral Transmission for Astronomy," 23.

82. For more on Muʾayyadzādah's career, see Pfeiffer, "Teaching the Learned."

83. Morrison, "Mūsā Cālīnūs' *Treatise*." On Ahī Çelebī's identity as Muḥammad b. Kamāl al-Tabrīzī, see İzgi, *Osmanlı Medreselerinde İlim*, 2:23. See "Ahî Çelebi, Mehmed," Islam Ansiklopedisi, https://islamansiklopedisi.org.tr/ahi-celebi-mehmed for how Ahī Çelebī emerged from retirement to be Sultan Selim's chief physician.

84. Corazzol, "Gli ebrei," 212. See also Capsali, *Seider Eliyahu Zuṭa*, 2:205. Here, Capsali laments the death of his teacher and uncle, Judah Delmedigo, during the plague in 1523.

85. Corazzol, "Gli ebrei," 212.

86. Cassuto, *I manoscritti*, 25. See Cassuto, 35 for how the statement of sale of Vat. MS Ebr. 201 to the Fuggers' agent was lost. See Cassuto, 33 on how the Fuggers' agents acquired the codex with *Urim wᵉ-ṯummim*, owned at one time by Moses Cohen Ashkenazi, from a Moses b. Judah, who most likely was Galeano/Jālīnūs.

87. Corazzol, "Gli ebrei," 213.

88. ASV, DDC 34, folder 13, 429b ff. Spagnola paid more taxes than Galeano/Jālīnūs.

89. ASV, DDC 34bis, folder 20, 56a–57a and 64a–b. See Corazzol, "Gli ebrei," 206, 214–5 on how Galeano/Jālīnus and his family departed the island for an unspecified destina-

tion, perhaps Tripoli in the Levant, where Galeano/Jālīnūs's wife hoped to repatriate (*repatriar*).

90. Lauer, "Venice's Colonial Jews," 382, 387. As evidence of the family's connection to Iberian Jewry, Profayṭ Duran wrote a letter to a Sheʾaltiʾel Gratian contained in Paris BnF Ebr. 1048.

91. Kozodoy, *The Secret Faith*.

92. ASV, NDC 17, folder 2, 66b–67a.

93. ASV, NDC 15bis, folder 6, 48a. A Jacob Gratian was listed as well in this 1521 document.

94. Richler, *Hebrew Manuscripts*, 71

95. See Corazzol, "Gli ebrei," 52, on how Sheʾaltiel was sometimes named Saltin.

96. ASV, DDC 33, folder 1a, 563a.

97. On the manuscripts connected to the Casanis that were sold to the Fuggers, see Richler, *Hebrew Manuscripts*, 27 (Vat. MS Ebr. 39), 184 (Vat. MS Ebr. 247, as a scribe), and 332 (Vat, MS Ebr. 336).

98. Lauer, *Colonial Justice*, 176.

99. Lauer. 166.

100. ASV, NDC 17, folder 1, 87a.

101. ASV, DDC 34bis, folder 20, 56a.

102. ASV, DDC 34, folder 13, 429b. For more on Leo Gratian, see Corazzol, "Gli ebrei," 214n910.

103. Lauer, *Colonial Justice*, 40–41. Lauer recounted the Astrucs's economic success on Crete and their intermarriage with Romaniots. See also Marcus, "The Composition," 70.

104. Lauer, "Venice's Colonial Jews," 437.

105. Richler, *Hebrew Manuscripts*, 450. That manuscript is now Vatican MS Ebr. 530.

106. There is an extensive discussion of the Nomicos found in Corazzol, "Gli ebrei," 160–205. David Capsali purchased the codex Vat. MS Ebr. 249, which contained texts owned by the Nomicos, in 1511. On that, see Richler, *Hebrew Manuscripts*, 186.

107. Richler, *Hebrew Manuscripts*, 184–86.

108. Lauer, *Colonial Justice*, 56.

109. Lauer, 94.

110. Corazzol, "Gli ebrei," 183–84.

111. See Lauer, *Colonial Justice*, 97.

112. ASV, NDC 17, folder 2, 54a. On the *messitaria*, a tax of 3 percent on commercial transactions, see Panagiotakes, *El Greco*, 34.

113. ASV, DDC 33bis, folder 6, 397a–b.

114. ASV, DDC 33, folder 2/2, 66b.

115. See the critique of the term "trade diaspora" in Aslanian, *From the Indian Ocean*, 7–12.

116. See Markovits, *Global World*, 33 on how some nineteenth- and twentieth-century networks were not created by immigration or emigration.

117. Markovits, *Global World*, 25. See also Aslanian, *Indian Ocean*, 13–14 for the reference to Markovits's work.

118. Goldberg, *Trade and Institutions*, 300, 311–12.

119. Aslanian, *Indian Ocean*, 14.

120. Cf. Goldberg, *Trade and Institutions*, 128–37 for a description of a formal relationship of friendship (*ṣuḥba*) between merchants.

121. ASV, DDC 32, folder 89, 133b–134a.

122. ASV, DDC 32, folder 101, 315b.

123. ASV, DDC 33bis, folder 6a, 153a.

Chapter 2

1. Svachula, "Like a Virgo."

2. Roberts, "Not a Slave to the Zodiac."

3. Wakin, "Pluto."

4. On the distinction between natural and judicial astrology, see Cooper, "Astrology," 381–82.

5. Barton, *Ancient Astrology*, 20–21. See Rochberg, *The Heavenly Writing*, 288 on how Babylonian celestial divination contained both the origins of horoscopic astrology and elements of what falls under the modern category of religion.

6. Cooper, "Astrology," 400–401.

7. On computing the houses in Islamic astrology, see Casulleras and Hogendijk, "Progressions, Rays and Houses." See also Casulleras, "Mathematical Astrology."

8. Thorndike, *History of Magic*, 4:413–37. See Hayton, *The Crown*, 2, on Maximilian's use of astrological predictions as a type of political propaganda. See also Azzolini, "Reading Health." On astrology at the Ottoman court, see Şen, "Astrology in the Service."

9. Shefer-Mossensohn, *Science among the Ottomans*, 51.

10. Azzolini, "Are the Stars Aligned?"

11. Westman, *The Copernican Question*, 66.

12. Westman, 68.

13. Westman, 66.

14. Şen, "Astrology in the Service," 173–76, 211.

15. Şen, 215–16.

16. ʿAbd al-Raḥmān b. Abī Yūsuf, *Kitāb Ḥifẓ al-ṣiḥḥa*, 6a.

17. ʿAbd al-Raḥmān, 68b.

18. The text (*Seiper ha-goralot* in Paris BnF MS Hébreu 1073, 31a–38b, 73a–77b) was attributed to Moses b. Elijah Galeano, but the name of the author's father was given in the manuscript as Y^eʾudah, i.e., Judah, not Elijah.

19. Galeano/Jālīnūs, *Seiper ha-goralot*, 31a. See also Savage-Smith, "Geomancy," 998–99.

20. Melvin-Koushki, "Geomancy in the Islamic World," 789.

21. Melvin-Koushki, 789. See also Melvin-Koushki, "In Defense of Geomancy." In the *Muqaddimah*, Ibn Khaldūn (d. 1382) opposed both geomancy and astrology, noting that astrologers were no longer trained to assess comprehensively astronomical data.

22. Galeano/Jālīnūs, *Seiper ha-goralot*, 31a–b. See also Rodríguez-Arribas, "Divination according to Goralot," 250.

23. Galeano/Jālīnūs, *Seiper ha-goralot*, 33b. But cf. Rodríguez-Arribas, "Divination according to Goralot," 254. There, she quotes from a fifteenth-century geomancy text that counseled the importance of choosing a propitious time to consult the lots.

24. Galeano/Jālīnūs, *Seiper ha-goralot*, 35a.

25. Galeano/Jālīnūs, 35a. For more on al-Ṭarābulsī, see Savage-Smith and Smith, "Islamic Geomancy," 213.

26. Galeano/Jālīnūs, *Seiper ha-goralot*, 36a–b. See also Fahd, *La divination*, 201–2. On the availability of al-Zanātī's work in Hebrew, see Steinschneider, *Die hebräischen Übersetzungen*, 855–56.

27. Galeano/Jālīnūs, *Seiper ha-goralot*, 35a.

28. Galeano/Jālīnūs, 31b.

29. On Ḥaydar-i Remmal, see Fleischer, "Seer to the Sultan," 296.

30. Galeano/Jālīnūs, *Seiper ha-goralot*, 33a.

31. Langermann, "A Compendium," 287. I read a zayin where Langermann read a waw, hence a one-year discrepancy in the year of completion of *Puzzles*.

32. Galeano/Jālīnūs, *Taʿalumot ḥokmah*, 65a.

33. On Galeano/Jālīnūs's sale of the text to the agents, see Steimann, "The Story," 33.

34. Cohen Ashkenazi, *Urim wᵉ-ṯummim*, 162a, 163b. See Langermann, "Science in the Jewish Communities," 450.

35. Ibn Ezra, *The Book of the World*, 88–89, 164–67.

36. Cohen Ashkenazi, *Urim wᵉ-ṯummim*, 162a.

37. Nesitorisi may have been Nestor-Iskender (b. ca. 1438), a chronicler of the fall of Constantinople, although I have no evidence that Nestor-Iskender practiced or wrote on astrology. For more on Nestor-Iskender, see Historians of the Ottoman Empire, "Nestor-Iskender," https://ottomanhistorians.uchicago.edu/en/historian/nestor-iskender.

38. Şen, "Astrology in the Service."

39. See Şen, "Reading the Stars," 571.

40. Westman, *Copernican Question*, 62–63.

41. Mercier, *An Almanac*.

42. Cohen Ashkenazi, *Urim wᵉ-ṯummim*, 162a. For the almanac, see Anon., *Taqwīm* (1450), 5a. The text reads: "Ikinjī iqlīm keh āftābʾa mawḍūʿdir."

43. Cohen Ashkenazi, *Urim wᵉ-ṯummim*, 162a. For the passage in the almanac, see Anon., Taqwīm (1450), 5a.

44. Cohen Ashkenazi, *Urim wᵉ-ṯummim*, 180a.

45. On *subaşı* as a military term, see Ilgürel, "Subaşı," 447–48.

46. Galeano/Jālīnūs, *Taʿalumot ḥokmah*, 76a.

47. At the North Pole, half of the degrees of the zodiac will never rise and half will never set.

48. See Augustine, *The City of God*, 146–49.

49. Galeano/Jālīnūs, *Taʿalumot ḥokmah* 65a–b.

50. Ibn Ezra, *Abraham Ibn Ezra's Introductions to Astrology*, 803. Ibn Ezra explained that with Hermes's balance or Enoch's rectification, "the position of the Moon at the moment of birth is the ascendant at the moment of conception, and vice versa."

51. Ibn Ezra, *Abraham Ibn Ezra on Nativities*, 41–45, 92–95, 230–32.

52. Galeano/Jālīnūs, *Taʿalumot ḥokmah*, 65b.

53. This may in fact be the philosopher Kamāl Pāsha Zādah (1468/9–1534). See Ménage, "Kemāl Pa<u>sh</u>a-Zāde." His grandfather was Kemal Paşa, one of Fatih Sultan Mehmed's ministers. See also Saraç, *Şeyhülislam Kemal Paşazade*, 18n25. Saraç determined a death date of 1469–70 for Kemal Paşa, meaning that Galeano/Jālīnūs probably

never met him but could have met his son. See Öçal, *Kemal Paşazade'nin Felsefi ve Kelami Görüşleri*, 18 on how Kamāl Pāshā Zādah was known to Galeano/Jālīnūs's patron ʿAbd al-Raḥmān Mu'ayyadzādah, who ensured Kemal Paşazade's appointment as a teacher at the Taşlık Medrese in Edirne.

54. Fahd, "Ḥurūf."

55. Melvin-Koushki, "Toward a Neopythagorean Historiography."

56. Many of the logical fallacies that Galeano/Jālīnūs identified in *Puzzles* were mentioned in Aristotle's *Sophistical Refutations*. On the fallacy of mistaken premises, see Averroës, *Talkhīṣ al-Safasṭa*, 52.

57. Galeano/Jālīnūs, *Taʿalumot ḥoḵmah*, 54a–b.

58. Galeano/Jālīnūs, 54b.

59. Şen, "Astrology in the Service," 13–14.

60. Şen, 190–91.

61. Anon., *Taqwīm* (1450), 3a.

62. Anon., 3a.

63. Pico , *Disputationes*, 1:312.

64. See also Vanden Broecke, *The Limits of Influence*, 57 where Vanden Broecke writes, "The *Disputations* criticized the practice of judicial astrology in late fifteenth-century Italy, and not so much its theoretical soundness. Accordingly, I claim that Pico's *Disputations* proposed a revision of late medieval astrological practice." See Hasse, *Success and Suppression*, 251–52, for a characterization of the *Disputationes* as "multi-layered."

65. Saliba, "Cometary Theory," 105. On astrology as a dimension of Christian thought, see Smoller, *History, Prophecy, and the Stars*, 122. On the integration of natural philosophy into religious thought in the Renaissance, see Vescovini, "The Theological Debate," 136.

66. Fahd, *La divination*, 81–90.

67. Van Ess, *Frühe muʿtazilitische Häresiographie*, 110 in German and 123 in Arabic.

68. Adamson, "Al-Kindī and the Muʿtazila."

69. Brethren of Purity, *On "Astronomia,"* ch. 29–32. See Marquet, "La Détermination astrale," 129–30 on the Brethren of Purity's view of the role of the stars in transformations of matter and 143 for the role of the planets in individuals' lives.

70. Saliba, "The Development of Astronomy."

71. Melvin-Koushki, "Powers of One."

72. See Griffel, *Al-Ghazālī's Philosophical Theology*, 147–73 on how Ghazālī did not intend his rebuttal of intermediate causality to be a denial.

73. Vesel, *Les Encyclopédies Persanes*, 36–37. See also Melvin-Koushki, "Powers of One," 146–47.

74. Al-Nīsābūrī, *Gharāʾib al-Qurʾān*, 30:16.

75. Morrison, *Islam and Science*, 66–67.

76. Gutas, *Greek Thought*, 16.

77. Şen, "Astrology in the Service."

78. Leicht, "The Planets," 274.

79. On the principle that Judaism spared one from the influences of the stars, see Neusner, *A History of the Jews in Babylonia*, 5:192. On the rabbinic rejection of the earlier, Hellenistic idea that the stars might determine the fate of Israel, see Toepel, "Yonton Revisited."

80. Sela, *Abraham Ibn Ezra*, 157.

81. Sela, 149, 184. See also Ibn Ezra, *Abraham Ibn Ezra on Nativities*, 84–85, 328.

82. As quoted in Sela, *Abraham Ibn Ezra*, 173. See also Ibn Ezra, *Abraham Ibn Ezra on Nativities*, 89.

83. Sela, *Abraham Ibn Ezra*, 187.

84. Comment on Ex. 23:25 ("and I will remove sickness from your midst"). See Alhatorah.org, https://mg.alhatorah.org/Full/Shemot/23.25#e0n6. This site provides editions of Ibn Ezra's scriptural commentaries.

85. Sela, *Abraham Ibn Ezra*, 178–82.

86. Langermann, "Some Astrological Themes," 56. Ibn Ezra argued that God's intervention in Jews' lives could occur through astral causes.

87. See Maimonides, "Letter." See also Marx, "Additions and Corrections," 493–94. The "Letter" was not translated into Latin, but the Latin translation of the *Guide*, entitled *Dux neutrorum*, was available by the 1240s. See Di Segni, "Early Quotations," 203–7.

88. Maimonides, "Letter," 232–33.

89. Maimonides, 234–35.

90. Maimonides, 234.

91. On Pico's knowledge of Levi's work in astronomy, see Goldstein, "Levi ben Gerson and the Cross Staff Revisited," 376–77 and Goldstein, *The Astronomy of Levi ben Gerson*, 11–13.

92. Langermann, "Gersonides on Astrology," 516–18.

93. Goldstein and Pingree, "Levi ben Gerson's Prognostication."

94. Goldstein, "Levi Ben Gerson's Astrology," 296.

95. Langermann, "Gersonides on Astrology," 511–12.

96. Maimonides, *Guide*, 2:578 (III.45). I often cite the *Guide* both by volume:page and part.chapter.

97. Langermann, "Gersonides on Astrology," 513.

98. Berzon, *Classifying Christians*, 110–12. See also Hippolytus, *Refutatio omnium haeresium*, 97.

99. Augustine, *City of God*, 146–51.

100. Ferrari, "Augustine and Astrology."

101. Lemay, *Abu Maʿshar and Latin Aristotelianism*, xxvi. For the Arabic text and translation, see Abū Maʿšar, *The Great Introduction*.

102. Lindberg, *Beginnings*, 279–80. On medical astrology, see Siraisi, *Medieval and Early Renaissance Medicine*, 22.

103. Lindberg, *Beginnings*, 277–79.

104. Caudano, "Astronomy and Astrology," 225.

105. Magdalino, "Astrology," 212–13.

106. On the problems of historical astrology, see Vanden Broecke, *The Limits of Influence*, 60.

107. On Pico's rejection of historical astrology, see Pico, *Disputationes*, 1:574–95. *De redemptione* was not among the works of Ibn Ezra found in Smithius, "Abraham Ibn Ezra's Astrological Works."

108. Sela, *Abraham Ibn Ezra*, 102–3. On historical astrology in *Scroll of the Revealer* and Bar Ḥiyya's messianic concerns, see Rodríguez-Arribas, "The Terminology," 12–32.

109. Von Stuckrad, "Christian Qabbalah," 10. See also Lawee, *Isaac Abarbanel's Stance*, 129–30.

110. See Pico, *Disputationes*, 1:592.

111. Pico , 1:594–97. Pico added that failed predictions harmed Jewish morale. Cf. Maimonides, "Letter," 235 for a similar concern for the impact of failed predictions on morale.

112. Guidi, "L'Astrologia," 47–51.

113. Lesley, "The Place," 82–83. See also Hughes, "Transforming the Maimonidean Imagination," 464. See Guidi, "L'Astrologia," 40 on how Isaac's son Judah (a.k.a. Leone Ebreo) may have written *De Coeli harmonia*, which is unfortunately no longer extant, at the request of Pico or, more likely, his nephew Gianfresco.

114. Galeano/Jālīnūs, *Taʿalumot ḥokmah*, 34a–b.

115. Pico, *Disputationes*, 1:136.

116. Pico, 1:180.

117. Pico, 1:232. The "aforementioned" was astrology.

118. Pico, 2:56.

119. Pico, 2:358. See Westman, *Copernican Question*, 87 on Pico's citations of Biṭrūjī and Averroës in his discussion of the order of the planets.

120. Pico, 2:414.

121. Langermann, "A Compendium," 286.

122. Vanden Broecke, *The Limits*, 74. On Abraham Ibn Ezra's preservation of the parameter of Ibn al-Zarqāl for the length of the tropical year, see Goldstein, "Astronomy and Astrology," 16–17.

123. Smithius, "Abraham Ibn Ezra's Astrological Works."

124. Sela, "Abraham Ibn Ezra-Peter of Limoges," 36–50, esp. 49. See also Juste and Burnett, "A Newly Discovered Treatise."

125. Sela, "Abraham Ibn Ezra-Peter of Limoges," 39–40.

126. Richler, *Hebrew Manuscripts*, 24–27, 31, 58.

127. Richler, 72–76.

128. On the vocalization of the surname, see Kumaṭiano, *Ḥibbur bᵉ-matemaṭiqah*, 13a where the name was transcribed in Arabic script as *qūmātiyānū*. On his death date, see Kumaṭiano, *Ḥibbur bᵉ-matemaṭiqah*, 60b.

129. Attias, *Le commentaire*, 78.

130. Kumaṭiano, *Ancient Commentary on "Yᵉsod Moraʾ"* (Heb.), 31.

131. As quoted from Kumaṭiano's biblical commentary in Kumaṭiano, *Ancient Commentary on "Yᵉsod Moraʾ,"* 33–34.

132. Attias, *Le commentaire*, 78.

133. Attias, 78n83.

134. Kumaṭiano, *Ancient Commentary on "Yᵉsod Moraʾ,"* 32.

135. Greenberg, "Urim and Thummim."

136. See Fahd, *La divination*, 138–42 on parallels with the *eiḇod* and the *urim* and *tummim* in pre-Islamic Arabia.

137. Ibn Ezra's long commentary on Ex. 28:30 can be found at Alhatorah.org, https://mg.alhatorah.org/Full/Shemot/28.30#e0n6.

138. Kumaṭiano, *Ancient Commentary on "Y^{e}sod Moraʾ,"* 146–47.

139. Kumaṭiano, 147–48. Likewise, the squares of two and eight are four and sixty-four respectively, and the squares of one and nine are one and eighty-one respectively. That is: 1, 4, 9, (1)6, (2)5, (3)6 (4)9, (6)4, (8)1 for 1^2, 2^2, 3^2, 4^2, 5^2, 6^2, 7^2, 8^2, and 9^2.

140. Kumaṭiano, in his commentary on Abraham Ibn Ezra's *Seip̱er ha-eḥad*, referred to mathematical rules underpinning the astrological principle of aspects. See Kumaṭiano, *Peirush "Seip̱er ha-eḥad,"* 88b.

141. Kumaṭiano, *Ancient Commentary on "Y^{e}sod Moraʾ,"* 196. In this appendix, Schwartz has edited portions of Kumaṭiano's scriptural commentary.

142. See Kumaṭiano, *Ancient Commentary on "Y^{e}sod Moraʾ,"* 26–39 for Schwartz's discussion of the intellectual context of the commentary.

143. Langermann, "Science in the Jewish Communities," 449. See also Richler, *Hebrew Manuscripts*, 341 on Moses Cohen Ashkenazi's location.

144. Richler, 341. There was a notation on folio 1a that an agent of the Fuggers purchased from Moses Judah, probably Moses b. Judah Galeano/Mūsā Jālīnūs.

145. Cohen Ashkenazi, *Urim w^e-ṯummim*, 121a–122a.

146. Cohen Ashkenazi, 138a. Cohen Ashkenazi's determination that amulets were illicit resembled Maimonides's position, on which see Twersky, *Introduction*, 481.

147. Cf. B^{e}reishit Rabbah, 10.

148. See Cohen Ashkenazi, *Urim w^e-ṯummim*, 122a. Other verses referenced by Cohen Ashkenazi were Deuteronomy 32:1 and 33:26.

149. Cohen Ashkenazi, 205b. Cohen Ashkenazi cited Deuteronomy 30:19 ("Choose life") in support.

150. Cohen Ashkenazi, 123a.

151. Cohen Ashkenazi, 205b.

152. Cohen Ashkenazi, 205b–206a.

153. On the derivation of *Tummim* from *tamm*, see Brown, *A Hebrew and English Lexicon*, 1070.

154. Cohen Ashkenazi, *Urim w^e-ṯummim*, 123b–124a.

155. In support, Cohen Ashkenazi cited Joshua 1:7: "But you must be very strong and resolute to observe faithfully all the Teaching that My servant Moses enjoined upon you. Do not deviate from it to the right or to the left, that you may be successful wherever you go."

156. Cohen Ashkenazi, *Urim w^e-ṯummim*, 205b–206a.

157. Cohen Ashkenazi, 206a.

158. Langermann, "Science in the Jewish Communities," 451.

159. Cohen Ashkenazi, *Urim w^e-ṯummim*, 124b.

160. Cohen Ashkenazi , 137a. Other astrology texts that Cohen Ashkenazi listed here were Ptolemy's *Tetrabiblos* and *Centiloquium*; Ibn Ezra's *Book of the Luminaries*, *Book of the World*, and *Book of Elections*; Māshāʾallāh's work; and Bālīnūs's (Ps. Apollonios of Tyana) *Introduction*. He included *The Book of Hermes* (*Seip̱er Hormos*) and the *Picatrix* (*Taḵlit he-ḥaḵam*) as texts on talismans.

161. Cohen Ashkenazi, *Urim w^e-ṯummim*, 125a. See Langermann's discussion in "Science in the Jewish Communities," 450.

162. Cohen Ashkenazi, 125b.

163. Cohen Ashkenazi, *Urim w^{e}-ṯummim*, 126a.

164. For the translation, see Maimonides, *Mishneh Torah*, 66a.

165. Cohen Ashkenazi, *Urim w^{e}-ṯummim*, 126b.

166. Cohen Ashkenazi, 126b.

167. Babylonian Talmud, *Mo'ed Qaṭan*, 28a.

168. Babylonian Talmud, *Shabbat*, 156a.

169. Cohen Ashkenazi, *Urim w^{e}-ṯummim*, 126b–127a. Other scholars mentioned were Abraham Ibn Ezra and Levi b. Gerson. The sixth chapter of *Pirqei de-Rabbi Eliezer* contains many statements in favor of astrology. See, e.g., Anon., *Pirke de Rabbi Eli'ezer*, 31. See also Keim, "*Pirqei de Rabbi Eliezer*," 250–51 on the sparse comments on astrology in the text.

170. Cohen Ashkenazi, *Urim w^{e}-ṯummim*, 127r. On *M^{e}gillat ha-m^{e}galleh*, see Langermann, "Gradations of Light." On Bar Ḥiyya's idea that Israel's exile is due largely to astral forces, see Töyrylä, *Abraham Bar Hiyya*, 315–17.

171. Töyrylä, 308–9.

172. Maimonides, *Guide*, 2:275. See also Cohen Ashkenazi, *Urim w^{e}-ṯummim*, 127b.

173. Cohen Ashkenazi, 127b–128a.

174. Cohen Ashkenazi, 130a.

175. Cohen Ashkenazi, 129a.

176. Cohen Ashkenazi, 130b–131a.

177. Langermann, "Science in Byzantium," 450.

178. On Galenic anatomy and its reception in medieval Europe, see Siraisi, *Taddeo Alderotti*, 186–87. On Hippocrates, see Lonie, *The Hippocratic Treatises*, 76.

179. Cohen Ashkenazi, *Urim w^{e}-ṯummim*, 131b.

180. Cohen Ashkenazi, 132a. Cohen Ashkenazi classified Maimonides's contradictions using the categories listed in Maimonides, *Guide*, 1:17–18.

181. Cohen Ashkenazi, *Urim w^{e}-ṯummim*, 133a.

182. Langermann, "Science in Byzantium," 450–51.

183. Cohen Ashkenazi, *Urim w^{e}-ṯummim*, 133a–b.

184. Cohen Ashkenazi, 134a.

185. Ibn Ezra's comment can be found online at Alhatorah.org, https://mg.alhatorah.org/Full/Shemot/23.26#e0n6.

186. Cohen Ashkenazi, *Urim w^{e}-ṯummim*, 134a.

187. Cohen Ashkenazi, 135a.

188. Ibn Ezra, *Elections, Interrogations, and Medical Astrology*, 189.

189. Sela, *Asṭrologyah* (Heb.), 188–89.

190. Cf. Ibn Ezra, *Ibn Ezra on Elections*, 93–94.

191. Cohen Ashkenazi, *Urim w^{e}-ṯummim*, 134a.

192. His argument was prompted by Deuteronomy 20:7, which reads, "Lest he die in a battle."

193. Cohen Ashkenazi, *Urim w^{e}-ṯummim*,134b.

194. Cohen Ashkenazi, 135b.

195. See Langermann, "Some Astrological Themes," 55 for Ibn Ezra's comment that God helps people evade a preordained flood by putting "the *thought* of going out of the city into the minds of the populace."

196. Cohen Ashkenazi, *Urim we-tummim*, 138a. See also Maimonides, *Guide*, 2:448–56 (III.13).

197. Anon., *Taqwīm* (1445), 13a. Cf. al-Bayḍāwī, *Tafsīr al-Bayḍāwī*, 2:34.

198. Cohen Ashkenazi, *Urim we-tummim*, 163a. Cohen Ashkenazi recounted, "Rabbi Abraham Ibn Ezra said in *The Book of the World* that King Doranius said that he found in the *Book of Secrets of Enoch* . . ." For the statements, see Ibn Ezra, *The Book of the World*. Van Bladel (*Arabic Hermes*, 172–73) ascribed Hermes's status as a prophet of philosophy and science to "the combination of ancient Judaean lore about Enoch with a Hellenistic astrological tradition."

199. On the patronage relationship between Galeano/Jālīnūs and Muʾayyadzādah, see Samsó, "Abraham Zacut," 83.

200. For more on Muʾayyadzādah's career, see Pfeiffer, "Teaching the Learned."

201. Al-Ījī, *al-Mawāqif*, 3:89–91.See also Şen, "Astrology in the Service," 89–90. On the significance of Jurjānī's *Sharḥ al-Mawāqif* for Ottoman *kalām*, see Özervarlı, "Theology," 572–76.

202. See Augustine, *City of God*, 144–49.

203. Muʾayyadzādah, *Ḥawāshī*, 11a.

204. Muʾayyadzādah, 10a. For Jurjānī's position, see al-Ījī, *al-Mawāqif*, 3:90–91.

205. Anon., *Taqwīm* (1445), 13a. The text reads, "God's custom is the external cause so that everything comes in its place."

206. Muʾayyadzādah, *Ḥawāshī*, 8b. See also Şen, "Astrology in the Service," 92–93.

207. Pico, *Syncretism*, 529.

208. See Garin, *Astrology in the Renaissance*, 99–104.

209. Bullard, "Inward Zodiac," 695–97.

210. Engel, *Elijah Delmedigo*, 7.

211. Vescovini, "The Theological Debate," 128.

212. Vescovini, 113. See also Zambelli, "Astrologers' Theories," 26–27.

213. See Garin, *Astrology*, 13 for the same ideas in the thought of Pomponazzi. See also Vescovini, "Peter of Abano," 29–30. Vescovini commented that scholars believed that God could suspend the effects of the planets.

214. Vescovini, "The Theological Debate," 100.

215. Bullard, "Inward Zodiac," 689, 698.

216. This is most likely Abraham Ibn Ezra.

217. This is most likely the third century CE Jewish scholar Samuel Yarḥinaeh.

218. Ficino, *Book of Life*, 167.

219. Bullard, "Inward Zodiac," 702–3.

220. Ficino, *Book of Life*, 90.

221. Ficino, 181–82. Ficino summarized, "This is what the Hebrew astrologer, Samuel, seems to mean, supported by the authority of David, his fellow astrologer, that the ancients clearly were makers of images and statues that would predict the future. He says that the harmony of the heavens was arranged in these."

222. Guidi, "L'Astrologia," 39.

223. Guidi, 56–61.

224. Lesley, "The Place," 82–83. See also Hughes, "Transforming the Maimonidean Imagination," 464.

225. Hughes, 474.
226. As translated in Hughes, 475–76.
227. Ebreo, *Dialogues*, 105.
228. Ebreo, *Dialogues*, 117.
229. See Harari, "Some Lost Writings" (Heb.).
230. Hughes, "Transforming the Maimonidean Imagination."

Chapter 3

1. See Grafton, *Cardano's Cosmos*, 10.
2. Tihon, "Astronomy," 195–96.
3. Gigante, "A Medieval Islamic Astrolabe." See also Goldstein and Saliba, "A Hispano-Arabic Astrolabe." For more astrolabes with multilingual labeling, see Pérez, *Catálogo razonado*.
4. Glasner, *Gersonides*, 59–60.
5. Goldstein, *The Astronomy of Levi ben Gerson*, 13–14.
6. Goldstein, "Levi ben Gerson and the Cross Staff Revisited," 378.
7. John of Gmünden, *Marʾit ha-koḵaḇim*, 125b. On the identification of the author as John of Gmünden, see Steinschneider, *Mathematik*, 199.
8. John of Gmünden, *Marʾit ha-koḵaḇim*, 118a–b. The translator did not connect himself specifically to earlier translation enterprises. See Gutas, *Greek Thought* on how, in the eighth and ninth centuries, scholars in the Abbasid Caliphate justified their translations of Greek, Persian, and Sanskrit texts as the recovery of lost knowledge.
9. John of Gmünden, *Marʾit ha-koḵaḇim*, 125b.
10. Richler, *Hebrew Manuscripts*, 75–76.
11. Anon., *Qoḇeṣ ba-madaʿim*, 6.
12. See Gurland, "The History of R. Mordechai Kumaṭiano" (Heb.), 8 on how Kumaṭiano's grandson Joseph grew up in Istanbul and married Rachel, the daughter of the merchant of knowledge Elijah Delmedigo.
13. Kumaṭiano, "Tiqqun k^{e}li ṣap̱iḥa," 173b. See 176b on the probable composition date of the 17th of Tevet in 5223, or July 14, 1463.
14. See Charette, *Mathematical Instrumentation*, 99 for the completion date of Ibn al-Zarqāl's text on his universal astrolabe plate.
15. Museum with No Frontiers, "Discover Islamic Art," https://islamicart.museumwnf.org/database_item.php?id=object;EPM;gr;Mus21;44;en
16. Kumaṭiano, *Ancient Commentary on "Y^{e}sod Moraʾ,"* 25.
17. See Lévy, "The Establishment," 442n30.
18. Kumaṭiano, "Seiper Tiqqun k^{e}li ha-n^{e}ḥoshet," 195b–196a.
19. Afendopolo, *Seip̱er K^{e}li roḇaʿ ha-shaʿot.*
20. Afendopolo, 80a
21. Afendopolo, 75a.
22. Ibn al-Zarqāl, "Iggeret," BnF hébreu 1021/8 (70b–88b). I have compared with BnF hébreu 1031 (Ibn Tibbon's translation), 12b–24b and BnF hébreu 1030 (also Ibn Tibbon's translation), 29a–41b. Galeano/Jālīnūs's translation of the introduction exists only in BnF hébreu 1021/8, 70b–88b. See also Langermann, "A Compendium," 288 on this translation. The Hebrew text, including the introduction translated by Galeano/Jālīnūs, has been edited

in Millás, *Tractat de l'assafea*, 1–40. See Millás, xlvi on the translation of the introduction. The text by Ibn al-Zarqāl that Ibn Tibbon translated was not precisely the same as the one present in MS Escorial 962, studied and translated by Puig in *Los Tratados de Construcción*.

23. Ibn al-Zarqāl, "Iggeret," 70b.

24. Ibn al-Zarqāl, 72a. The annotator explained that the stars discussed in the text were known in Arabic as *al-kawākib al-thābita* (the fixed stars).

25. Zonta, "Medieval Hebrew Translations," 67–72.

26. This text is sometimes described as a translation. For example, see Munk, *Manuscrits hebreux de l'Oratoire*, no. 182. There are, however, a number of places where Galeano/Jālīnūs inserted his own comments. See Langermann, "A Compendium," 289n5 for a description of the text as "a translation or Hebrew paraphrase."

27. See De Young, "Sibṭ al-Māridīnī," 1058.

28. On the attribution of the Arabic original to al-Sunbāṭī, see Zonta, "Medieval Hebrew Translations," 72. Cf. Munk, *Catalogues des manuscrits Hébreux*, 184. At the end of the treatise, the author encouraged the interested reader to consult the long treatise (*iggeret*).

29. al-Māridīnī, *Ḥibbur ʿal ha-robaʿ*, 17b. Galeano/Jālīnūs used the Arabic term *ẓill al-zawāl* (spelled as *zab̲āl* in Hebrew characters) for the shadow at noon. His misspelling (b̲ for *w*) suggests oral exchange.

30. al-Māridīnī, 18b.

31. al-Māridīnī, 20b

32. al-Māridīnī, 15b.

33. al-Māridīnī, 14b.

34. Presuming that Zonta ("Medieval," 72) read זיג (astronomical handbook with tables) for גיוב (sines), this text may be the one that he identified as "On the Construction of Astronomical Tables." The two codices are St. Petersburg RNL MSS Evr. 350–51, 22b–26a and St. Petersburg IOS B446, 87a–90a.

35. Anon., *Peirush ʿal ha-juyub*, 24a–b.

36. Gutas, *Greek Thought*, 113.

37. Parameters are measurements, such as the sizes of orbs and the distances between their centers, that shape the values shown in tables.

38. See al-Muthannā, *Ibn al-Muthannā's commentary*, 8–12. The medieval Latin translation was made directly from the Arabic.

39. Alfonso's connection to the tables has been questioned by Poulle, "The Alfonsine Tables," 99.

40. Kremer, Husson, and Chabás, "Introduction," 8.

41. Swerdlow, "The Derivation," 426.

42. On the use of the Alfonsine Tables in Ilkhanid Iran, see Samsó, "Alfonso X," 33. Samsó noticed that a parameter for the obliquity of the ecliptic, found in texts patronized by Alfonso, also appeared in a fourteenth-century Arabic text produced in China. See also Comes, "The Possible."

43. Despite the fact that the thirteenth-century Alfonsine Tables transformed the Toledan Tables, there was a fourteenth-century Greek translation of the Toledan Tables. On that Greek translation, see Neugebauer, *History of Ancient Mathematical Astronomy* (*HAMA* hereafter), 1:12.

44. Langermann, "The Scientific Writings," 15–20.

45. Langermann, 41. See Langermann, 8–11 et passim for Finzi's contacts with Christian scholars. See Lacerenza, "A Rediscovered Autograph Manuscript," 318 on how Finzi, in his Hebrew translation of Ibn al-Zarqāl's treatise on the universal astrolabe plate, was assisted by an astrologer and mathematician of the Gonzagas.

46. Swerdlow, "Nature, Experiment, and the Sciences," 166.

47. Chabás and Goldstein, *The Astronomical Tables*, 2.

48. On Copernicus's reliance on Bianchini's tables, see Goldstein and Chabás, "Ptolemy, Bianchini, and Copernicus," 453. There, Goldstein and Chabás reported a copy of Bianchini's tables for planetary latitudes that were in Copernicus's handwriting.

49. Chabás and Goldstein, *The Astronomical Tables*, 22. See also Anon., *Qoḇeṣ ba-asṭronomyah*, 27a–b, 29a–b for a Hebrew version of some tables of Bianchini. On Hebrew translations of the tables of Bianchini, see Steinschneider, *Die hebräischen Übersetzungen*, 626–29.

50. Anon., *Luḥot*, 346b–369a. Bianchini was named on 360a, 366b, and 367a. The codex that contained *Luḥot* was copied for Israel Cornelius Adelkind, a Jewish scholar born in Padua who spent part of his career (1524–44) as an editor in Venice in the publishing house of Daniel Bomberg. On Adelkind's career in Venice, see Raz-Krakotzkin, "Censorship," 152. Gershom Soncino, a contact of Galeano/Jālīnūs, made the acquaintance of Bomberg. See Marx, "Gershom (Hieronymus) Soncino's Wanderyears," 474.

51. On Finzi's citation of Bianchini's tables, see Langermann, "The Scientific Writings," 20.

52. Chabás, "The Astronomical Tables," 283.

53. Tihon, "L'Astronomie Byzantine," 253. George Plethon, in his *Manual of Astronomy*, extracted information from *Six Wings*. See Tihon and Mercier, *Georges Gémiste Pléthon*. On the Slavic translation of *Six Wings*, see Gardette, "Judaeo-Provençal Astronomy."

54. Chabás, "Interactions," 153.

55. Chabás and Roca, "Early Printing," 124, 128–32.

56. The title was an allusion to Isaiah 6:2, where there was a description of angels with six wings. On the method, see Chabás and Goldstein, *Essays*, 50–51.

57. Emmanuel b. Jacob, *Sheish k^{e}naḇayim*, 5b–6b.

58. Solon, "Six Wings," 15. Though Emmanuel Bonfils explained that he depended on the tables of Battānī (d. 929), Solon found, in his study of *Six Wings*, evidence of either some independent observations by Bonfils or the existence of some other source.

59. Solon, "Six Wings," 1.

60. Tihon, "L'Astronomie," 253–54.

61. Solon, "Six Wings," 1.

62. Emmanuel b. Jacob, *Sheish k^{e}naḇayim*, 40b.

63. Solon, "Six Wings," 15.

64. Tihon, "Astronomy," 193.

65. See Chabás and Goldstein, *Essays*, 340–41.

66. Goldstein and Chabás, "Isaac Ibn al-Ḥadib." Flavius Mithridates's father was a student of Isaac Ibn al-Ḥadīb in Sicily.

67. Goldstein and Chabás, 148.

68. Tihon, "Science," 196–97.

69. See, e.g., Saliba, "The Astronomical Tradition of Maragha."

70. For a sample of scholarship on *zījes* at Marāgha, see Saliba, "An Observational Notebook"; Mozaffari, "Muḥyī al-Dīn al-Maghribī's Lunar Measurements"; and Mozaffari, "Muḥyī al-Dīn al-Maghribī's Measurements."

71. None of the authors of the aforementioned Jewish sources for syzygy computation used the Islamic method favored by Kumaṭiano.

72. Ragep, "New Light."

73. See, e.g., Nicol, *The Last Centuries*, 139–40.

74. Ragep, "New Light," 236–37.

75. See Ragep, 238–43 for the authorship of the *Schemata*.

76. Paschos and Sotiroudis, *Schemata*, 16. On the relationship between the *Schemata* and the *zījes*, see Paschos and Sotiroudis , 12.

77. See Ragep, "New Light," 244–45.

78. Leichter, "The Zīj al-Sanjarī," 7.

79. See Bardi, "Persiche Astronomie," 17 n. 49 for a citation of the edited introduction to the *Persian Syntaxis*. Bardi edited the *Paradosis* and translated it into German.

80. Mercier, "The Greek 'Persian Syntaxis,'" 36–37. Anne Tihon ("Astronomy," 192) agreed with Mercier.

81. Pingree, "In Defence," 437–38; and Leichter, "The Zīj al-Sanjarī," 9. See also Pingree, *The Astronomical Works*, vol. 1, part 1, 16. See Pingree, 158–59 for Chioniades's mention of a translation of *Zīj-i Īlkhānī*.

82. Ragep, "New Light," 238.

83. On Solomon ben Elijah Sharbiṭ ha-Zahab, see Gardette, "Judaeo-Provençal Astronomy," 196. Chrysococcès acknowledged the role of Shams al-Dīn al-Bukhārī in transmitting the *Persian Tables* (see Mercier, "Shams al-Dīn al-Bukhārī," 1047–1048).

84. Sharbiṭ ha-Zahab, *Luḥot*, 34a.

85. Sharbiṭ ha-Zahab , 37a. See also Tihon, "Astronomy," 116 where she remarked, "The fifteenth century was a period of intense intellectual exchanges, which left their traces in manuscripts, so that one finds notes in Latin or Hebrew in Greek astronomical manuscripts of the period."

86. Mercier, "The Astronomical Tables," 126–27. See Neugebauer, *HAMA*, 1:13 on how Nicholas of Cusa (d. 1464) produced a separate partial translation of the *Persian Syntaxis* into Latin after seeing the text in Basel.

87. Tihon, "L'Astronomie Byzantine," 246.

88. See Sharbiṭ ha-Zahab, *Luḥot*, 45b for how the Hebrew translation perpetuated Chrysococcès's misunderstanding.

89. Tihon, "L'Astronomie Byzantine," 249.

90. For the dating, see Kumaṭiano, *Peirush Luḥot Paras*, 1b. In a calculation, Kumaṭiano mentioned that 5,184 years have elapsed since creation according to the Jews. The year 5184 was 1424. On the date of the commentary, see also Attias, *Le commentaire*, 21.

91. Kumaṭiano, *Peirush Luḥot Paras*, 15b.

92. Sharbiṭ ha-Zahab, *Luḥot*, 16b. He wrote, "If you want to know this from the tables of Chrysococcès (Heb. Korshoqoqi) . . ."

93. See Kumaṭiano, *Peirush Luḥot Paras*, 8a. See also Sharbiṭ ha-Zahab, *Luḥot*, 41a for Sharbiṭ ha-Zahab's value of 23;35° for the obliquity of the ecliptic. See al-Ṭūsī, *The*

Memoir, 394 for how many Islamic sources reported the obliquity to be 23;35° but how Ṭūsī gave the value of 23;30° in the *Zīj-i Īlkhānī*.

94. Toomer, *Ptolemy's "Almagest,"* 252.

95. See Saliba, "Theory and Observation" on how Ibn al-Shāṭir proposed a new solar model on the basis of new observations of the apparent size of the sun.

96. Kumaṭiano, *Peirush Luḥot Paras*, 23a.

97. See al-Ṭūsī, *The Memoir*, 460.

98. al-Khāzinī, *al-Zīj al-Sanjarī*, 114a. Values for the solar diameter found in the Arabic *al-Zīj al-Sanjarī* range from 0;31,28° to 0;33,58°. See also Pingree, *The Astronomical Works*, vol. 1, part 2, 175–76. In Chioniades's translation of *al-Zīj al-ʿAlāʾī*, we find a range of values for the apparent diameters of the sun: from 0;31,21° to 0;34,4°.

99. Leichter, "The Zīj al-Sanjarī," 104–5. See also al-Khāzinī, 38a on computing the diameters of the luminaries through the tables.

100. Kumaṭiano, *Peirush Luḥot Paras*, 33a. For more references to Argyros, see Kumaṭiano, 31a–b.

101. Pingree, "The Astrological School," 198.

102. Bardi, "Persiche Astronomie," 199.

103. Kumaṭiano, *Peirush Luḥot Paras*, 3a. My recomputation yields one degree every 70.59 years.

104. Pingree, *The Astronomical Works*, vol. 1, part 2, 35–36.

105. Kumaṭiano, *Peirush Luḥot Paras*, 31a. See also Bardi, "Paradosis," 251–52 on how Bardi found no evidence that Argyros was a supporter of Ptolemaic astronomy against Islamic astronomy. Bardi added that Argyros's authorship of the *Paradosis* is unclear. To that end, see also Bardi, "Bessarione," 34 on how there was a reference in one MS of the *Paradosis* to tables that were calculated for 1425/26, after Argyros died. See also Bardi, "The Reception," 174 where Bardi remarked that the third book of Meliteniotes's (d. 1393) *Tribiblos Astronomike* "is, more specifically, a commentary on the same set of tables commented in the *Paradosis*, and it constitutes a refined and enriched version of the *Paradosis*."

106. For the Greek text of the *Paradosis* that Kumaṭiano paraphrased, see Bardi, "Persiche Astronomie," 154, ll. 50–55. For the German translation of the *Paradosis*, see Bardi, 155, ll. 16–21.

107. Kumaṭiano, *Peirush Luḥot Paras*, 31a.

108. For the Greek text of the *Paradosis* that Kumaṭiano paraphrased, see Kumaṭiano, 154, l. 66–156, l. 72. For the German translation of the *Paradosis*, see Kumaṭiano, 155, l. 31–157, l. 3.

109. Kumaṭiano, 31a.

110. Bardi, "Scientific Interactions," 349.

111. See Tihon, "L'Astronomie Byzantine," 259. She noted that there was no trace of the critiques that Kumaṭiano attributed to Argyros in Argyros's surviving writings. See Bardi, "Paradosis," 252–54 on Meliteniotes's (d. 1393) knowledge of Ṭūsī's astronomy. See also Bardi, "Persiche Astronomie," 103 on evidence for Argyros's favorable impression of Islamic astronomy.

112. Frank, "Karaite Exegetical and Halakhic Literature," 535. See de Lange, "Abra-

ham Ibn Ezra," 188 for how Elijah Bashyatchi, a Qaraite, maintained "that according to that [Qaraite] tradition, Ibn Ezra was a disciple of the great Qaraite master Yefet ben ʿAli." See also Bashyatchi, *Adderet Eliyahu*, 11 for his assertion that Ibn Ezra and Maimonides's criticism of the Qaraites arose from the need to say something negative about the Qaraites for public consumption.

113. Lasker et al., "Karaites."

114. Bowman, *The Jews of Byzantium*, 144–45. See also Bowman, 148 for how Ashkenazic and Sephardic Jews came to Adrianople in the years before the conquest.

115. Morrison, "Tables."

116. Lasker, "Medieval Karaism and Science," 432.

117. Cf. Samuel b. Benjamin, "ʿIr qᵉṭanah," 4a–5b. The entire treatise by Samuel was about the determination and sanctification of the new month. For a survey of the Qaraites' earlier rejection of calendar computations, see Goldstein, "Astronomy and the Jewish Community," 32–48.

118. Morrison, "Tables," 170.

119. Bashyatchi, *Adderet Eliyahu*, 17a. See also Morrison, "Tables," 166.

120. Morrison, "Tables," 174; and Yazdi, "Naṣīr al-Dīn al-Ṭūsī," 233–34.

121. Morrison, "Selective Appropriation."

122. Steinschneider, *Mathematik bei den Juden*, 196. I cannot find a reference in *Adderet Eliyahu* to support Steinschneider.

123. Goldstein, "Survival," 38.

124. See Goldstein, *The Astronomical Tables*, 75–76 on how Abraham copied in Istanbul in 1510 a manuscript of Levi b. Gerson's *The Wars of the Lord*.

125. Goldstein, "Astronomy in Hebrew in Istanbul."

126. Tihon and Mercier, *Manuel d'Astronomie*, 250–63, esp. 255–60. For Tihon's discussion of Plethon's Jewish sources, see Tihon and Mercier, 10–13.

127. Tihon and Mercier, 6–7. It is possible that the Elisha in question, Plethon's putative Jewish informant, authored a text called *The Key of Medicine*. For more on that Elisha, see Langermann, "Science in the Jewish Communities," 443–44.

128. Bardi, "Bessarione," 21–23.

129. Shank, "Regiomontanus and Astronomical Controversy," 89–90.

130. Shank, 89–93. On Bessarion's completion of his copy of the *Paradosis* in the first half of the fifteenth century, see Bardi, "Bessarione," 9. The codex with Bessarion's copy of the *Paradosis* contains another text by Argyros as well as a series of Islamic astronomical tables.

131. Bardi, "Paradosis," 243.

132. Şen, "Reading the Stars," 583–85. Şen found that Ottoman scholars drew heavily on *The Verified Ilkhanid Tables* (i.e., the *Zīj-i Muḥaqqaq*) by Wābkanawī/Shams al-Dīn Bukhārī, which Chioniades did not translate.

133. Chabás and Goldstein, *Astronomy in the Iberian Peninsula*, 1. See also Chabás, "Interactions," 152. Zacut did not formally attend the University of Salamanca.

134. Chabás, "Interactions," 153.

135. Gómez-Aranda, "Science and Jewish Identity," 165.

136. Cantera Burgos, *Abraham Zacut*, 38.

137. On Zacut's position as royal astronomer, see Goldstein, "Abraham Zacut's Signature," 159. See also Cantera Burgos, *Abraham Zacut*, 39.

138. Chabás and Goldstein, *Essays*, 64. On the tables for 1513, see Goldstein and Chabás, "New Evidence," 33–48. On the tables of de Heybech, see Chabás and Goldstein, "Transmission of Computational Methods."

139. Chabás and Goldstein, *Astronomy in the Iberian Peninsula*, 49–52.

140. Chabás and Goldstein, 141–43.

141. Chabás and Goldstein, 8.

142. Chabás and Goldstein, 53.

143. Chabás and Goldstein, 2.

144. Chabás and Goldstein, 95–98.

145. Chabás and Goldstein, 161.

146. Chabás, "Astronomy for the Court," 183–84. Chabás reported that Ludwik Birkenmaijer (*Stromata Copernicana* [Cracow: Polish Academy, 1924], 353) was the first to identify Alfonso of Cordoba, the editor of the *Almanach perpetuum*, as the astronomer whose value for the length of the year was cited by Copernicus in the *Commentariolus*.

147. On Zacut's lack of participation in the production of the printed *Almanach perpetuum*, see Goldstein and Chabás, *Astronomy in the Iberian Peninsula*, 3, 98. See also Samsó, "Abraham Zacut," 82.

148. Van der Werf, "History and Critical Analysis," 212.

149. Cantera Burgos, *Abraham Zacut*, 33–36. Goldstein and Chabás (*Astronomy in the Iberian Peninsula*, 2, 9–11) were skeptical.

150. For the dedication to Muʾayyadzādah, see Jālīnūs/Galeano, *Kitāb al-Zīj*, 1b. This Arabic translation was one of at least two Arabic translations of Zacut's work. For a preliminary study about translations of the *Almanach perpetuum*, including the two into Arabic, see Samsó, "Abraham Zacut." For a comprehensive study, see Parra, "Estudio."

151. Pfeiffer, "Teaching the Learned," 289. See also Pfeiffer, "Emerging from the Copernican Eclipse." The latter publication contains a list of titles in Muʾayyadzādah's library.

152. King, *In Synchrony with the Heavens*, 2:783–96. One astrolabe includes Persian in the inscriptions and was made by Shukrallāh Mukhliṣ Shirwānī (King, 790–91). The second is made by al-Aḥmar al-Nujūmī al-Rūmī (King, 796), whom King hypothesized was a Turk from Central Anatolia. King has also proposed that Galeano/Jālīnūs produced a ca. 1480 Ottoman spherical astrolabe. See King, "Spherical Astrolabes."

153. Parra, "Estudio," 85–86.

154. Parra, 85.

155. Parra, 86, 100.

156. Samsó, "Abraham Zacut," 83.

157. Cf. Goldstein and Chabás, *Astronomy in the Iberian Peninsula*, 96–97. There is a better correspondence between the order of the tables in Escorial MS 966 and the tables of contents that Goldstein and Chabás provided. Aspects of the table of contents would seem to follow the Castilian headings more closely rather than the Latin. See, e.g., Jālīnūs/Galeano, *Kitāb al-Zīj*, 9r. See also Goldstein and Chabás, *Astronomy in the Iberian*

Peninsula, 163 for how the canons of the Arabic version were adapted from the *Almanach perpetuum*.

158. See Parra, "Estudio," 41 for the suggestion that Galeano/Jālīnūs acquired the 1502 printing of the *Almanach perpetuum*.

159. Arbel, "Jews in International Trade," 85.

160. See Ravid, "The Legal Status," 192. Officially, Jews could spend only fifteen days in Venice, after which they had to be absent for four months, or, after 1496, for a year. Jews also had to wear a yellow beret.

161. Akasoy, "Mehmed II," 254.

162. Jālīnūs/Galeano, *Kitāb al-Zīj*, 1b–2a.

163. Cf. Gómez-Aranda, "Science and Jewish Identity," 164. Note that Gómez-Aranda and Parra ("Estudio," 37) read "Tīrawī" as "Yatrawī" (i.e., physician). I have retranslated the section that Gómez-Aranda had translated in order to modify some of his readings.

164. Cf. the translation of this paragraph in Gómez-Aranda, "Science and Jewish Identity," 164.

165. I am reading *yusannid* for *tusayyid* (*Allāh arkānahā*).

166. Jālīnūs/Galeano, *Kitāb al-Zīj*, fol. 1b–2a.

167. Jālīnūs/Galeano, 2a.

168. Chabás and Goldstein, *Astronomy in the Iberian Peninsula*, 58–59.

169. Jālīnūs/Galeano, *Kitāb al-Zīj*, 30b.

170. Chabás and Goldstein, *Astronomy in the Iberian Peninsula*, 113–15.

171. Jālīnūs/Galeano, *Kitāb al-Zīj*, 81a.

172. Chabás and Goldstein, *Astronomy in the Iberian Peninsula*, 115.

173. Jālīnūs/Galeano, *Kitāb al-Zīj*, 3a. See also Parra, "Estudio," 283n16.

174. Jālīnūs/Galeano, *Kitāb al-Zīj*, 2b. A few of the comments are sewn into the binding of the manuscript and, thus, are too hard for me to interpret.

175. Jālīnūs/Galeano, *Kitāb al-Zīj*, 33b–34a.

176. Necipoğlu, Kafadar, and Fleischer, *Treasures of Knowledge*.

177. Langermann, "Medieval Hebrew Texts," 34–35.

Chapter 4

1. Renan, *Averroès*, 22–23.

2. Renan, 250. Renan wrote, "But in general, the Middle Ages had no hesitation in asking for lessons in philosophy from those whom faith required be damned."

3. On Avicenna's denial of astrology, see Mehren, *Vues d'Avicenne* and Michot, *Avicenne*.

4. Ivry, "Jewish Philosophy," 797–801.

5. Ivry, 803. Isaac Israeli was also influenced by Aristotelianism.

6. de Lange, "Hebrew Scholarship," 32. On how there are more Byzantine manuscripts of Ibn Ezra's work than of Maimonides, see de Lange, "Abraham Ibn Ezra and Byzantium," 188.

7. Langermann, "Abraham Ibn Ezra."

8. Smithius, "Abraham Ibn Ezra's Astrological Works," 240–43.

9. Frank, "Karaite Exegetical and Halakhic Literature," 535. See de Lange, "Abraham

Ibn Ezra," 188 for how Elijah Bashyatchi, a Qaraite, maintained that "according to that [Qaraite] tradition, Ibn Ezra was a disciple of the great Qaraite master Yefet ben ʿAli."

10. Lasker, "Karaites."

11. Cf. Attias, *Le commentaire*, 70.

12. On Kumaṭiano's reputation, see Bowman, *The Jews of Byzantium*, 149. Neither Moses Capsali nor his successors were in charge of a centralized community. On the definition of Moses Capsali's post, see Shmuelevitz, *The Jews of the Ottoman Empire*, 19–22. For more on Moses Capsali's status as the leading rabbi, rather than the chief rabbi, see Rozen, *A History*, 70–77. See Paudice, *Between Several Worlds*, 43–44 on how Moses Capsali at one point found himself in a disagreement over Jewish marriage law with Rabbi Joseph Colon of Pavia, in Lombardy. Because Colon's participation was solicited by other prominent rabbis in Istanbul, Capsali must have been in a relatively weak position.

13. Kupfer, "Comtino."

14. Grätz, *History of the Jewish People*, 6:235.

15. Gurland, "The History of R. Mordechai Kumaṭiano" (Heb.), 6.

16. Attias, *Le commentaire*, 71–72. See Attias, 45–48 on how Kumaṭiano's biblical commentary contained anti-Qaraite polemics, though directed against historical figures.

17. Hacker, "Mizraḥi, Elijah." See also Galante, *Histoire des juifs de Turquie*, 1:97 on how Qaraites were in Constantinople before 1453, had a synagogue, and hoped to study at the (Rabbanite) Yeshivah.

18. Mizraḥi, *Teshuḇot*, 93b.

19. See also David, "New Information" (Heb.), 180 on Afendopolo's death in 5283.

20. See Lasker, *From Judah Hadassi*, 106 for Bashyatchi's position that one needs to have learned science and philosophy to understand the Torah. Bashyatchi also possessed a copy of Kumaṭiano's commentary on the Torah that is now preserved as Paris BnF MS Hébreu 265.

21. Richler, *Hebrew Manuscripts*, 286.

22. Richler, 291.

23. Kumaṭiano, *An Ancient Commentary on "Yesod Moraʾ*," 26.

24. Kumaṭiano, 63–65.

25. Kumaṭiano, 65–68. For Maimonides's method of computing lunar crescent visibility, see Maimonides, *Sanctification*.

26. For Kumaṭiano's position on the role of science and philosophy in understanding Jewish law, see Kumaṭiano, *An Ancient Commentary on "Yesod Mora'*," 68–71.

27. Kumaṭiano, 67. For more on the place of astronomy in the debates between Rabbanites and Qaraites, see Goldstein, "Astronomy and the Jewish Community," 31–45. For the claim, shared by Qaraites and Rabbanites, that Israel's sages knew the rational sciences but that this knowledge had been lost to Jews, see Bashyatchi, *Adderet Eliyahu*, 21a. See also Maimonides, *Sanctification*, 73.

28. Kumaṭiano, *Ancient Commentary on "Yesod Moraʾ*," 65. For the text, see Kumaṭiano, *Peirush Luḥot Paras*, esp. 10b.

29. Kumaṭiano, *Ancient Commentary on "Yesod Moraʾ*," 101–2. See Kumaṭiano, 104–7 on how theories of medicine were incorporated into Kumaṭiano's comments on Deuteronomy 4:19. For the passage in the *Canon*, see Avicenna, *The Canon*, 218 (§356).

30. Kumaṭiano, *Ancient Commentary on "Yesod Moraʾ*," 116.

31. An earlier discussion of the hylic intellect is found in Averroës and Narboni, *Epistle on the Possibility*, 2–3. See Kumaṭiano, *An Ancient Commentary on "Y^{e}sod Mora',"* 118n239 and 119n245 on how Narboni's commentary was an important source for Kumaṭiano.

32. Kumaṭiano, 124.

33. Kumaṭiano, 44.

34. Averroës, *Middle Commentary*, 111. More information about how the human intellect was both eternal and incarnated will follow.

35. See Anon., *Taqqanot Qandiya*, 62 for Balbo's signature on a decree on the sale of kosher meat written between 1406 and 1424. On the dating of the document, see Lauer, "Cretan Jews," 133.

36. Anon., *Taqqanot Qandiya*, 36–39.

37. Delmedigo mentioned both the *Guide* and Abraham Ibn Ezra's works in his 1482 *D^{e}rushim*, 150b.

38. See Rosenthal, "From Arabic Books and Manuscripts," 20 for a description of a manuscript of the *Guide* in Arabic script, in a codex in the Millet Library in Istanbul, mostly copied between 1477 and 1480. On this codex, see Langermann and Kasher, "The Critical Notes" (Heb.), 249–51. Langermann and Almog transcribed and translated al-Nihmī's praise of Maimonides for his defenses of Aristotle.

39. See Attias, *Le commentaire*, 25 on the copying date being the date of composition. On the Cambridge MS of Kumaṭiano's commentary on the *Guide*, see Kumaṭiano, *An Ancient Commentary on "The Guide,"* 30–31. See also Schwartz, "To Understand Something" (Heb.), 128.

40. Shabb^{e}tai b. Moses also copied a manuscript of Kumaṭiano's scriptural commentary. See Kumaṭiano, *Peirush "ha-Torah,"* 198b.

41. Raby, "Mehmed the Conqueror's Greek Scriptorium," 17–18.

42. Maimonides, *Guide*, 1:206.

43. Maimonides, 1:207.

44. For the importance of Taftāzānī and Jurjānī to Ottoman theologians, see Özervarlı, "Theology," 572.

45. For Taftāzānī's position, see Taftāzānī, *Sharḥ al-Maqāṣid*, 1:480. For Jurjānī's position, see Hasan, "Foundations of Science," 156–57.

46. See Attias, *Le commentaire*, 32–33 on Kumaṭiano's reference to Greek, Arabic, and Romance (*la'az*) terms in his biblical commentary.

47. See Eisenmann, "Scientific" (Heb.), 102–4 for a complete list of Kumaṭiano's philosophical and scientific sources. See also Stroumsa, *Maimonides*, 67–70, 73–75 on how Maimonides was influenced by Averroës's *Bidāyat al-mujtahid* as well as by his *Decisive Treatise* and *Exposition of the Traditions of Proofs regarding Muslim Beliefs*.

48. Maimonides, *Guide*, 1:210.

49. For the commentary and the reference to Heron, see Kumaṭiano, *An Ancient Commentary on "The Guide,"* 235. On the absence of a Hebrew version of Heron, see Zonta, "Medieval Hebrew Translations." Zonta did not list any text by Heron, though Heron's texts were available in Arabic.

50. Schub, "A Mathematical Text," 57–58.

51. Schwartz, "To Understand," 146–47.

52. Kumaṭiano, *Ancient Commentary on "The Guide,"* 198–99. See also Endreß, "Averroes'

De caelo," 39 on how the orbs in Averroes's cosmos were alive. For the Latin text, see Averroës, *Averrois Cordubensis commentum magnum*, 2:391–92.

53. For the presence of manuscripts of Averroës's *Incoherence of the Incoherence* (Ar. *Tahāfut al-Tahāfut*) in Ottoman libraries, see Özervarlı, "Arbitrating," 388n47. For evidence that part of Averroës's long commentary on Aristotle's *De caelo* was known in the Ottoman Empire between 1508 and 1524, see Morel and Aouad, "Un fragment," 203.

54. Hasse, "*Averroica Secta*," 322.

55. Adamson, *Philosophy in the Islamic World*, 189.

56. Merlan, *Monopsychism*, 86.

57. Bland, "Elijah Del Medigo, Unicity," 6.

58. Bland, 15.

59. As translated in Averroës, *Long Commentary on the "De anima"* (*LCDA* hereafter), 303–4.

60. See also Zonta, "Un'ignota versione ebraica," 7–8 on how a Hebrew version of the *LCDA* was available by the end of the fifteenth century.

61. Averroës, *LCDA*, 304.

62. Averroës, 305.

63. Cf. Adamson, *Philosophy in the Islamic World*, 193 where Adamson observed that Averroës overlooked the possibility that individuals could have particular thoughts about universal facts.

64. Averroës, *LCDA*, 307–9.

65. Averroës, 311. See also Davidson, *Alfarabi*, 283 for how the disposition of which Alexander of Aphrodisias conceived was an epiphenomenon of matter.

66. Averroës, *LCDA*, 308. On Themistius, see Hasse, "Arabic Philosophy," 116. See also Bland, "Elijah Del Medigo, Unicity," 8, and Davidson, *Alfarabi*, 261.

67. I.e., the agent intellect.

68. I.e., the agent intellect.

69. Averroës, *LCDA*, 322.

70. Merlan, *Monopsychism*, 86.

71. Davidson, *Alfarabi*, 355–56. See also, regarding Plato's influence on Averroës's doctrine of the unity of the intellect, Ogden, *Averroes on Intellect*, 113.

72. Hasse, "Arabic Philosophy and Averroism," 117. Hasse explained that Paul held that "the intellect is united to the body as its substantial form."

73. Bland, "Elijah Del Medigo, Unicity," 14.

74. I have been translating this word as "agent."

75. Bland, 15. He added, "It is the mark of coherence in the mature philosophy of Averroes: One single mind, and therefore one unified, intelligible cosmos all the way down."

76. On the date of the printed edition, see Engel, *Elijah Delmedigo*, 123.

77. Carpi, *L'individuo*, 220. See also Cassuto, "Gli ebrei," 284 on how Delmedigo was not recorded as a professor in university archives.

78. Bartòla, "Eliyahu Del Medigo," 257–58.

79. Engel, *Elijah Delmedigo*, 7. Other Christian students of Delmedigo were Antonio Pizzamanno, Domenico Grimani, and Hieronymus Donatus.

80. Engel, *Elijah Delmedigo*, 35–42.

81. Engel, 95.

82. Engel, 106.

83. Engel, 103–7.

84. Bartòla, "Eliyahu del Medigo," 265n46. On Pico's increasing interest in Plato in the 1480s, see Cassuto, "Gli ebrei," 287.

85. Bland, "Elijah Del Medigo, Unicity," 10. See also Averroës, *Avveroè: Parafrasi della "Repubblica."*

86. On the continued contact between Pico and Delmedigo, see Bartòla, "Eliyahu Del Medigo," 258.

87. Pico, *Syncretism*, 253.

88. For Delmedigo's expressions of frustration with Jean de Jandun, see Delmedigo, *D^e^rushim*, 102b, 131b.

89. Pico, *Syncretism*, 113.

90. Zonta, *Hebrew Scholasticism*, 13–14. On thirteenth-century antecedents of Hebrew scholasticism in the Iberian Peninsula, see Chabás, "Interactions," 148–49.

91. Zonta, *Hebrew Scholasticism*, 8–9. For example, the Hebrew term *ep̱shariyyot ʿatidiyyot* (future possibilities) was a calque for the Latin *contingentia futura*.

92. Zonta, 29–30.

93. Hartig, *Die Gründung*, 256.

94. Tamani, "Le traduzioni," 613–14. See Tamani, 619–20 for texts that De Balmes translated into Latin, such as Ibn al-Haytham's *Kitāb al-Hayʾa*, Ghazālī's *Incoherence*, and Geminos's *Introduction to Astrology*, that did not originate with Averroës. See Langermann, *Ibn al-Haytham's "On the Configuration,"* 41 on how the translation of Ibn al-Haytham's *On the Configuration* was at the request of Grimani.

95. Tamani, "Le traduzioni," 614–15. We read that De Balmes's grandfather Abraham ben Mosheh de Balmes was the personal physician of Ferdinand of Aragon.

96. Conforte, *Qoreiʾ ha-dorot*, 39a. The translation of Bricot's work in the 1490s complicates the identification.

97. The Latin title was *Textus abbreviatus in cursum totius physices et metaphysicorum Aristotelis.*

98. Ben Shushan, *Tol^e^dot Adam*, 4b. New York JTSA MS 5475 of the same text was copied in Salonika in 1504.

99. See Zonta, *Hebrew Scholasticism*, 20.

100. Delmedigo, "Annotationes," in Jean de Jandun, *Quaestiones Joannis de Janduno de physico auditu*, 136b. "Per imaginationem autem intelligit apprehensionem seu cognitionem simplicem que vocat apud latinos simplicium apprehensio. Et iste fuit modus loquendi Arabum et Hebreorum." For the Hebrew, see Averroës, *ha-Beiʾur he-arok̠*, 5a. A universal (*koleil*) was simple because it was not differentiated.

101. For similar though not identical wording, see Averroës, *ha-Beiʾur he-arok̠*, 340a.

102. Delmedigo, "De primo motore," in Jean de Jandun, *Quaestiones*, 126a.

103. Glasner, *Averroës' "Physics,"* 13–14.

104. On the weaknesses of Christian theology, see Delmedigo, *D^e^rushim*, 150a. See also Bland, "Elijah Del Medigo, Unicity," 20–21. See also Bland, "Elijah del Medigo's Averroist Response," 24.

105. On humans' ability to understand the order and rationality of the universe, see

Bland, "Elijah Del Medigo, Unicity," 9–10. See also Ivry, "Remnants," 256 for how Delmedigo excluded the possibility that God disrupts nature, God's expression of perfection.

106. Montada, "Eliahu del Medigo," 165.

107. Delmedigo, "De primo motore," in Jean de Jandun, *Quaestiones*, 122a.

108. Bland, "Elijah Del Medigo, Unicity," 15.

109. See Di Segni, "Early Quotations," 203–7 on how the Latin translation of the *Guide*, entitled *Dux neutrorum*, was available by the 1240s, though the precise date of the translation is unknown.

110. Delmedigo "De primo motore," in Jean de Jandun, *Quaestiones*, 123b. For Maimonides's statement of his position, see Maimonides, *Guide*, 2:258–59: "It cannot be true that the intellect that moves the highest sphere should be identical with the necessary of existence."

111. Delmedigo, "De primo motore," in Jean de Jandun, *Quaestiones*, 123b.

112. Delmedigo, 131a. The rest of this paragraph draws on this page of *De primo motore*.

113. Delmedigo, *Bei'ur "ʿEṣem ha-galgal"* (Heb.), 48b. For the Latin, see Delmedigo, *Expositio Averrois "De substantia orbis,"* 33b.

114. Pico, *Syncretism*, 467.

115. See Hyman, *Averroes' "De substantia orbis,"* 113–15. Averroës thought that the orbs were moved by the desire for a more excellent object, i.e., the prime mover, but it is unclear, in *De substantia*, whether there can be more than one object of desire. See Davidson, *Alfarabi*, 226n33 for Davidson's interpretation that this passage from *De substantia* meant that all orbs, except the outermost, could have two objects of desire.

116. Niphus, *Commentationes in librum Averrois "De substantia orbis,"* 56.

117. Niphus, 68.

118. Hasse, *Success and Suppression*, 231. See Hasse, 210–14 for clues that Nifo still acceded to parts of the unicity thesis in later life.

119. Delmedigo, *Dᵉrushim*, 91b–92a. Cf. Engel, *Elijah Delmedigo*, 43–44.

120. Delmedigo, *Dᵉrushim*, 96a–96b.

121. See Adamson, "On Knowledge of Particulars" for how, according to Avicenna, humans, like God, could know a given eclipse as a subset of a universal. Cf. Avicenna, *Metaphysics of the Healing*, 288–90. For Averroës's use of eclipses to talk about God's knowledge, see Belo, "Averroes on God's Knowledge."

122. Belo, 97a. See Adamson, "On Knowledge of Particulars," 275 for how, from Avicenna's perspective, God's self-knowledge was the knowledge of a particular.

123. Delmedigo, *Dᵉrushim*, 93b.

124. Hyman, *Averroës' "De substantia orbis,"* 92. In the Hebrew text of *De substantia*, *ʿib̲uy* was the word for "density." See also Glasner, "The Question of Celestial Matter," 317. Glasner noted that Averroës, in *De substantia orbis*, "does not consider the celestial body to be composed of matter and form, and insists that it is a simple body. It is difficult, however, to define the ontological status of this simple body."

125. See also Averroës, *Kitāb al-Āthār al-ʿulwiyya*, 21–22.

126. For the dates of the Hebrew and Latin versions of *De substantia orbis*, see Engel, *Elijah Delmedigo*, 124. The Hebrew version was a translation, with insertions, of the Latin original.

127. Delmedigo, *Bei'ur "ʿEṣem ha-galgal,"* 44b, and Delmedigo, *Expositio*, 30b–31a. The use of "spongy" (*s^epogi*) in other Hebrew sources in the scholarly nexus is attested in Mizrahi, *Peirush seiper al-Magisṭi*, 163b.

128. Delmedigo, *Bei'ur "ʿEṣem ha-galgal,"* 45b, and Delmedigo, *Expositio*, 31b. See also Averroës, *De substantia orbis*, 93.

129. Delmedigo, *Bei'ur "ʿEṣem ha-galgal,"* 44b, and Delmedigo, *Expositio*, 31a.

130. Pico, *Syncretism*, 385. This is thesis §2>41. "No part of heaven differs from another in being bright or not bright, but in being more or less bright."

131. . Averroës, *De Substantia*, 93–94.

132. Delmedigo, *Bei'ur "ʿEṣem ha-galgal,"* 46b, and Delmedigo, *Expositio*, 32b. Cf. Averroes, *De substantia orbis*, 94n75 for a citation of Jean Jandun's comment.

133. Delmedigo, *Bei'ur "ʿEṣem ha-galgal,"* 46b, and Delmedigo, *Expositio*, 32b. See also Delmedigo, *Bei'ur "ʿEṣem ha-galgal,"* 48a, and Delmedigo, *Expositio*, 33a. There, Delmedigo's reference to a translation (his words: "I have translated for you") was most likely to his Latin translations of Averroës's commentaries on the *Meteorology* of Aristotle.

134. Delmedigo, *Bei'ur "ʿEṣem ha-galgal,"* 48a, and Delmedigo, *Expositio*, 33a.

135. Delmedigo, *Bei'ur "ʿEṣem ha-galgal,"* 54b, and Delmedigo, *Expositio*, 38b.

136. Delmedigo, *Bei'ur "ʿEṣem ha-galgal,"* 53b–54a, and Delmedigo, *Expositio*, 37a.

137. For an outline of Pico's engagement with natural philosophy in the *Disputationes*, see Rutkin, "Astrology," 371–72.

138. See Rutkin, "Astrology," 375 on Pico's repetition of the Aristotelian analogy between the heavens and an animal that Delmedigo and Averroës made.

139. Pico, *Disputationes*, 1:56.

140. Pico, 1:238.

141. Pico, 1:242.

142. Rutkin, "Astrology," 384.

143. Delmedigo, *Bei'ur "ʿEṣem ha-galgal,"* 48b–49a, and Delmedigo, *Expositio*, 33b–34a. Delmedigo commented that Averroës allowed epicycles and eccentrics in his *Ikhtiṣār al-Majisṭī* because he wrote it during his youth. The earth is farther from the sun in the summer, but the angle of the earth's axis accounted for summer heat in the northern hemisphere.

144. Achillini, *De orbibus libri quattouor*, 19a–b.

145. Galeano/Jālīnūs, *Taʿalumot ḥokmah*, 85a–b.

146. Galeano/Jālīnūs's remark about "satisfaction with ignorance" probably pertains to the following sentence from chapter 8 of *Sh^emoneh P^eraqim*: "There is no necessity that we fathom His wisdom and comprehend why He chose to punish a particular person in this fashion and not in another." See Maimonides, *Pirkei Avot*, 54, 95 in the Hebrew.

147. Maimonides expounded upon God's knowledge of particulars in his commentary on *Pirqei Aḇot* (Chapters of the Fathers), an order of the Mishnah, and in *Sh^emoneh P^eraqim* (Eight Chapters), which was his introduction to *Pirqei Aḇot*.

In chapter 8 of *Sh^emoneh P^eraqim*, Maimonides disagreed with theologians (50, 86 in the Hebrew: *m^edabb^erim*) who held that God's "desire is invested in every entity at all times. Our faith differs, maintaining that God's will [was determined at the time of] the six days of creation. [At that time, He willed that] all entities continue to function according to their nature." In the eighth chapter (57–60, 104–10 in the Hebrew), Maimonides also addressed the question of God's knowledge of the future. First, Maimonides argued that God knew in a way that was different from the way in which humans knew. Perfect knowledge for humans meant knowing the cause, and because God was uncaused, humans could not fully know God.

148. See Zonta, "Medieval Hebrew Translations," 26 for how Aristotle's *De Interpretatione* was translated into Hebrew as *Seiper ha-Meliṣa* (cf. Ar. *Kitāb al-ʿIbāra*) by an anonymous translator. Galeano/Jālīnūs's transcription of the Greek title, פריארמוניאה, is curious; Zonta transcribed the Hebrew as ביריאמיננאס. In al-Fārābī, *Qiṣṣur mi-kol Melʾeket ha-higgayon*, Paris BnF hébreu 898, 42a, the title was transcribed as בארי ארמנניאס. In al-Fārābī, *Qiṣṣur mi-kol Melʾeket ha-higgayon*, Paris BnF hébreu 917, 134a the title was transcribed as ביריאמינאס or, possibly, ביריארמינאס, as there is an error notation (three dots in a triangle) between the *yod* and the *alep̱*.

149. The subsequent paragraphs tracked Fārābī's commentary on *De Interpretatione*. See al-Fārābī, *Al-Fārābī's Commentary*, 94–96.

150. Galeano/Jālīnūs, *Taʿalumot ḥoḵmah*, 87a.

151. Galeano/Jālīnūs, 88b. Cf. al-Fārābī, *Fārābī's Commentary*, 96.

152. Zonta, "Al-Fārābī's Commentaries," 224–25. See also Zonta, "Fonti antiche," 528n38 on Hebrew translations of passages from Fārābī's *Long Commentary*. Ḥizqiyyah bar Ḥalafta cited the *Long Commentary* in another work in 1320, but the best evidence for the existence of a partial or complete Hebrew translation of Fārābī's *Long Commentary* was how a text from Ṭodros Ṭodrosi (fourteenth century) preserved a fragment of the introduction from Fārābī's *Long Commentary on "De Interpretatione."* See also Manekin, "Logic in Medieval Jewish Culture," 118.

153. On the Latin translations, see Zonta, "Al-Fārābī's Commentaries," 220n3.

154. In addition, in Galeano, *Ha-Dibbur*, 1a, a text attributed to Galeano/Jālīnūs, we read, "Discourse about what is included in the *Book of Categories*, called *Categorias* in Latin and Greek."

155. Manekin, "Logic in Medieval Jewish Culture," 114. With the exception of Maimonides's *Maqāla fī ṣināʿat al-manṭiq* and the work of Ibn Kammūna, "texts on logic composed by Jews in the Islamic world are virtually nonexistent." Manekin (133) reported on Galeano/Jālīnūs works on logic but did not locate him in the Islamic world.

156. Galeano/Jālīnūs, *Melʾeḵet ha-higgayon*, 5b–6a.

157. Galeano/Jālīnūs, 16b–17b.

158. Galeano/Jālīnūs, 19a. This quotation is a verbatim translation of Rāzī's *Manṭiq al-Mulakhkhaṣ*, 37. See also Galeano/Jālīnūs, *Melʾeḵet ha-higgayon*, 17a–b. This passage from *Melʾeḵet ha-higgayon* was a verbatim translation of al-Rāzī *Manṭiq al-Mulakhkhaṣ*, 26.

159. Al-ʿAṭūfī, *Kitāb al-kutub*, 172a. See also Necipoğlu et al., *Treasures of Knowledge*, 1:899.

160. On a manuscript of Rāzī's *Commentary on the Difficulties of the "Qānūn"* in Hebrew characters, see Langermann, "Criticisms of Authority," 268n35. A portion of Rāzī's *al-Mabāḥith al-mashriqiyya* exists in Judeo-Arabic in New York, JTSA MS 9108, a sixteenth-century manuscript.

161. Galeano/Jālīnūs, *Melʾeḵet ha-higgayon*, 26a, 32b for Burley and 18b for Marsilio.

162. For a reference to the Thomist-Scotistic sect, see Galeano/Jālīnūs, *Melʾeḵet ha-higgayon*, 16b. For other references to these philosophers in Hebrew texts, see also Zonta, *Hebrew Scholasticism*, 166, 202–8.

163. Manekin, "Logic in Medieval Jewish Culture," 128–29. Manekin added that Shᵉmariah ben Elijah of Negroponte, the Cretan (fourteenth century), produced the earliest Hebrew translation of Peter of Spain's *Tractatus* on logic.

164. Galeano/Jālīnūs, *Melʾeḵet ha-higgayon*, 17a–b.

165. Erismann, "Logic in Byzantium," 380.

166. Morrison, "The Role of Oral Transmission," 11–12.

167. Langerman, "Compendium," 286. See also Langermann, "From My Notebooks," 354 for some evidence in *Puzzles* that Galeano/Jālīnūs composed another text, now lost, in which he raised interdisciplinary or transdisciplinary questions. In *Puzzles* (*Taʿalumot ḥoḵmah*), 93b–94a, amid remarks on methods of bloodletting, Galeano stated that he addressed relevant questions "in my medical questions in *Seiḇer ha-shorᵉshim* (The Book of Principles)." Although Langermann interpreted the reference to "my medical questions" as a reference to *Seiḇer ha-shorᵉshim*, it is also possible that the medical questions were but a single section of *Seiḇer ha-shorᵉshim* because Galeano/Jālīnūs also referred to *Seiḇer ha-shorᵉshim* as an astronomy text in *Puzzles* (*Taʿalumot ḥoḵmah*), 68a.

Chapter 5

1. Scholem et al., "Kabbalah."

2. Margoliouth, *Catalogue*, part 2, 190.

3. Katz, "The Qabbalah," 193.

4. Scholem, *Origins*, 265.

5. Anon., *The Zohar*, 1:107–9. See also Scholem, *On the Kabbalah*, 103.

6. Idel, "Mysticism," 1554.

7. Scholem et al., "Kabbalah," 596.

8. Anon., *The Zohar*, 1:113–14.

9. Yisraeli, "The Mezuzah," 149–51.

10. Scholem, "The Beginnings," 24–27. See also Dan, "Christian Kabbalah," 194. Dan explained, "The Christian kabbalah can be characterized as a Christian acceptance of a Jewish claim that was denied, often vehemently, by ancient and medieval Christian theologians: that Jewish non-biblical traditions contain ancient, universal truth that, because of its antiquity and divine origin, must contain the essential Christian message." See Coudert, "Christian Kabbalah," 163 on the writings of Raymond Lull (d. 1315) as the first evidence of Christian *Qabbalah*.

11. See, e.g., Reifenberg and Schwabe, "A Judaeo-Greek Amulet," on how Greek-speaking Jews had been using amulets well before *Qabbalah* developed.

12. Yisraeli, "The Mezuzah," 138.

13. Scholem, "Gilgul."

14. Ogren, *Renaissance*, 11–12.

15. Ḥalamish, *An Introduction*, 286.

16. See Scholem, *Origins*, 458 on how the doctrine of metempsychosis facilitated the interpretation that Job suffered for sins committed in an earlier life.

17. Scholem, 191–2. See also Scholem, "Gilgul," and Ḥalamish, *Introduction*, 286–87, 296.

18. Ogren, *Renaissance*, 59.

19. Ogren, 41–42. Cf. Goetschel, "Elie Ḥayyim de Genazzano," 98 for Genazzano's (fl. late fifteenth and early sixteenth century) view that the evidence for *gilgul* was scriptural. See also Deuteronomy 25:5–10 for the prooftext for levirate marriage.

20. Ogren, *Renaissance*, 60–62.

21. See Anon., *Seip̱er ha-Zohar.*

22. Gottlieb, "The Metempsychosis Controversy" (Heb.), 47–48, and Ravitzky, "The God of the Philosophers," 119–120. See also Ogren, *Renaissance*, 42.

23. Balbo, "Iggeret ʿal ha-gilgul," 44b.

24. Ogren, *Renaissance*, 43.

25. Malachi, "The Balbo Family" (Heb.), 256. Philosophy texts that Balbo studied were, inter alia, Narboni's commentary on the *Guide*, Ghazālī's *Intentions of the Philosophers*, portions of other commentaries on the *Guide*, and the introduction to Aristotle's *Theology*.

26. Ogren, *Renaissance*, 47.

27. Idel, *Mystical Experience*, 8.

28. Wolfson, "The Doctrine of Sefirot (Part 2)," 80–84. Wolfson emphasized that he was "not suggesting that Abulafia adopted a theurgic interpretation of the divine hypostases akin to the theosophic kabbalists" (82). See also Wolfson, "The Doctrine of Sefirot."

29. Ogren, *Renaissance*, 47. See also Idel, *Mystical Experience*, 138.

30. Ogren, *Renaissance*, 43–44.

31. Gottlieb, "The Metempsychosis Controversy," 50. Michael Balbo supported Capsali in his debate with Rabbi Joseph Colon. See Rabinowicz, "Joseph Colon." On Michael Balbo's dirge for the fall of Constantinople, see Bowman, *The Jews of Byzantium*, 178, 342–43.

32. Malachi, "Balbo Family," 263.

33. Vatican MS Ebr. 105 also contains Kumaṭiano's commentary on *Yᵉsod Moraʾ* and Balbo's epistolary exchange with Yedidyah Rak̠ of Rhodes. On that exchange, see Elior, "Rabbi Yedidyah Rakh."

34. Kumaṭiano, *An Ancient Commentary on "Yᵉsod Moraʾ,"* 184–85.

35. Halbertal, *Maimonides*, 279. For Maimonides's own view, see Halbertal, 285.

36. Idel, "Maimonides' *Guide*," 198–99.

37. On this book, which was not a commentary on the *Guide*, see Hames, "A Seal," 163. For the passage in which the author alluded to the *Guide*, see Idel, *Mystical Experience*, 109.

38. Maimonides, *Guide*, 2:429. As quoted in Idel, *Mystical Experience*, 111.

39. Kumaṭiano, *An Ancient Commentary on "The Guide,"* 177.

40. Kumaṭiano, 73 apud *Guide* 1:19.

41. Kumaṭiano, 18. See Louth, "Knowing," for the influence of *Qabbalah* on the emergence of hesychasm.

42. Russell, "The Hesychast Controversy," 494, 496.

43. Maimonides, *Guide*, 1:80.

44. Kumaṭiano, *Ancient Commentary on "The Guide,"* 99.

45. Kumaṭiano, 19. On *Guide* I.35 (Maimonides, *Guide* 1:79–81), in which Maimonides

discussed God's incorporeality, Narboni commented, "we should choose the most noble concomitant of bodies and say that He is light." As translated in Shemesh, "Averroes as Intertext," 184. For the background of Narboni's comment in the works of Averroës, see Shemesh, 185–93.

46. Cohen Ashkenazi, *Wikkuaḥ*, 29b; Balbo, *Wikkuaḥ*, 213b. For the translation, see Maimonides, *Mishneh Torah*, 34b. See also Gottlieb, "The Metempsychosis Controversy," 61 on the opening of the twentieth argument.

47. Cohen Ashkenazi, *Wikkuaḥ*, 29b; Balbo, *Wikkuaḥ*, 213b.See also Maimonides, *Guide*, 1:221.

48. Cohen Ashkenazi, *Wikkuaḥ*, 30a; Balbo, *Wikkuaḥ*, 214a.

49. Cohen Ashkenazi, 30b; Balbo, 214b.

50. Cohen Ashkenazi, 31a; Balbo, 215a. Here, Cohen Ashkenazi described Ghazālī as a *ḥasid elohi* (a divinely righteous person).

51. See Ghazālī, *Maqāṣid al-falāsifa*, part 3, 54. For the quote, see Cohen Ashkenazi, *Wikkuaḥ*, 30b; Balbo, *Wikkuaḥ*, 214b.

52. Clark, "Neoplatonism."

53. See Ravitzky, "The God of the Philosophers," 140–41.

54. Cohen Ashkenazi, *Wikkuaḥ*, 72b; Balbo, *Wikkuaḥ*, 243a. See Davidson, *Alfarabi*, 228–29 on how the concept of emanation was unsuitable to incorporeal entities.

55. On the separate intellects necessary to move epicycles and eccentrics, see Janos, "Moving the Orbs," 199.

56. Cohen Ashkenazi, *Wikkuaḥ*, 43a–b; Balbo, *Wikkuaḥ*, 223a. See also Maimonides, *Guide*, 2:323.

57. Balbo cited Isaac b. Laṭīf (d. 1280), who made the same argument against Ptolemaic astronomy in *Shaʿar shamayim* (Gate of Heaven), 13a.

58. Cohen Ashkenazi, *Wikkuaḥ*, 44b; Balbo, *Wikkuaḥ*, 224a.

59. Cohen Ashkenazi, 45a; Balbo, 224a.

60. Averroës, *Tafsīr Mā baʿd aṭ-ṭabīʿa*, 3:1649–50.

61. Cohen Ashkenazi, *Wikkuaḥ*, 46b, marginal comment. The passage quoted is not from *The Intentions of the Philosophers* but is instead found in the margin of the translation of Ghazālī, *Kawwanot ha-pilosopim*, 110b. See Davidson, *Alfarabi*, 230 on Averroës's acceptance of emanation in his earlier *Epitome of the "Metaphysics."*

62. Cohen Ashkenazi, *Wikkuaḥ*, 71b–72b; Balbo, *Wikkuaḥ*, 242b–243a. See also Glasner, *Gersonides*, 47–50 for Levi b. Gerson's dismissal of the need for intellects to move the orbs. See also Shemesh, "Averroes as Intertext," 22–23.

63. Cohen Ashkenazi, 74a–b; Balbo, 244a.

64. This much follows Friedlander, *Pirḳê De Rabbi Eliezer*, 22.

65. The quote terminates here in Balbo, *Wikkuaḥ*, 243b.

66. The description of the camps of the angels parallels the description of the camps of the tribes of Israel found in Numbers 2:17ff.

67. I.e., from a single cause.

68. The quote terminates here in Cohen Ashkenazi, *Wikkuaḥ*, 73b.

69. Cohen Ashkenazi, 74a; Balbo, 244a.

70. Cf. Ravitzky, "The God of the Philosophers," 143–44.

71. For an exploration of the relationship between *Qabbalah* and astronomy, see

Chajes, "Spheres." See also Chajes, *The Kabbalistic Tree*, 210–11 for a diagram in a manuscript of *Qabbalah* in which the planets are associated with the *s*ᵉ*pirot*. See Cohen Ashkenazi, *Wikkuaḥ*, 44a; Balbo, *Wikkuaḥ*, 223b for how one possible difference between the soul of the orbs and the human soul was that the human soul, but not the soul of the orb, ought to be governed by Jewish law (*dīn*; also the name of a *s*ᵉ*pira*). The two records of the polemic are not identical here.

72. Cf. Ravitzky, "The God of the Philosophers," 145–48.

73. See also Ogren, *Renaissance*, 43–45, 48 on how philosophical argumentation, for Balbo, was a means to an end. Ogren cited Balbo, *Wikkuaḥ*, 201b.

74. Bland, "Elijah del Medigo's Averroist Response," 25n8 listed many secondary sources in which this view was advanced. For more on Delmedigo's later life, see also Geffen, "Insights," 77–79.

75. See Fraenkel, "Considering the Case," 224 for an argument that the *Examination* did not propound the double truth theory.

76. Delmedigo, *Examen*, 77–78.

77. Delmedigo, 79.

78. Delmedigo, 78–79.

79. Delmedigo, 89.

80. Ivry, "Remnants," 251.

81. Delmedigo, 92.

82. Engel, *Elijah Delmedigo*, 7.

83. On Delmedigo's acknowledgment of *Qabbalah*, see Bland, "Elijah del Medigo's Averroist Response," 28–29, 40–41. For a critique of Bland, see Ogren, *Renaissance*, 38n91.

84. For the dates of the Hebrew and Latin versions of the *Commentary on "De substantia orbis,"* see Engel, *Elijah Delmedigo*, 124.

85. Delmedigo, *Beiʾur "ʿEṣem ha-galgal"*, 41a–b.

86. That is, any way of saving Aristotle's explanation that the east winds were warmer than the west winds, aside from the animation of the heavens, was invalid.

87. Transcribed in Kieszkowski, "Les rapports," 70, from Delmedigo, *Epistolae*, 74a. I did not have time to include the findings of Licata, *Secundum Avenroem* in this book.

88. This question arose in Averroës, *Kitāb al-Āthār al-ʿulwiya*, 37.

89. Transcribed in Kieszkowski, "Les rapports," 71–72, 74b from Delmedigo, *Epistolae*.

90. On Pico's access to the *Qabbalah* of Abulafia, see Wirszubski, *Pico della Mirandola's Encounter*, 60–61, and Ogren, *Renaissance*, 215–16 on Pico's access to Abraham Abulafia's work through the Latin translations of Flavius Mithridates.

91. Transcribed in Kieszkowski, "Les rapports," 73, and Delmedigo, *Epistolae*, 75b.

92. For the corresponding Hebrew passage, which does not track the Latin letter perfectly, see Delmedigo, *Beiʾur "ʿEṣem ha-galgal,"* 41a–b. The relevant lines are indicated in the margin of the manuscript with a line that is sometimes squiggly. In the Hebrew commentary on *De Substantia*, instead of *ipsi enim opinantur* (for these scholars suppose), Delmedigo wrote that "it is apparent from the words of the Qabbalists of the people of our nations (*umoteinu*)." See Delmedigo, *Beiʾur "ʿEṣem ha-galgal,"* 41a. Instead of *dei gloriosi*, the Hebrew reads *he-ʿasiri* (i.e., the tenth, outermost orb). The Latin *infinitum* was rendered in Hebrew as *ein sop̱*.

93. This sentence is not reflected in the Hebrew version of Delmedigo's commentary on *De substantia orbis*.

94. There is an illegible word in Delmedigo, *Epistolae*, 75b; Kieszkowski ("Les rapports," 74) transcribed X. Because the Latin letter is not a direct translation of a passage from the Hebrew translation of the commentary, the Hebrew translation does not help us reconstruct the illegible word.

95. Kieszkowski, "Les rapports," 73–74.

96. See Idel, *Abraham Abulafia's Esotericism*, 66, 105, 197 for how Abulafia took the same view of the *s^e^pirot* as Delmedigo did.

97. Pico, *Syncretism in the West*, 549.

98. I.e., Elijah Delmedigo.

99. For more on the letter from Ficino, see Ruderman, *The World of a Renaissance Jew*, 40–43.

100. Ficino, *Letters*, 7:26.

101. Pico's curiosity extended to the Qur'ān, and he eventually acquired Flavius's 1481 transliteration of the Qur'ān into Hebrew characters featuring an interlinear Latin translation. On the transliteration of the Qur'ān into Hebrew characters, see Grévin, "Flavius Mithridate," 30, 45–46. The extensive Latin marginalia in the manuscript, some of which were due to Pico, have led to this text (now MS Vat Ebr. 357) being characterized as the first commentary on the Qur'ān written in Latin. But cf. Burman, *Reading the Qur'ān*, 133–45 on how Flavius Mithridates's errors of philology indicate shortfalls in comprehension. Burman suggested that Flavius's translation of the Qur'ān was more of a patronage gift than a philological exploration of terms.

102. Andreatta, "Subverting Patronage," 192.

103. See Andreatta, 176 on how Flavius Mithridates began his translation as early as 1485, and Bartòla, "Eliyahu Del Medigo," 270 on how Pico and Delmedigo were in contact through 1486.

104. See Novak, "Giovanni Pico," 131–32.

105. Andreatta, "Subverting Patronage," 185.

106. Cassuto, "Gli ebrei," 291. For the reference to Recanati in the letter (*el Ricanato adesso*), see Kieszkowski, "Les rapports," 74, and Delmedigo, *Epistolae*, 75b. See also Dukas, *Recherches*, 65–66. On the character of Recanati's *Qabbalah*, see Idel, *Kabbalah in Italy*, 117–27.

107. Murano, *La biblioteca*, 81.

108. On the *terminus post quem* of the translation of Flavius Mithridates's translation of Recanati's commentary, see Wirszubski, *Pico della Mirandola's Encounter*, 15. On Pico's esteem of Recanati, see Recanati, *Commentary on the Daily Prayers*, 1:20.

109. Wirszubski, *Pico della Mirandola's Encounter*, 20–56, 59–64. See also Pico, *Syncretism in the West*, 344–45 (notes). See also Kieszkowski, "Les rapports," 60 on the unmistakable similarities between Pico's views and Flavius's Latin translation of Recanati's work.

110. On Pico, *Qabbalah*, and astrology, see Pico, *Syncretism in the West*, 541. See also Rutkin, "Astrology," 262–63. See also Fornaciari, "Elementi," 25–37.

111. Pico, *Syncretism in the West*, 529.

112. The thesis reads: "Every Hebrew Cabalist, following the principles and sayings

of the science of the Cabala, is inevitably forced to concede, without addition, omission, or variation, precisely what the Catholic faith of Christians maintains concerning the Trinity and every divine Person, Father, Son, and Holy Spirit" (Pico, *Syncretism in the West*, 523).

113. On the Latin translation produced for Pico, see Fenton, "Joseph Ibn Waqār et sa tentative." 339.

114. Fenton, 331. Fenton concluded: "Ibn Waqār is convinced of the preeminence of qabbalistic gnosis. Nevertheless, he hardly denies the value of philosophy. He thinks that the ultimate goal of the first does not differ radically from that of philosophical speculation. . . . For him, the particular method of Qabbala leads to analogous but superior results."

115. Fenton, 330–31.

116. Bartòla, "Eliyahu Del Medigo," 274. See Geffen, "Insights," 72–73 for how Delmedigo composed a treatise for Donato in 1480, when Delmedigo was twenty. On Pizzamano, see Montana, "Eliahu Del Medigo," 185–86.

117. Cassuto, "Gli ebrei," 291–92. Delmedigo translated the text for Pico but did not keep a copy.

118. Lelli, "Cabbalà e aristotelismo." See also Tamani, "Le traduzioni," 613–14. See also Ogren, "Sefirotic Depiction."

119. Lelli, "Cabbalà e aristotelismo," 231, 234–36.

120. Lelli, 242.

121. Lelli, 236–37. See also De Balmes, *The Letter of Ten*, 1b.

122. De Balmes, 1b–2a. See also Lelli, "Cabbalà e aristotelismo," 236–37.

123. Hacker, "The Immigration,"150.

124. See also Cohen Ashkenazi, *Sh*ᵉ*ʾeilot*, 18a where Abravanel remarked that he did not have with him Averroës's *Epistle on the Possibility of Conjunction*, as he had sent it on to Salonika, believing that he would end up there.

125. Cohen Ashkenazi, 17a.

126. See Cohen Ashkenazi, 17a–b for remarks on the indirect speech in the passage.

127. Cohen Ashkenazi, 17b–18a.

128. Cohen Ashkenazi, 20b.

129. Ogren, *The Beginning*, 90. His father Isaac equated "the Forms to the *sefirot*."

130. Richler, *Hebrew Manuscripts*, 140. It is possible that the book that Galeano/Jālīnūs bought from Abraham Algazi was not Vatican MS Ebr. 201 but MS Ebr. 202, a jumbled, unbound codex of twenty-one texts. On that possibility, see Cassuto, *I manoscritti*, 36, and Morrison, "A Scholarly Intermediary," 56. Vatican MS Ebr. 202 still joined the Fuggers' collection. Kumaṭiano possessed a manuscript of Recanati's scriptural commentary, which is now Paris BnF MS Hébreu 830.

131. See Scholem, *Origins*, 386–87, where he commented that a secret was really something that could be interpreted by Qabbalists.

132. Described in Ez. 1:5–11.

133. Galeano/Jālīnūs, *Taʿalumot ḥoḵmah*, 22a.

134. Schwartz, "Conceptions," 199–200.

135. Schwartz, "From Theurgy to Magic."

136. Galeano/Jālīnūs, *Taʿalumot ḥokmah*, 22a.

137. Though Galen's commentary on the Hippocratic *Epidemics* was not available in Hebrew, the Arabic text was known to some Jews. See Pormann, "Case Notes," 264.

138. Galeano/Jālīnūs, *Taʿalumot ḥokmah*, 5a. Galeano/Jālīnūs transcribed many Arabic technical terms into Hebrew characters.

139. Galeano/Jālīnūs, 22a–b.

140. Anon., *Peirush ha-Torah*, 137b. See 145a and 155b for more marginal comments on the secret of sacrifices.

141. Anon., 206a. See also 66a and 70b for a marginal note with a reference to a *sᵉpirah*.

142. Anon., 60a. There are two lines through this comment. Nevertheless, the connection between the *sᵉpirot*, the scriptural passage, and the *minḥah* prayer must have seemed plausible at one point.

143. Anon., 112b.

144. Anon., 216a.

145. Anon., *Seiper Tana dᵉbei Eliyahu*, 65, 117, 122. Here, the author of the midrash explained that the utmost heights were the upper heavens. The "land of treasure" is a reference to the Israelites.

146. From Babylonian Talmud Baba Qama 42a7; the expression meant something that was not possible.

147. Lelli, "'Prisca Philosophia,'" 60–61. See also Lelli, 67: "What Ficino could find in Platonic works was traced by both Alemanno and Pico back to the supposedly more ancient tradition of kabbalah."

148. Pico, *Oration*, 109. See note 3 on the same page for how there is no settled scholarly consensus on the identity of ʿAbd Allāh. He may have been Ibn al-Muqaffaʿ; ʿAbdallāh b. Salām (d. 663–4), an early Jewish convert to Islam; or ʿAbd Allāh b. al-Thāmir (d. 523), a Christian martyr of Najrān.

149. Pico, 225.

150. Pico, 243.

151. Pico, 213. See also note 252 about how Pico believed the available translations of the Arabic and Greek originals were poor.

152. Saliba, "Arabic Science," 128–29. On Almuli, see Kuntz, *Guillaume Postel*, 25. Kuntz specified that Postel received Aramaic books on *Qabbalah* from Almuli.

153. See Chajes, "Spheres," 232 on Postel's interest in diagrams of the *sᵉpirot* and the cosmology of *Qabbalah*. See also Chajes, *The Kabbalistic Tree*, 1.

154. Fleischer, "Seer to the Sultan," 292–93.

155. Saliba, "Arabic Science," 135–36. On Postel's translation of the *Zohar*, see Saliba, 137–38.

156. See Anon., *The Book of Bahir*, 118–19.

157. Saliba, "Arabic Science," 139–40. Postel acquired the *Tadhkira* before arriving in Istanbul. On Postel's relationship with the noted printer of Hebrew books, Daniel Bomberg, see Bobzin, "Guillaume Postel (1510–1581)," 59.

158. Kuntz, *Guillaume Postel*, 78.

159. Kuntz, 34. Kuntz added: "It is not surprising that Postel was so enthusiastic about the Arabic language; he states that it is closely related to the Hebrew language

and that almost two-thirds of the world use this language. Therefore, Arabic must be studied by scholars and taught in the schools" (39).

160. Kuntz, 25n74.

161. McLean, *The "Cosmographia,"* 16–18.

162. Münster, *Compendium aritmetices*, 1.

Chapter 6

1. See also Barker, "Copernicus," 351 on the connection between Averroist psychology and Averroist ideas about the structure of the heavens.

2. On the observational practices that led to these critiques of Ptolemy, see Munns, "The Challenge."

3. Michael Shank has published extensively on Regiomontanus's interest in homocentric astronomy. See Shank, "Regiomontanus as a Physical Astronomer," 326. See also Shank, 327, where Shank discussed "the inaugural lecture of his [Regiomontanus's] course on al-Farghānī at the University of Padua in 1464, in which he mentioned wistfully Averroes's unsuccessful efforts to construct a concentric astronomy." See also Shank, "The 'Notes on al-Biṭrūjī,'" 15 where Shank wrote, "And yet, paradoxical though it may seem, Regiomontanus was very interested in the homocentric tradition, in spite of the fact that he was an exceptionally competent mathematical astronomer." See also Shank, "Regiomontanus and Astronomical Controversy," 86–87, 96. See Shank, "Regiomontanus versus George of Trebizond," 364 on Regiomontanus's proposal of a concentric orb for the sun that revolved nonuniformly.

4. Shank, "Regiomontanus and Astronomical Controversy," 83.

5. On Fracastoro's appointment as a professor of logic at the University of Padua, see *Archivio Antico dell' Università*, vol. 669, 3a.

6. Fracastoro, *Homocentrica*, 60a–b.

7. Shank, "Geometrical Diagrams," 52n25.

8. See Barker, "The Reality," 17 on how Averroists did not accept partial orbs as calculating devices.

9. Robinson, "The First References."

10. Copernicus, *Nicolaus Copernicus Gesamtausgabe*, 2:17. Copernicus cited Biṭrūjī's finding that Venus was above the sun and that Mercury is below the sun.

11. Sylla, "The Status," 48–55.

12. Barker, "The Reality," 18.

13. Sylla, "The Status."

14. Swerdlow, "Regiomontanus's Concentric-Sphere Models," 2. See also Mancha, "Ibn al-Haytham's Homocentric Epicycles" for the argument that Ibn al-Haytham's (eleventh-century) homocentric epicycles are transmitted to Europe by the fourteenth century.

15. For more technical information, see Morrison, "A Scholarly Intermediary," 45.

16. Regiomontanus was not the only Renaissance astronomer to theorize a reciprocating mechanism. See Swerdlow, "Aristotelian Planetary Theory," 38–41 for how Amico hypothesized a reciprocating mechanism.

17. Swerdlow, "Regiomontanus's Concentric-Sphere Models," 14–15.

18. Swerdlow, 17.

19. Ibn Naḥmias, *The Light of the World*, 1.

20. For more on the relationship between these two versions of *The Light of the World*, see Ibn Naḥmias, *The Light of the World*, xiii–xiv.

21. Morrison, "An Astronomical Treatise," 40 for the reference to the sultan's court and 404–6 for the passage from *The Light of the World*.

22. See Ibn Naḥmias, *The Light of the World*, 42 on the chronology of the career of Galeano/Jālīnūs.

23. Bartolocci, *Bibliotheca magna rabbinica*, 4:501. The report came via a certain Petrus Rivier who was associated with the Collegium Neophytorum (where Bartolocci had also been a professor) and was a convert from Judaism (Bartolocci, 4: 228). Petrus Rivier was likely a contemporary of Bartolocci (Bartolocci, 4: 229). On the founding of the Collegium Neophytorum in 1543, see Wilkinson, *Orientalism*, 42n45.

24. Regiomontanus, "ʿEsrim wᵉ-shmoneh," 387/84. Vatican Ebr. 379, Parma Palatina Library Cod. Parma 2637, and Budapest, Kaufmann A 508 contain more extensive Hebrew translations from tables produced by Regiomontanus.

25. Galeano/Jālīnūs, *Taʿalumot ḥokmah*, 9a.

26. Marx, "Gershom," 427n1.

27. See Marx, 441–42n23 on how Soncino became acquainted in Venice with Francesco Giorgio, a churchman and philosopher who read *Qabbalah* and the *Talmud*; Zaccaria Dolfino, to whom Pope Gregory dedicated his *Dialogues*; and the Venetian senator Marco Tiepolo.

28. Sandal, "I libri scolastici." See also Marx, "Gershom," 492, where he wrote, "The influence of the Renaissance education was too great for the printers, who had constant close and friendly associations with educated Italians, to have escaped it."

29. Matsen, *Alessandro Achillini*, 23–26. For the relevant primary sources documenting Achillini's time in Padua and his return to Bologna, see Matsen, "Alessandro Achillini (1463–1512)."

30. Shank, "Setting up Copernicus?," 294–95.

31. Shank, 313–14. See also Barker, "Copernicus," 346: "First, it was the Averroists, not Copernicus, who were the main threat to Ptolemy in the sixteenth century. . . . For Copernicus and his contemporaries the only technically accurate mathematical astronomy was the Ptolemaic tradition. Hence, the Averroist attack on Ptolemy questioned the legitimacy of mathematical astronomy in general." See Goddu, *Copernicus*, 241 for his skepticism of direct influence between Achillini and Copernicus.

32. Achillini, *De orbibus libri quattuor*, 15b. Achillini made these comments in the course of his critical analysis of the Ptolemaic lunar model.

33. Achillini, 15b. The use of the term "spiral" was due to a mistranslation. On that, see Endreß, "Averroes' *De caelo*," 45. See Hasse, *Success and Suppression* on the study of Arabic, however limited, by Renaissance humanists.

34. Achillini, *De orbibus libri quattuor*, 16b.

35. Achillini, 22a. For the Arabic, see Averroës, *Tafsīr*, 2:563. My translation is: "And the things, animal or divine, including their parts, which are composed of bodies." For Bessarion's translation from the Greek coupled with the Latin version of Averroës's *Long Commentary*, see Aristotle, *Aristotelis Metaphysicorum Libri XIIII*, 117b–118a.

36. Averroës, *Tafsīr*, 3:1675. Averroës clarified that the solution for the motion of the

zodiac could be adapted for the motion of other planets: "Nothing impossible appears from theorizing this hypothesis and it is possible that all that appears corresponds to it. This spiral motion is a motion found in the heavens by the coincidence of the diurnal motion and the motion of the planets in their orbs."

37. Niphi, *Expositiones*, 583.

38. al-Biṭrūjī, *On the Principles of Astronomy*, 1:9–10.

39. Delmedigo, "De primo motore" in Jean de Jandun, *Quaestiones*, 131a. Delmedigo mentioned that these Spanish scholars made astrolabes from which increases and decreases in the rate of precession, the slow, west-to-east motion of the fixed stars, followed.

40. For more details about the three proposals found in *The Light of the World*, see Ibn Naḥmias, *The Light of the World*, 246–47 (§B.1.II.26/X.7–8) and the associated technical commentary.

41. Ibn Naḥmias, 248 (§B.1.II.26/X.10).

42. Mathematical analysis of this hypothesis is found in Ibn Naḥmias, 368–75.

43. Ibn Naḥmias, 248 (§B.1.II.26/X.11–13).

44. Ibn Naḥmias, 253–54 (§B.1.II.26/X.30).

45. Ibn Naḥmias, 368–70.

46. For more on the history of this model found in the Hebrew recension of *The Light of the World*, see Ibn Naḥmias, 19–22.

47. On the presence of the Ṭūsī couple in the astronomy of Copernicus, see al-Ṭūsī, *The Memoir*, 1:57–58.

48. Ragep, "From Tūn," 171–72. See also Ṭūsī, *The Memoir*, 216–17. See Ragep, "From Tūn," 182 for Ragep's inclusion of Ibn Naḥmias's *The Light of the World* in his catalog of appearances of the Ṭūsī couple in Europe before 1543 and as a source of Europeans' knowledge of the Ṭūsī couple.

49. Ibn Naḥmias, *The Light of the World*, 245 (§B.1.II.26/X.3).

50. For more on the Ṭūsī couple in the Judeo-Arabic original of *The Light of the World*, see Ibn Naḥmias, 134–39 and the associated commentary.

51. On Levi b. Gerson's (d. 1344) knowledge of the authentic and pseudo Ṭūsī commentaries on Euclid's *Elements*, see Lévy, "Gersonide," 90–91. On exchange of instruments, see Comes, "The Possible Scientific Exchange." See also Samsó, *On Both Sides*, 814–16 on knowledge in the Maghrib of a *zīj* of al-Maghribī (d. 1283), an astronomer at Marāgha.

52. Perhaps the first observation of the presence of a Ṭūsī couple in Amico's work is Ruths, "Das homozentrische Sphärensystem," 54–55. Ruths noted that Fracastoro mentioned the Ṭūsī couple only when discussing the work of Callippos and Eudoxos.

53. Ibn Naḥmias, *The Light of the World*, 134–39 (§B.1.II.13–26). Ragep, "From Tūn," 162 explained that Ibn Naḥmias's two-sphere Ṭūsī couple was a truncated version of the three-sphere version that Amico used. Copernicus preferred the truncated two-sphere version.

54. Di Bono, *Le sfere*, 11.

55. Fracastoro, *Homocentrica*, 61b–62a. See also di Bono, "Copernicus, Amico," 143–44.

56. Ragep, "From Tūn," 183.

57. Swerdlow, "Aristotelian Planetary Theory," 41.

58. Bartolocci, *Bibliotheca magna*, 4:501.

59. Ibn Naḥmias, *The Light of the World*, 139 (§B.1.II.26). See Ibn Naḥmias, 244 for a suggestion that the Ṭūsī couple be applied to the lunar model.

60. Swerdlow, "Aristotelian Planetary Theory," 42–44.

61. Ibn Naḥmias, 306.

62. Ibn Naḥmias, 139.

63. On Amico's erroneous equating of the planar and spherical Ṭūsī couples, see Swerdlow, "Aristotelian Planetary Theory," 40. See Ragep, "From Tūn," 186 for how Amico provided no evidence that he knew of these mathematical imprecisions in the Ṭūsī couple he used. Naṣīr al-Dīn Ṭūsī was aware of such a displacement. For mathematical analysis of the displacement, see Kennedy and Saliba, "The Spherical Case." See Ibn Naḥmias, *The Light of the World*, 306–9 for how Ibn Naḥmias's Ṭūsī couple was mathematically equivalent to the spherical Ṭūsī couple.

64. Di Bono, *Le sfere*, 65–66.

65. Barker, "Copernicus," 345–46.

66. Granada and Tessicini, "Copernicus and Fracastoro," 458–64.

67. Granada and Tessicini, 437–47, esp. 447.

68. Ragep, "From Tūn," 183.

69. On the dates of the Hebrew and Latin versions of Delmedigo's commentary on *De substantia orbis*, see Engel, *Elijah Delmedigo*, 124–25.

70. Delmedigo, *Bei'ur "ʿEṣem ha-galgal,"* 48b and Delmedigo, *Expositio*, 33b.

71. Zonta, "Medieval Hebrew Translations," 29. The *Compendium* was translated into Hebrew between 1231 and 1235 by Jacob Anatoli. See Lay, "*L'Abrégé*," 47–48 on Ibn al-Haytham's *Shukūk* and 48 for a discussion of Averroës's acknowledgement of eccentrics and epicycles in the *Compendium*.

72. On the relationship between Copernicus and Albert of Brudzewo, see Malpangotto, "The Original Motivation," 365. Sylla ("The Status of Astronomy," 70) wrote, "If Brudzewo did not lecture on Peurbach's *Theoricae novae planetarum* with Copernicus in attendance, he had taught such a course in previous years, and whoever taught Peurbach with Copernicus in attendance had likely heard Brudzewo himself earlier." For a twelfth-century European attempt at non-Ptolemaic theoretical astronomy, see Grupe, "Stephen of Pisa's Theory."

73. Malpangotto, "The Original Motivation," 373–75.

74. As translated in Malpangotto, 374.

75. Barker, "Albert of Brudzewo's *Little Commentary*," 137–39. See also Malpangotto, "The Original Motivation," 378.

76. Ibn Naḥmias, *The Light of the World*, 164. Ibn Naḥmias added that the repugnancies attached to the first two lunar anomalies differed from the other repugnancies of Ptolemy's astronomy.

77. Comments on other difficulties of Ptolemaic astronomy, e.g., the equant, may exist in the chapters of *The Light of the World* on the planets, chapters which are mostly missing from the manuscript of the Judeo-Arabic original of *The Light of the World* and completely missing from the manuscript of the Hebrew recension.

78. Duran, *Peirush al-Magisṭi*, 12a–b.

79. Kozodoy, *The Secret Faith*, 61–62. Duran responded to *The Light of the World* and criticized how Ibn Naḥmias explained away, rather than explained, observations of planets'

varying distances. For Duran's text and translation, see Ibn Naḥmias, *The Light of the World*, 393–98. See Kozodoy, *The Secret Faith*, 89–90 for how Profayṭ Duran, following Ibn Naḥmias's own claims, recognized *The Light of the World* as an attempt to solve, unsuccessfully in Duran's view, Maimonides's "true perplexity."

See Monfasani, *Collectanea Trapezuntiana*, 677–78 for the suggestion that Duran conveyed his concerns about Ptolemy to George of Trebizond (d. 1486) in a now-lost Latin version of Jābir ibn Aflaḥ's commentary on Averroës's *Compendium*. George of Trebizond knew of Jābir's criticisms of the *Almagest*, which were available only in Hebrew, even though George did not know Hebrew. See Akasoy, "Mehmed II," 254 on how George attempted to get Mehmed II to be the dedicatee of his new Latin translation of the *Almagest*.

80. Kumaṭiano, *An Ancient Commentary on "The Guide,"* 342. Cf. Maimonides, *Guide*, 2:326 (II.24).

81. Kumaṭiano, in his commentary, used *t*^{e}*ḵunah* to mean model in the sense of "hypothesis"; the eccentric and the epicycle were both examples of a *t*^{e}*ḵunah*.

82. Kumaṭiano, *An Ancient Commentary on "The Guide,"* 345.

83. Maimonides, *Guide*, 2:326.

84. Maimonides, *Guide*, 2:307.

85. Kumaṭiano, *Ancient Commentary on "The Guide,"* 326.

86. Kumaṭiano, 329.

87. Eisenmann, "Scientific," (Heb.) 112.

88. Kumaṭiano, *Ancient Commentary on "The Guide,"* 328–29. Cited also in Eisenmann, "Scientific," (Heb.) 113.

89. In *Correction*, Jābir improved Ptolemy's methods of proof and located Mercury and Venus above the sun. See Bellver Martinez, "El Lugar," 84 on how Jābir's mathematical criticisms of Ptolemy had theoretical implications. See Zonta, "Medieval Hebrew Translations," 58 on how *Qiṣṣur al-Magisṭī* was the title of the Hebrew translation of Jābir's *Iṣlāḥ al-Majisṭī*.

90. Jābir b. Aflaḥ, *Qiṣṣur al-Magisṭi*, 106b.

91. Jābir b. Aflaḥ, 84a. See also Jābir b. Aflaḥ, 37a, 60a for comments about where a word/phrase was missing in the Arabic version.

92. Jābir b. Aflaḥ, 89b.

93. Jābir b. Aflaḥ, 117a.

94. Afendopolo, *Peirush ʿal ha-Aritmeṭiqah*, 1a.

95. Freudenthal and Mancha edited and translated the Hebrew and Latin versions of this chapter in Freudenthal and Mancha, "Levi ben Gershom's Criticism." Other innovative models of Levi are found later in book 5, part 1 of *Wars*.

96. On David's reputation in Salonika, see Tirosh-Rothschild, *Between Worlds*, 84. Elijah Mizraḥi, Galeano/Jālīnūs's teacher and a contemporary of Afendopolo, recognized David b. Judah Messer Leon's comprehensive erudition.

97. For the statement of sale, see Levi b. Gerson, *Milḥamot ha-Shem*, 257b. David b. Judah Messer Leon arrived in Constantinople in 1495–96 and left for Salonika before 1504. On those dates, see Tirosh-Rothschild, *Between Worlds*, 55. The sale of this MS to Afendopolo was David's last known action in Istanbul before departing for Salonika.

98. On the ascription of the commentary to Mizraḥi, see Morrison, "A Scholarly Intermediary," 42.

99. Benayahu, *Rabbi Elijah Capsali* (Heb.), 88. As evidence for the connection, Benayahu pointed to Mizraḥi's kind words on Judah Mintz's death that Capsali recorded. See Benayahu, 151 for Capsali's reference to Mizraḥi's responsa.

100. The author of the colophon noted on 214a that the writer (*koteib̲*) of the commentary was not a single person; I take this to mean that the St. Petersburg MS was produced by multiple hands. See Langermann, "Science in the Jewish Communities," 446 on the copyist's reference to strange letters (*otiyot meshonot*) in the MSS upon which he depends; these references could be to Greek and/or Arabic script. After all, Mizraḥi cited Greek and Arabic MSS of the *Almagest.*

101. See Mizraḥi, *Peirush Seip̲er al-Magisṭi*, 85b, 86b, 89b, 130b–131a for references to the Arabic *Almagest* (*nusḥa ʿarab̲it*). On 86a, Mizraḥi compared two Hebrew manuscripts of the *Almagest*. Mizraḥi referenced other manuscripts to sort out difficult passages in the lunar model. See, e.g., Mizraḥi, 121a. On Mizraḥi's references to Ibn Rushd and Jābir b. Aflaḥ's writings, see Langermann, "Science in the Jewish Communities," 447.

102. See Ibn al-Haytham, *al-Shukūk*, 19 on how the Ptolemaic lunar epicycle had a single mover that moved it in two opposite directions at the same time.

103. Mizraḥi, *Peirush Seip̲er al-Magisṭi*, 163a.

104. Mizraḥi, 163b. On Quṭb al-Dīn al-Shīrāzī's (d. 1311) use of the Ṭūsī couple in such a demonstration, see Morrison, "Quṭb al-Dīn al-Shīrāzī's Hypotheses," 91.

105. Mizraḥi, *Peirush Seip̲er al-Magisṭi*, 163b. See also Langermann, "Science in the Jewish Communities," 448.

106. Mizraḥi, *Peirush Seip̲er al-Magisṭi*, 164a.

107. On an earlier exploration of how an outer concentric orb could move an inner concentric orb, see Saliba, "Early Arabic Critique."

108. Mizraḥi, *Peirush Seip̲er al-Magisṭi*, 163b.

109. Mizraḥi, 163b. See Langermann, "A Compendium," 301–2 for Galeano/Jālīnūs's use of these terms. See Langermann, 303–4 for his use of the term "mathematical body" to denote the heavens.

110. Grant, *Planets*, 324–70.

111. Mizraḥi, *Peirush Seip̲er al-Magisṭi*, 167a.

112. Mizraḥi, 164a. Mizraḥi explained, "It follows necessarily from this that the center of the epicycle approaches and distances from the earth."

113. On the relationship between Galeano/Jālīnūs and Elijah Galeano, see Langermann, "From My Notebooks," 376.

114. Ragep, *Jaghmīnī's "Mulakhkhaṣ,"* 20.

115. For commentaries on Jaghmīnī and Qushjī's summary astronomy texts, see İzgi, *Osmanlı Medreselerinde İlim*, 1:334. For Ṭāsh Kubrī Zādah's classification of Jaghmīnī's *Mulakhkhaṣ*, see Ṭāsh Kubrī Zādah, *Miftāḥ al-saʿāda*, 1:349.

116. İzgi, *Osmanlı Medreselerinde İlim*, 1:371. See now Ragep, *Jaghmīnī's "Mulakhkhaṣ,"* 25. Istanbul MS Laleli 2141/43 is the oldest manuscript of the *Mulakhkhaṣ* in a Turkish library.

117. Ragep, *Jaghmīnī's "Mulakhkhaṣ,"* 15n61.

118. İzgi, *Osmanlı Medreselerinde İlim*, 1:373. On the relationship between Dawānī and Muʾayyadzādah, see Pfeiffer, "Teaching the Learned."

119. On Ṭāsh Kubrī Zādah's biography, see Fleming et al., "Ṭās͟h köprüzāde."

120. Ṭāsh Kubrī Zādah, *Miftāḥ al-saʿāda*, 1:348. On the connection between ʿUrḍī and Copernicus, see Saliba, *Islamic Science*, 202–4.

121. Ṭāsh Kubrī Zādah, *Miftāḥ al-saʿāda*, 1:348–49. Parts of *Nihāyat al-idrāk* have been studied by Savadi, "The Historical and Cosmographical Context" and Nikfahm-Khubravan, "The Reception of Ptolemy's Latitude Theories," 423–509.

122. See Galeano/Jālīnūs, *Taʿalumot ḥokmah*, 21a for the reference to Elijah Galeano as *ziqni* (my elder).

123. Morrison, "The Role." See Ragep, *Jaghmīnī's "Mulakhkhaṣ*," 290–91 on glosses on the *Mulakhkhaṣ* in Persian and Turkish.

124. Morrison, "The Role," 13.

125. On Maḥmūd Çelebi, see Adıvar, *Osmanlı Türklerinde İlim*, 26–27.

126. See Morrison, "The Role," 11–12 for more details on the activities of Mevlānā Aḥmet.

127. Morrison, 15–17.

128. Halevi, *The Kuzari*, 126. There, Halevi proclaimed, "From the mouths of scholars, but not from the mouth of books."

129. This translation comes from Morrison, "The Role," 11.

130. Arnaldo da Villanova, "Panim ba-mishpaṭ," 77a.

131. For evidence that the text was studied in those locales, see the list of commentaries in İzgi, *Osmanlı Medreselerinde İlim*, 1:373–76. For more on science and religion at Samarqand, see Trigg, "From Samarqand to Istanbul."

132. Among Bernard Goldstein's many publications on Gersonides's astronomy, one might begin with *The Astronomy of Levi ben Gerson* and "Levi ben Gerson's Contributions to Astronomy."

133. The translation comes from Langermann, "A Compendium," 291.

134. Isaac b. Samuel Abū al-Khayr, *Peirush al-Fargani*, 2b.

135. Eliyahu al-Faji, *Mik͟tab Eliyahu*. I follow Steinschneider, "Die mathematischer Wissenschaften bei den Juden, 1441–1500," 73 to date this text to around 1500.

136. See the collection of responsa from Turkey in Anon., *Leqeṭ tᵉshuḇot posqim*, 34a. G. Margoliouth, in his *Catalogue of the Hebrew and Samaritan Manuscripts* (3:345–46) identified al-Faji with the Elijah ben David al-Faji who signed a responsum (*tᵉshuḇah*) for Elijah Mizraḥi in 1518.

137. Al-Faji, *Mik͟tab Eliyahu*, 2b.

138. Cf. Langermann, "A Compendium," 290–91.

139. See al-Faji, *Mik͟tab Eliyahu*, 3b where al-Faji mentioned Levi b. Gerson's proof against the existence of the epicycle.

140. Al-Faji, 2b. For more on al-Faji, see Morrison, "A Scholarly Intermediary," 40–42.

141. Langermann ("A Compendium," 295n17) mentioned this reference from *Mik͟tab Eliyahu*.

142. Langermann, 290.

143. See Langermann, 287 on the completion of the text in the 1530s.

144. For more details on Ibn Naḥmias's lunar theory, see Ibn Naḥmias, *The Light of the*

World, 161–66 and the associated commentary. For the lunar model of Ibn al-Shāṭir, see Roberts, "The Solar and Lunar Theory."

145. Swerdlow, "The Most Complex Homocentric Astronomy," 360. See also Nikfahm-Khubravan and Ragep, "The Mercury Models," 7. They wrote, "It may also be the case that Copernicus, when writing the *Commentariolus*, was under the influence of the Paduan Averroists and saw Ibn al-Šāṭir's models, with their eschewing of eccentrics and the potential of a return to [a] single, Aristotelian center, as a way to achieve a 'quasi-homocentricity.'"

146. İzgi, *Osmanlı Medreselerinde İlim*, 1:362–63. See also Fazlıoğlu, "Osmanlı felsefe-biliminin arkaplanı." An English translation is "The Samarqand Mathematical-Astronomical School: A Basis for Ottoman Philosophy and Science," in *Journal for the History of Arabic Science* 14 (2008): 3–68, https://islamsci.mcgill.ca/Fazlioglu.pdf.

147. Ibn al-Shāṭir, *Nihāyat al-sūl*, 1a. See al-Ḥamawī, *Muʿjam al-buldān*, 5:372 on Warsanīn. On other known manuscripts of *Nihāyat al-sūl*, see "Nihāyah al-suʾl fī taṣḥīḥ al-uṣūl (Ibn al-Shāṭir)," Islamic Scientific Manuscripts Initiative, https://ismi.mpiwg-berlin.mpg.de/index.php/text/105844.

148. See Morrison, "An Astronomical Treatise," 386–88 for more on Galeano/Jālīnūs's knowledge of the contents of *The Light of the World* despite the confusion about the identity of the author. One of the few surviving MSS of Biṭrūjī's *On the Principles* exists in the Topkapı Palace Library, in the same codex (MS Ahmet III 3302) as Mūsā Jālīnūs's *Dhikr baʿḍ al-maḥallāt*.

149. Galeano/Jālīnūs, *Taʿalumot ḥokmah*, 76a.

150. Galeano/Jālīnūs, 33a. Galeano/Jālīnūs wrote, "Ptolemy's hypotheses (*shoreshim*) can yield true effects but, nevertheless, not be true due to their conflict with physics."

151. Galeano/Jālīnūs, 33a.

152. Galeano/Jālīnūs, 18b.

153. Galeano/Jālīnūs, 18b. Galeano/Jālīnūs added an example from arithmetic wherein the presence of the expected remainder should not lead one to presume that the arithmetic operation is necessarily carried out correctly.

154. Galeano/Jālīnūs, 83a.

155. The related matter of whether Euclid demonstrated that parallel lines must be in the same plane is the topic of an intervention at Samarqand by Fatḥ Allāh Shirwānī (d. 1486). See Fazlıoğlu, "Osmanlı Felsefe," 48–50.

156. Euclid, *The Thirteen Books of The Elements*, 1:155.

157. Maimonides, *Guide*, 1:210. Although Pines stated in his footnote that Maimonides referred to book 2, proposition 13, this was a reference to book 2, theorem 14 of the *Conics* in which an asymptote approaches the hyperbola so that the distance between them becomes "less than any assignable length." See Apollonios of Perga, *Treatise on Conic Sections*, 61. See Profayṭ Duran, "Beiʾur shenei ha-qawwim she-einam nipgashim," 67a for Profayṭ Duran's exegesis of Maimonides's reference to the *Conics* in the *Guide*. Paris 1021 is also the codex that contains Galeano/Jālīnūs's Hebrew translation of Ibn al-Zarqāl's Arabic treatise on the quadrant. Maimonides, "Notes to Some Propositions in the Conics" exists in Arabic script in Manisa İl Halk Kütüphanesi MS Genel 1706, 36b–33b. On that manuscript, See Hogendijk, *Ibn al-Haytham's "Completion,"* 129.

158. Maimonides, *Guide*, 1:210–11.

159. Some of Philoponus's philosophy was known to Islamic and Jewish philosophers, though there is no trace of a Hebrew version of his work. See Davidson, "John Philoponus." See also Maimonides, *Guide*, 1:177.

160. Galeano/Jālīnūs, *Taʿalumot ḥokmah*, 83a.

161. Langermann, "A Compendium," 305–9, esp. 307–9. The passage I quote supports Langermann's inference that Galeano/Jālīnūs disagreed with Apollonios.

162. Langermann, "A Compendium," 306–8. Langermann translated and studied this passage. He wrote, "I must confine my remarks here to Galeano's interesting contextualizations of the issues involved. The first of these is his reference to the 'calculatori,' the only explicit mention of this school known to me in Hebrew letters."

163. Bellosta and Heyberger, "Abraham Ecchellensis," 195.

164. Fazlıoğlu, "Hendese," 201.

165. Lévy, "The Hebrew Mathematics Culture," 164–65. See also Lévy, "The Establishment," 438–39 on the availability of texts about Apollonios's mathematics in Hebrew.

166. Shefer-Mossensohn, *Science among the Ottomans*, 164.

167. Feldhay and Ragep, "Introduction," 7.

168. Goddu, "Birkenmajer's Copernicus," 203.

169. On the date of the composition of the *Commentariolus*, see Swerdlow, "Copernicus' Derivation," 34. For the possibility of a date as early as 1508, see Shank, "Regiomontanus and Astronomical Controversy," 108. Copernicus, in his later, better-known *De revolutionibus orbium coelestium* (On the Revolutions of the Celestial Orbs) presented heliostatic models that also resembled Islamic models, though not Ibn al-Shāṭir's.

170. The first publication to mention the connection between Copernicus and Ibn al-Shāṭir was Roberts, "The Solar and Lunar Theory," 428. See also Kennedy and Roberts, "The Planetary Theory of Ibn al-Shāṭir," 227. For a summary of the parallels between Copernicus and Islamic astronomers, see Saliba, *Islamic Science*, 196–209.

171. Swerdlow and Neugebauer, *Mathematical Astronomy*, 47.

172. Westman, *The Copernican Question*, 100, 200–201. But see Swerdlow, "Copernicus and Astrology."

173. Swerdlow, "Copernicus and Astrology," 365–66, 373. See also Swerdlow, 360 ("Pico himself says nothing, not one word, about 'the order of the planets and the assignment of elemental qualities.'")

174. Goddu, *Copernicus*, 389–91.

175. For a reaction, see Barker and Vesel, "Goddu's Copernicus," 327–32. See Ragep, "From Tūn," 186–87 for a critique of Goddu's explanation for how Copernicus devised a Ṭūsī couple without knowing of Ṭūsī's work. Goddu, in any case, did not explain how Copernicus devised double-epicycle models identical to those of Ibn al-Shāṭir. Albert Brudzewo's (d. 1497) models (cf. Goddu, *Copernicus*, 484) were not identical to those of Ibn al-Shāṭir.

176. See Vesel, *Copernicus*, 354–56. In that respect, Vesel agreed with Goldstein, "The Origins."

177. Goddu, "Birkenmajer's Copernicus," 203.

178. Vesel, *Copernicus*, 343–45. Vesel suggested that the exchange most likely transpired in Padua or Venice.

179. See Westman, *Copernican Question*, 531n136 for a reference to Swerdlow and Neu-

gebauer's finding of the identity between the lunar models of Copernicus and Ibn al-Shāṭir.

180. Ragep, "Ibn al-Shāṭir and Copernicus."

181. Wilson, "Rheticus," 32–35. On p. 32, Wilson acknowledges the possibility of Arabic sources for the epicycle-epicyclet hypothesis.

182. Goddu, "Ludwik Antoni Birkenmajer," 241, 246. Birkenmaijer argued that once Copernicus came upon the epicycle-epicyclet hypothesis, he proposed a heliostatic arrangement to replace the large Ptolemaic epicycles.

183. Swerdlow, "Copernicus' Derivation."

184. Goddu, "Ludwik Antoni Birkenmajer," 241. Birkenmaijer "did not assert categorically that either Brudzewo or Copernicus invented his own model."

185. Neugebauer, *HAMA*, 1:11, 3:1035.

186. Blåsjö, "A Critique." See Blåsjö, 190 for a "natural continuation." See also Blåsjö, "A Rebuttal."

187. See Ragep,"ʿAlī Qushjī," for a persuasive argument that Regiomontanus's (d. 1476) demonstration for converting Mercury and Venus's epicycles to eccentrics was due to ʿAlī Qushjī (d. 1474), an astronomer who finished his career in the Ottoman Empire.

188. Bloom, *The Anxiety of Influence*, 27. I would like to thank Professor Harriet Murav of the University of Illinois for suggesting this book.

Chapter 7

1. See Morrison, "Mūsā Cālīnūs' *Treatise*."

2. On the place of logic in the medical curriculum at Padua, see Ruderman, *Jewish Thought*, 107.

3. Langermann, "A Compendium," 285.

4. On the absence of interest of premodern Hebrew writers in mechanics, see Langermann, "From My Notebooks," 367.

5. See Langermann, "A Compendium," 315.

6. Galeano/Jālīnūs, 11a.

7. Galeano/Jālīnūs, 39a. This vignette has been described in Langermann, "From My Notebooks," 365.

8. This text was known as the *Pantegni* of Haly Abbas to physicians working in Latin.

9. Galeano/Jālīnūs, *Taʿalumot ḥokmah*, 80b.

10. Galeano/Jālīnūs, 80b.

11. Galeano/Jālīnūs, 60b.

12. This vignette was added, probably by Galeano/Jālīnūs, after his student Abraham Algazi copied the extant manuscript of *Puzzles* in 1539. For more on the history of the composition of *Puzzles*, see Langermann, "A Compendium," 289–90.

13. See, e.g., Brentjes, "Courtly Patronage"; Brentjes, "Patronage of the Mathematical Sciences"; Goldstein, "Astronomy among Jews"; and Chabás, "Interactions."

14. Eamon, "Court," 26.

15. Findlen, "The Economy," 13. See also Westfall, "Science and Patronage."

16. Findlen, "The Economy," 24.

17. Galeano/Jālīnūs, *Taʿalumot ḥokmah*, 9a.

18. Galeano/Jālīnūs, 6a–10a. Folios 7 and 8 are missing. Strangely, the text reads

smoothly from folio 6a to 8b. For a report of a similar feat of memory, see Galeano/Jālīnūs, 77b.

19. See Langermann, "A Compendium," 295–96 for an initial discussion of this device.

20. In the figure in the manuscript (10a), at the end of the spokes corresponding with Aries, Cancer, Libra, and Capricorn are circles. These are the signs that correspond with the beginning of each season.

21. Galeano/Jālīnūs, *Taʿalumot ḥoḵmah*, 8b–9a.

22. Galeano/Jālīnūs, 9a.

23. Galeano/Jālīnūs, 8b.

24. Galeano/Jālīnūs, 9b.

25. Cf., Truitt, *Medieval Robots*, 32. With regard to a Christian perspective on devices made by Muslims, Truitt found "that although how automata *work* can be explained in familiar mechanical terms, how automata are *made* requires invoking mysterious, occult knowledge of questionable moral legitimacy." On the connection between mathematics and the occult sciences, see Melvin-Koushki, "Powers of One."

26. Tirosh-Rothschild, *Between Worlds*, 71–73 and 105. See also Tirosh-Rothschild, 76 where she commented, "Renaissance Jewish humanism was a distinct Italian phenomenon which could exist only within the Appenine Peninsula. In Italy, and there only, Jewish intellectuals such as Judah Messer Leon and his son R. David blended a unique syncretism of Ashkenazic legalism, Maimonidean rationalism, medieval scholasticism and Renaissance humanism. This amalgam of seemingly contradictory trends was challenged when Italian Jews exported it to their new homes in the Ottoman Empire."

27. On the compass, see Galeano/Jālīnūs, *Taʿalumot ḥoḵmah*, 4b. For another mention of a compass, see Capsali, *Seider Eliyahu Zuṭa*, 2:255. See also Dietrich and Wiedemann, "Maghnāṭīs."

28. Galeano/Jālīnūs, *Taʿalumot ḥoḵmah*, 28b. The last half of the second sentence was in the hand responsible for the later additions to the text.

29. Galeano/Jālīnūs, 61a–b. It is plausible that Galeano/Jālīnūs was describing the conflicts between the Ottomans and the Habsburgs in the 1520s.

30. The word in Hebrew is רושקופיטירוש, which I believe is a transliteration of *rochepeter*. On *rochepeter* as a synonym for *saltpetre*, see Partington, *A History of Greek Fire*, 318.

31. Galeano/Jālīnūs, *Taʿalumot ḥoḵmah*, 11a–b.

32. On Barbarossa's approach to Minorca while flying Christian standards, see Markham, *The Story of Majorca*, 237–39.

33. Langermann, "A Compendium," 287–90. In Galeano/Jālīnūs, *Taʿalumot ḥoḵmah*, 95b, I read a zayin (7) where Langermann read a waw (6), hence the discrepancy of a year in the date of completion of *Puzzles*.

34. Galeano/Jālīnūs, *Taʿalumot ḥoḵmah*, 80a.

35. Langermann, "A Compendium," 287, 290.

36. See Olivieri, *Pietro d'Abano*, 40 on the impact of d'Abano's studies in Paris on Averroism at Padua.

37. Olivieri, 149.

38. Galeano/Jālīnūs, *Taʿalumot ḥoḵmah*, 29b. Here and elsewhere, Galeano/Jālīnūs transcribed Arabic technical terms into Hebrew characters.

39. See de Abano, *Expositio*, 1:217 (XXIV.1). See Langermann, "From My Notebooks,"

370–71 for another place in *Puzzles* where he suspected that Galeano/Jālīnūs cited d'Abano's commentary. For the corresponding passage in the *Expositio* that Galeano/Jālīnūs cited, see de Abano, *Expositio*, 2:268 (XXV.59).

40. For more, see Zonta, *La filosofia*, 167–69.

41. Langermann, "From My Notebooks," 368.

42. Galeano/Jālīnūs, *Taʿalumot ḥokmah*, 21a.

43. Manekin, "Logic in Medieval Jewish Culture," 129.

44. Bloodstone, a variety of jasper, began to be used as a material for amulets in Late Antiquity. See Kotansky, "The Magic 'Crucifixion Gem.'" See also Albertus Magnus, *The Book of Minerals*, 88–89.

45. Galeano/Jālīnūs, *Taʿalumot ḥokmah*, 21b. See also Langermann, "From My Notebooks," 376.

46. From Lat. *commissura*, joining together. Langerman translated "comosure."

47. The last phrase could be translated as such: "which is where the place where its [viz. the coin's] quality and effect passes to the brain."

48. I have adopted Langermann's translation from Langermann, "From My Notebooks," 375–77.

49. Galeano/Jālīnūs, *Taʿalumot ḥokmah*, 26a. Cf. Maimonides, *Guide*, 2:545 (III.37). Galeano/Jālīnūs concluded by citing *Psalms* 49:14: "This is the way of them that are foolish."

50. Galeano/Jālīnūs, *Taʿalumot ḥokmah*, 45a.

51. Galeano/Jālīnūs, 45a–b.

52. Galeano/Jālīnūs, 24b. Galeano/Jālīnūs has discussed events in Exodus 17 and Numbers 33.

53. Galeano/Jālīnūs, 22a.

54. Heb. *sharbīn*. In Persian, *sharbīn* is a tree, perhaps the *oxycedrus*, from which liquid pitch flows.

55. The classical Greek δᾴδινος means "made of pine" and is the origin of the Latin *taeda*.

56. Cf. Latin *taeda*.

57. Galeano/Jālīnūs, 68b.

58. See Hasse, *Success and Suppression*, 137–78 on Renaissance humanists' profound preoccupation with materia medica. See also Brentjes, *Travellers*.

59. Another early example of Turkish scientific literature is Ahī Çelebī's translation of *al-Mūjaz fī al-ṭibb* into Turkish. See İhsanoğlu, *Osmanlı Tıbbi Bilimler*, 1:108–9. See Langermann, "Science in the Jewish Communities," 443 on Elisha the Greek, who worked at the sultan's court and whose Hebrew text *Mapteiaḥ ha-rᵉpuʾah* depended on Ibn al-Nafīs's *Mūjaz fī al-ṭibb* and on Kāzarūnī's commentary on the *Mūjaz*. Langermann identified Elisha with a Jew named Elissaos.

60. Morrison, "Mūsā Cālīnūs' *Treatise*," 79.

61. García-Ballester et al., "Jewish Appreciation," 90. See also Hacker, "Local Patriotism" (Heb.), on a localized intellectual patriotism among the Sephardic *emigrés* in the Ottoman Empire. See also Hacker, "The Intellectual Activity."

62. For the Latin sources, see Morrison, "Mūsā Cālīnūs' *Treatise*," 110n71.

63. Lorqi composed in Arabic, but only the Hebrew translation by Don Vidal Joseph b. Benveniste survives.

64. Ahī Çelebī, "Risāla fī al-ṭibb," 39b.

65. On Şāh Çelebi, see https://islamansiklopedisi.org.tr/muhyiddin-mehmed-sah. For a translation of Maimonides's treatise, see Maimonides, "On Poisons," 77–104.

66. Maimonides, "al-Maqāla al-Fāḍiliyya," 169b.

67. Galeano/Jālīnūs, *Taʿalumot ḥokmah*, 32a.

68. The text is unmentioned in Zonta, "Medieval Hebrew Translations."

69. Galeano/Jālīnūs, *Taʿalumot ḥokmah*, 32a. In the quoted portion, Galeano/Jālīnūs paraphrased Silvaticus, *Pandectarum*, 94b. The Latin reads: "Si medicus visitando infirmum ipsam in manu portaverit aegroto insciente. Et dicit aegroto qualiter stas. Si aeger dicit bene sanabitur. Et si dicet male morietur."

70. Barkai, "Between East and West," 53. The treatise has not been published.

71. Barkai, 57.

72. Barkai, 58. See Stearns, *Infectious Ideas*, 163 on how a significant minority of Muslim scholars believed that the plague was contagious, although their acceptance of contagion theory did not necessarily lead to a response. See Varlık, *Plague and Empire*, 232 on how Rabbi al-Mosnino and Taşköprüzadeh believed that plague could come from outside. Varlık found that both contagion and miasma could be combined to explain different aspects of the plague (e.g., its outbreak and spread).

73. Stearns, *Infectious Ideas*, 162. For a sixteenth-century European's perception that Ottoman attitudes to the plague changed, see Varlık, *Plague and Empire*, 245.

74. Varlık, 242. Hence, Ilyās was not the first to advocate flight.

75. Langermann, "From My Notebooks," 356–57. For the original text, see Galeano/Jālīnūs, *Taʿalumot ḥokmah*, 12a–12b. Both of the examples come from these pages.

76. See also David, "Taʿalumot ḥokmah" (Heb.), 1339; and Langermann, "A Compendium," 314.

77. Terzioğlu, "Un Traité Turc." For a reproduction of the manuscript and a transcription of the text into modern Turkish, see İzgöer, *16. Yüzyıl Osmanlı Tabibi*.

78. Benayahu, "Rabbi Joseph Hamon" (Heb.), 285. See Marcus and Bornstein-Makovetsky, "Hamon" on how Joseph Hamon's father Isaac was for a time the physician of King ʿAbdallāh of Granada.

79. Benayahu, "Rabbi Joseph Hamon," 283. See Kohen, *History of the Turkish Jews*, 27 on how Joseph's success brought accusations of poisoning Bayezit II.

80. Adamson, *Al-Kindī*, 161–65.

81. Adamson, 162–64.

82. Adamson, 164.

83. Langermann, "Another Andalusian Revolt?," 354.

84. Kindī argued that with an integer progression (1, 2, 3, 4), the problem was that the ratio between the integers was, in fact, constantly changing as 3:2 was not equal to 2:1.

85. Langermann, 353. The treatise on theriac (*Kitāb al-tiryāq*) has been published in Averroës, *Rasāʾil Ibn Rushd al-ṭibbiyya*, 387–422.

86. McVaugh, "Quantified Medical Theory," 398.

87. Morrison, "Mūsā Cālīnūs' *Treatise*," 110n71. There I mentioned a possible Latin source, Arnaldo's *De gradibus*, for Galeano/Jālīnūs's report.

88. Demaitre, *Doctor Bernard de Gordon*, 20, 50.

89. Morrison, "Mūsā Cālīnūs' *Treatise*," 110.

90. Morrison, 110.

91. Isaac Israeli (d. ca. 955) did not author a text called *Gerem ha-maʿalot*, but Lorqi made this very point in *Gerem ha-maʿalot*, 87b–88a. See Necipoğlu, Kafadar, and Fleischer, *Treasures of Knowledge*, 1:553 for a record of a copy of Israeli's Arabic *Kitāb al-Aghdhiya* (Book of Foods) in Bayezit II's library.

92. Galeano/Jālīnūs, *Taʿalumot ḥoḵmah*, 94a.

93. Galeano/Jālīnūs, 74a. He added, "Most of the error is in this area. Were it only that their stupidity was a deficiency, but it is also acquired."

94. Morrison, "Mūsā Cālīnūs' *Treatise*," 109.

95. Morrison, 112–13. See Morrison, 113n80 for the source of Galeano/Jālīnūs's references to Ptolemy.

96. Galeano/Jālīnūs, *Taʿalumot ḥoḵmah*, 30b. The ant descends the mountain because the ant's legs are to be considered a force that compels the ant.

97. Cf. al-Tirmidhī, *al-Jāmiʿ al-ṣaḥīḥ*, 2:316. *Ḥadīth* #453 reads: "Inna Allāh witr yuḥibb al-witr." The Arabic *witr* and the Hebrew *nip̱rad* both mean "odd" (viz. "not even"), "separate," and "indivisible."

98. Galeano/Jālīnūs, *Taʿalumot ḥoḵmah*, 30b–31a.

99. Park, *Doctors and Medicine*, 245.

100. See Cooper, "Galen and Astrology," 122 on how Galen included astrology in order to persuade the reader of the correctness of his analysis, meaning that Galen's theories could stand without astrology. See also Cooper, 124: "While Galen's medicine is consistent with *natural* astrology, it is not so with *judicial* astrology, and his attempt to present his theory of the critical days in Book II of the *Critical Days* in terms of judicial astrology is a practical failure, if a rhetorical success." For discussions of the doctrine of critical days in Hebrew texts, see Langermann, "The Astral Connections."

101. Schmitt, "Thomas Linacre," 55–61.

102. As quoted and translated in Siraisi, *Medieval and Early Renaissance Medicine*, 136. Siraisi found that even physicians less conversant in astrology drew on the principle of the critical days in their medical practice.

103. Seller, *Scientia Astrorum*, 105–6. On d'Abano's Latin translations of Abraham Ibn Ezra's works of astrology *Book of Reasons*, see Smithius, "Abraham Ibn Ezra's Astrological Works," 248–49.

104. Siraisi, *Medieval and Early Renaissance Medicine*, 68: "There can be no doubt that leading Latin medical writers of the thirteenth to fifteenth centuries were indeed the products of extensive training in logic and natural philosophy, and increasingly as time went on, in astrology as well. A few of them—Pietro d'Abano himself is an especially celebrated example—had a command of these subjects equal to that of any of their colleagues in other faculties, along with a pronounced interest in philosophical questions." See also Siraisi, "Two Models of Medical Culture," 81–83.

105. Galeano/Jālīnūs, *Taʿalumot ḥoḵmah*, 65b.

106. See Ibn Ezra, *Abraham Ibn Ezra's Introductions*, 488–91. On these pages from *The Book of Judgments*, Ibn Ezra explained that the seasons were marked by the passage of the sun through the zodiac.

107. Galeano/Jālīnūs, *Taʿalumot ḥoḵmah*, 64b–65a. The name Bernardo was modified to Barnaldo, a portmanteau of Bernard and Arnaldo that reflects the role of Bernard's

De Gradibus in spreading the teachings of Arnaldo's *De Gradibus*. On the medieval Hebrew translation of *De Gradibus* as *Seiper ha-madragot*, see Zonta, "Medieval Hebrew Translations," 48.

108. Galeano/Jālīnūs attributed the *Introduction* to Ptolemy. For the passage that Galeano/Jālīnūs referenced, see Evans and Berggren, *Geminos's Introduction*, 223.

109. Siraisi, *Medieval and Early Renaissance Medicine*, 64.

110. Siraisi, 67–68. Siraisi's examples for the importance of astrology at Padua came from the fourteenth and fifteenth centuries, before Galeano/Jālīnūs's career. Elsewhere, Siraisi showed that astrology was not a physician's preferred instrument. See Siraisi, "Medicine and the Renaissance World," 30. She explained, "The role of astrology in early modern medical practice may not have been quite as pervasive as is sometimes claimed; nevertheless, some sixteenth-century medical practitioners did indeed erect astrological figures for the onset of patients' illnesses, and probably almost all paid attention to the 'critical days' of illness supposed to depend on the phases of the moon."

111. Fracastoro, *Syphilis*, 40–41.

112. See Fracastoro, 146–47 for the translation of the passage on the causes and 104–5 for the original. Fracastoro acknowledged that these semences were subject to astral forces but found that astrology could not save someone from syphilis. Rather, a better knowledge of the disease was necessary so as not to be contaminated.

113. Siraisi, "Renaissance Readers," 221–24.

114. See Maclean, *Logic*, 304–6 on critiques of the use of astrology in medicine.

115. See Schmitt, "Science in the Italian Universities," 45. See also Maclean, *Logic*, 30 on how logic was central to medical education in Padua.

116. Cf. Long, "Trading Zones," 841–42. Long used "trading zone" to denote an arena in which scholars and artisans might productively interact.

117. Serjeant, "Notices on the 'Frankish Chancre,'" 241–44.

118. Fracastoro, *The Sinister Shepherd*, 77.

119. Varlık, *Plague and Empire*, 230–31.

120. Varlık, 231.

121. Corazzol, "Gli ebrei," 212.

Conclusion

1. Shefer-Mossensohn, *Science among the Ottomans*, 113.

2. Latour, *We Have Never Been Modern*, 93

3. Mufti, *Forget English!*, 29. I thank my colleague Nasser Abourahme for bringing this book to my attention.

BIBLIOGRAPHY

Archival Sources

Archivio di Stato di Venezia (ASV)

Duca di Candia (DDC), the records of the court of the Duke of Candia

Notai di Candia (NDC), the records of the Venetian notary

Manuscript Sources

ʿAbd al-Raḥmān b. Abī Yūsuf. *Kitāb Ḥifẓ al-ṣiḥḥa* (Book on the Preservation of Health). Istanbul Ayasofya, 3635.

Afendopolo, Caleb. *Peirush ʿal ha-Aritmeṭiqah shel Niqomaḵus* (Commentary on the *Arithmetic* of Nicomachus). Berlin, State Library Or. Qu. 760 (Steinschneider 226).

Afendopolo, Caleb. *Seiḇer Kᵉli roḇaʿ ha-shaʿot* (The Book of the Horary Quadrant). St. Petersburg Academy of Sciences, Institute of Oriental Studies, B313, 75a–100a.

Ahī Çelebī. "Risāla fī al-ṭibb" (Epistle on Medicine). Istanbul, Lala İsmail 701, 29b–46b.

Albo, Joseph. *Seiḇer ha-ʿIqqarim* (The Book of Roots). Vatican, BAV Ebr. 257.

Anonymous. *Leqeṭ tᵉshuḇot posqim mi-ḥoḵmei Turkyah wᵉ-ereṣ Yisraʾeil* (Anthology of the Responsa of the Adjudicators of the Scholars of Turkey and the Land of Israel). New York, Jewish Theological Seminary of America Rab. 1429.

Anonymous. *Luḥot* (Tables). Munich, Bavarian State Library Hebrew 31, 346b–369a.

Anonymous. *Peirush ʿal ha-juyub* (Commentary on Sines). Translation by Moses Galeano (?). St. Petersburg, Russian National Library, Evr. 351, 22b–26a.

Anonymous. *Peirush ha-Torah ʿal dereḵ ha-qabbalah* (Qabbalistic Commentary on the Torah). Vatican, BAV ebr. 201.

Anonymous. *Qoḇeṣ ba-asṭronomyah* (Collection in Astronomy). Parma, Palatina Library 2111.

Anonymous. *Qoḇeṣ ba-madaʿim* (Collection in the Sciences). Budapest, Library of the Hungarian Academy of Sciences, Kaufmann A 507.

Anonymous. *Seiper ha-Zohar.* Toronto, Thomas Fisher Rare Book Library Digital Collections, Friedberg MSS 5-015, https://collections.library.utoronto.ca/view/fisher2%3A115.

Anonymous. *Taqwīm* (Almanac for 1445). Paris, BnF Turc. 180.

Anonymous. *Taqwīm* (Almanac for 1450). Florence, Biblioteca Medicea Laurenziana, Or. 27, 1a–6b. http://mss.bmlonline.it/s.aspx?Id=AWOS4cmzI1A4r7GxMdhY&c=Tabulae%20Astronomicae#/oro/13.

Arnaldo da Villanova. "Panim ba-mishpaṭ" (Aspects in Judgment). Translation by Solomon b. Avigdor. Oxford, Bodleian Reggio 13 (Neubauer 2028), 77a–82b.

ʿAṭūfī, al-. *Kitāb al-kutub* (The Inventory of the Books). Budapest, Hungarian Academy of Sciences Török F59.

Averroës. *ha-Beiʾur he-arok̲ la-maʾamar ha-ḥamishi la-sheimaʿ ha-ṭib̲ʿi* (Long Commentary on Aristotle's "*Physics*"). Translation by Qalonymos b. Qalonymos. Paris, BnF hébreu 883.

Balbo, Michael. "Iggeret ʿal ha-gilgul" (Epistle on Metempsychosis). Paris, BnF hébreu 800, 44a–46b.

Balbo, Michael. *Wikkuaḥ ʿal ha-gilgul* (Polemic on Metempsychosis). Vatican, BAV ebr. 105, 116b–149a.

Ben Shushan, David. *Tolᵉdot Adam* (Genealogy of Man). Oxford, Bodleian Canonici Or. I 7 and New York, JTSA 5475.

Cohen Ashkenazi, Moses. *Urim wᵉ-t̲ummim*. Vatican, BAV ebr. 393.

Cohen Ashkenazi, Moses. *Wikkuaḥ ʿal ha-gilgul* (Polemic on Metempsychosis). Vatican, BAV ebr. 254.

Delmedigo, Elijah. *Beiʾur "ʿEṣem ha-galgal"* (Commentary on *De substantia orbis*). Paris, BnF hébreu 968, 1b–74b.

Delmedigo, Elijah. *Dᵉrushim kᵉ-p̲i shorᵉshei ha-pilosop̲im* (Investigations in Accordance with the Principles of the Philosophers). Paris, BnF hébreu 968, 79a–177a.

Delmedigo, Elijah. *Epistolae nonnullae autographae*. Paris, BnF Lat. 6508, 71a–81a.

Delmedigo, Elijah. *Expositio Averrois "De substantia orbis."* Vatican, BAV lat. 4553, 1a–51a.

Duran, Profayṭ. "Beiʾur shᵉnei ha-qawwim she-einam nip̲gashim" (Explanation of Two Lines That Do Not Meet). Paris, BnF hébr. 1021, 67–68b.

Duran, Profayṭ. *Peirush al-Magisṭi* (Commentary on the *Almagest*). Paris, BnF hébreu 1026.

Emmanuel b. Jacob. *Sheish kᵉnap̲ayim* (Six Wings). Paris, BnF hébreu 1080.

Faji, Elijah al-. *Mik̲tab̲ Eliyahu* (Elijah's Letter). London, British Museum Add. 15454 (Margoliouth 1017).

Fārābī, al-. *Qiṣṣur mi-kol melʾek̲et ha-higgayon* (Compendium of the Oeuvre of Logic). Paris, BnF hébreu 898, 917.

Galeano, Moses b. Judah. *Ha-Dibbur bᵉ-mah she-kᵉlalo seip̲er ha-maʾamarot* (A Discussion of What the Book of *The Categories* Includes). Cambridge, Cambridge University Library Mm. 6.32.2.

Galeano, Moses b. Judah. *Melʾek̲et ha-higgayon* (The Art of Logic). Cambridge, Cambridge University Library Mm. 6.32.1

Galeano, Moses b. Judah. *Seip̲er ha-goralot* (The Book of Lots). Paris, BnF hébreu 1073, 31a–38b and 73a–77b.

Galeano, Moses b. Judah. *Taʿalumot ḥok̲mah* (Puzzles of Wisdom). Cambridge, Cambridge University Library Add. 511, 1.

Galeano, Moses b. Judah. *See also* Jālīnūs, Mūsā.

Ghazālī, Abū Ḥāmid al-. *Kawwanot ha-pilosopim* (Intentions of the Philosophers). Translated and annotated by Isaac Albalag. Moscow, Russian State Library, Günzburg, 335.

Ibn al-Shāṭir. *Nihāyat al-sūl fī taṣḥīḥ al-uṣūl* (The Ultimate Quest in the Rectification of the Hypotheses/Principles). Istanbul, Kadizade Mehmed Ef. 339.

Ibn al-Zarqāl. "Iggeret ha-maʿaseh ba-luaḥ ha-niqraʾ ṣapiḥa" (Treatise on How to Use a Universal Astrolabe). Translation by Jacob b. Makir and Moses Galeano. Paris, BnF hébreu 1021, 70b–88b.

Isaac b. Laṭīf. *Shaʿar shamayim* (Gate of Heaven). London, British Library Add. 27051.

Isaac b. Samuel Abū al-Khayr. *Peirush al-Fargani* (Commentary on al-Farghānī). New York, Jewish Theological Seminary of America, 9825.

Jābir b. Aflaḥ. *Qiṣṣur al-Magisṭi* (Compendium of the *Almagest*). Translation by Jacob b. Makir Ibn Tibbon. Originally published as *Iṣlāḥ al-Majisṭī* (Correction of the *Almagest*). Paris, BnF hébreu 1024.

Jālīnūs, Mūsā. *Kitāb al-Zīj* (The Astronomical Handbook with Tables). San Lorenzo de El Escorial, Escorial árabe, 966.

John of Gmünden. *Marʾit ha-kokabim* (The Appearance of the Stars). Translation by David Kalonymos b. Maestro Jacob. Paris, BnF hébreu 1051, 118a–126b.

Khāzinī al-. *Al-Zīj al-Sanjarī* (The Sanjarī Astronomical Handbook with Tables). London, British Library Or. 6669.

Kumaṭiano, Mordechai. *Ḥibbur bᵉ-matemaṭiqah* (Composition on Mathematics). Oxford, Bodleian Neubauer 2774.

Kumaṭiano, Mordechai. *Peirush "ha-Torah"* (Commentary on the "Torah"). St. Petersburg, the National Library of Russia Evr. I 51.

Kumaṭiano, Mordechai. *Peirush Luḥot Paras* (Commentary on the Persian Tables). Paris, BnF hébreu 1085, 1a–33b.

Kumaṭiano, Mordechai. *Peirush "Seiper ha-eḥad."* Paris, BnF hébreu 681, 79b–101a.

Kumaṭiano, Mordechai. "Seiper Tiqqun kᵉli ha-nᵉḥoshet" (Repairing the Astrolabe). Munich, Bavarian State Library Hebrew 36, 195b–203b.

Kumaṭiano, Mordechai. "Tiqqun kᵉli ṣapiḥa" (Repairing the Universal Astrolabe). Munich, Bavarian State Library Hebrew 36, 164b–173b.

Levi b. Gerson. *Milḥamot ha-Shem* (Wars of the Lord). Paris, BnF hébreu 724.

Lorqi, Joshua. *Gerem ha-maʿalot* (The Cause of the Degrees). Vienna, Austrian National Library Hebrew 45.

Maimonides, Moses. "Al-Maqāla al-Fāḍiliyya fī al-ṭibb" (Treatise on Poison and Antidotes). Transcribed and annotated by Mūsā Jālīnūs. Istanbul, Nuruosmaniye 3590, 150b–170a.

Māridīnī, Sibṭ al-. *Ḥibbur ʿal ha-robaʿ* (Treatise on the Quadrant). Translation by Moses Galeano. St. Petersburg, Russian National Library, Evr. 350, 14b–22a.

Mizraḥi, Elijah. *Peirush seiper al-Magisṭi* (Commentary on the *Almagest*). St. Petersburg, Institute of Oriental Studies, C128.

Muʾayyadzādah, ʿAbd al-Raḥmān. *Ḥawāshī min qibal ʿilm al-kalām* (Glosses on Philosophical Theology). Istanbul, Aya Sofya, 2283.

Qāḍī Zādah. *Sharḥ al-Mulakhkhaṣ li-l-Jaghmīnī* (Commentary on the *Mulakhkhaṣ* of al-Jaghmīnī). Qatar National Library, 9545.

Rāzī, Fakhr al-Dīn al-. *Al-Mabāḥith al-mashriqiyya* (Eastern Investigations). New York, Jewish Theological Seminary of New York, MS 9108.

Regiomontanus. "ʿEsrim wᵉ-shmoneh maḥanot ha-lᵉbanah li-shnat 1464" (Twenty-Eight Houses of the Moon for 1464). Vatican, BAV ebr. 387, 22a.

Samuel b. Benjamin. "ʿIr qᵉṭanah" (Small City). St. Petersburg Academy of Sciences, Institute of Oriental Studies B313, 1a–7a.

Sharbiṭ ha-Zahab. *Luḥot Paras* (Persian Tables). Paris, BnF hébreu 1042.

Printed Sources

Abū Maʿšar. *The Great Introduction to Astrology*. Edition and translation by Keiji Yamamoto and Charles Burnett. 2 volumes. Leiden: Brill, 2019.

Achillini, Alessandro. *De orbibus libri quattuor*. Bologna: Benedictus Hectoris, 1498.

Adamson, Peter. *Al-Kindī*. Oxford: Oxford University Press, 2007.

Adamson, Peter. "Al-Kindī and the Muʿtazila: Divine Attributes, Creation and Freedom." *Arabic Sciences and Philosophy* 13 (2003): 45–77.

Adamson, Peter. "On Knowledge of Particulars." *Proceedings of the Aristotelian Society*, n.s., 105 (2005): 257–78.

Adamson, Peter. *Philosophy in the Islamic World: A History of Philosophy without any Gaps*, vol. 3. Oxford: Oxford University Press, 2018.

Adıvar, Abdülhak Adnan. *Osmanlı Türklerinde İlim*. İstanbul: Maarif Matbaası, 1943.

Akasoy, Anna. "Mehmed II as a Patron of Greek Philosophy." In *The Renaissance and the Ottoman World*, edited by Anna Contadini and Claire Norton, 245–56. Farnham, England: Ashgate, 2013.

Akkach, Samer. *ʿAbd al-Ghani al-Nabulusi: Islam and the Enlightenment*. Oxford: OneWorld, 2007.

Albertus Magnus. *The Book of Minerals*. Translation by Dorothy Wyckoff. Oxford: The Clarendon Press, 1967.

Andreatta, Michela. "Subverting Patronage in Translation: Flavius Mithridates, Giovanni Pico Della Mirandola, and Gersonides' *Commentary on the 'Song of Songs.'*" In *Patronage, Production and Transmission of Texts in Medieval and Early Modern Jewish Cultures*, edited by Esperanza Alfonso and Jonathan Decter, 165–98. Turnhout: Brepols, 2014.

Anonymous. *The Book of Bahir: Flavius Mithridates' Latin Translation, the Hebrew Text, and an English Version*. Edition, translation, and introduction by Saverio Campanini. Turin: Nino Aragno, 2005.

Anonymous. *Pirke de Rabbi Eli'ezer: A Critical Edition*. Edition by C. M. Horowitz. Jerusalem: Makor Publishing, 1972.

Anonymous. *Seiper Tana dᵉbei Eliyahu*. Lublin, 1959–60.

Anonymous. *Taqqanot Qandiya*. Edition by Elias Artom and Umberto Cassuto. 2 volumes. Jerusalem: Mᵉqiṣei Nirdamim, 1943.

Anonymous. *The Zohar: Pritzker Edition*. Translation and commentary by Daniel C. Matt. 12 volumes. Stanford, CA: Stanford University Press, 2004.

Apollonius of Perga. *Treatise on Conic Sections*. Edition by Thomas L. Heath. New York: Cambridge University Press, 1896.

Appuhn, Karl. "Tools for the Development of the European Economy." In *A Companion to the Worlds of the Renaissance*, edited by Guido Ruggiero, 259–78. Oxford, UK: Blackwell Publishing, 2002.

Arbel, Benjamin. "Jews in International Trade: The Emergence of the Levantines and Ponentines." In *The Jews of Early Modern Venice*, edited by Robert Davis and Benjamin Ravid, 73–96. Baltimore, MD: Johns Hopkins University Press, 2001.

Arbel, Benjamin. "Notes on the Delmedigo of Candia." In *Non solo verso Oriente*, edited by M. Del Bianco Cotrozzi, R. Di Segni, and M. Massenzio, 119–30. Florence: Leo Olschki, 2014.

Aristotle. *Aristotelis Metaphysicorum Libri XIIII cum Averrois Cordubensis in eosdem commentariis*. Venice: Junctas, 1562.

Aslanian, Sebouh D. *From the Indian Ocean to the Mediterranean: The Global Trade Networks of Armenian Merchants from New Julfa*. Berkeley: University of California Press, 2011.

Attias, Jean-Christophe. *Le commentaire biblique: Mordekhai Komtino ou l'herméneutique du dialogue*. Paris: Éditions du cerf, 1991.

Augustine. *The City of God*. Translation by Marcus Dods and introduction by Thomas Merton. New York: Random House, 1950.

Averroës. *Avveroè: Parafrasi della "Repubblica" nella traduzione latina di Elia del Medigo*. Edition by A. Coviello and P. E. Fornaciari. Florence: Olschki, 1992.

Averroës. *Averrois Cordubensis commentum magnum super libro De celo et mundo Aristotelis*. Edition by F. J. Carmody with Rüdiger Arnzen. Preface by Gerhard Endreß. 2 volumes. Leuven: Peeters, 2003.

Averroës. *Kitāb al-Āthār al-ʿulwiyya*. Edition by Suhayr Faḍl Allāh and Suʿād ʿAlī ʿAbd al-Rāziq. Annotated by Zaynab Muḥammad al-Khuḍayrī. Cairo: Supreme Council of Culture, 1994.

Averroës. *Long Commentary on the "De anima" of Aristotle*. Translation and introduction by Richard Taylor with Thérèse-Anne Druart. New Haven, CT: Yale University Press, 2009.

Averroës. *Middle Commentary on Aristotle's "De anima."* Edition and translation by Alfred Ivry. Provo, UT: Brigham Young University Press 2010.

Averroës. *Rasāʾil Ibn Rushd al-ṭibbiyya*. Edition by Georges Anawati and Said Zayid. Cairo: al-Hayʾa al-Miṣriyya al-ʿāmma li-l-kitāb, 1987.

Averroës. *Tafsīr Mā baʿd aṭ-ṭabīʿa* (Long Commentary on Aristotle's "Metaphysics"). Edition by Maurice Bouyges, S. J. 4 volumes. Beirut: Dar el-Machreq, 1990.

Averroës. *Talkhīṣ al-Safasṭa*. Edition by Muḥammad Salīm Sālim. Cairo: Maṭbaʿat Dār al-kutub, 1973.

Averroës, and Moses Narboni. *Epistle on the Possibility of Conjunction*. Edition, translation, and annotations by Kalman Bland. New York: The Jewish Theological Seminary, 1982.

Avicenna. *The Canon of Medicine of Avicenna*. Translation by O. Cameron Gruner. New York: AMS Press, 1973.

Avicenna. *Metaphysics of the Healing*. Translation by Michael Marmura. Provo: Brigham Young University Press, 2005.

Azzolini, Monica. "Are the Stars Aligned? Matchmaking and Astrology in Early Modern Italy." *Isis* 112, no. 4 (2021): 766–75.

Azzolini, Monica. "Reading Health in the Stars: Politics and Medical Astrology in Renaissance Milan." In *Horoscopes and Public Spheres: Essays on the History of Astrology*, edited by Günther Oestmann, H. Darrel Rutkin, and Kocku von Stuckrad, 183–205. Berlin: Walter de Gruyter, 2005.

Balıkçıoğlu, Efe. *Verifying the Truth on Their Own Terms: Ottoman Philosophical Culture and the Court Debate between Zeyrek and Ḫocazāde*. Venice: Edizioni Ca' Foscari, 2023.

Bardi, Alberto. "Bessarione a lezione di astronomia di Cortasmeno." *Byzantinische Zeitschrift* 111, no. 1 (2018): 1–38.

Bardi, Alberto. "The *Paradosis* of the Persian Tables: A Source on Astronomy between the Ilkhanate and the Eastern Roman Empire." *Journal for the History of Astronomy* 49, no. 2 (2018): 239–60.

Bardi, Alberto. "Persiche Astronomie in Byzanz." PhD Diss., Ludwig-Maximilians-Universität München, 2017.

Bardi, Alberto. "The Reception and Rejection of 'Foreign' Astronomical Knowledge in Byzantium." In *Finding, Inheriting or Borrowing? The Construction and Transfer of Knowledge in Antiquity and the Middle Ages*, edited by Jochen Althoff, Dominik Berrens, and Tanja Pommerening, 167–84. Bielefeld: Transcript Verlag; New York: Columbia University Press, 2019.

Bardi, Alberto. "Scientific Interactions in Colonial, Multilinguistic and Interreligious Contexts: Venetian Crete and the Manuscript *Marcianus latinus* VII.31 (2614). A Preliminary Study." *Centaurus* 63 (2021): 339–52.

Barkai, Ron. "Between East and West: A Jewish Doctor from Spain." In *Intercultural Contacts in the Medieval Mediterranean: Studies in Honour of David Jacoby*, edited by Benjamin Arbel, 49–63. New York: Routledge, 1996, reprinted 2012.

Barker, Peter. "Albert of Brudzewo's *Little Commentary on George Peurbach's 'Theoricae Novae Planetarum.'*" *Journal for the History of Astronomy* 44 (2013): 125–48.

Barker, Peter. "Copernicus and the Critics of Ptolemy." *Journal for the History of Astronomy* 30 (1999): 343–58.

Barker, Peter. "The Reality of Peurbach's Orbs: Cosmological Continuity in Fifteenth and Sixteenth Century Astronomy." In *Change and Continuity in Early Modern Cosmology*, edited by Patrick J. Boner, 7–32. Dordrecht: Springer, 2011.

Barker, Peter, and Matjaž Vesel. "Goddu's Copernicus: An Essay Review of André Goddu's *Copernicus and the Aristotelian Tradition*." *Aestimatio* 9 (2012): 304–36.

Bartòla, Alberto. "Eliyahu Del Medigo e Giovanni Pico Della Mirandola." *Rinascimento* 33 (1993): 253–79.

Bartolocci, Giulio. *Bibliotheca magna rabbinica de scriptoribus*. 5 volumes. Rome: Sacrae Congregationis de Propaganda Fide, 1675–94.

Barton, Tamsyn. *Ancient Astrology*. Oxford: Routledge, 1994.

Barzilay, Isaac. *Yoseph Shlomo Delmedigo, Yashar of Candia: His Life, Works and Times*. Leiden: Brill, 1973.

Bashyatchi, Elijah (with Caleb Afendopolo). *Adderet Eliyahu*. Ramleh: The Council of the Qaraite Jewish Community in Israel, 1966.

Bayat, Ali Haydar. "Ahī Çelebi, Mehmed." In *Türkiye Diyanet Vakfı İslâm Ansiklopedisi*, 1:528–29. İstanbul: Türkiye Diyanet Vakfı, İslâm Ansiklopedisi Genel Müdürlüğü, 1988.

Bayḍāwī, Nāṣir al-Dīn al-. *Tafsīr al-Bayḍāwī al-musammā anwār al-tanzīl wa-asrār al-taʾwīl*. 2 volumes. Beirut: Dār al-kutub al-ʿilmiyya, 1988.

Bellosta, Hélène, and Bernard Heyberger. "Abraham Ecchellensis et *Les Coniques* d'Apollonius: Les enjeux d'une traduction." In *Orientalisme, Science, et Controverse: Abraham Ecchellensis (1605–1664)*, edited by Bernard Heyberger, 191–201. Turnhout: Brepols, 2010.

Bellver Martinez, José. "El Lugar del Iṣlāḥ al-Maŷisṭī de Ŷābir b. Aflaḥ en la llamada 'rebelión andalusí' contra la astronomía ptolemaica." *Al-Qanṭara* 30, 1 (2009): 83–136.

Belo, Caterina. "Averroes on God's Knowledge of Particulars." *Journal of Islamic Studies* 17, no. 2 (2006): 177–99.

Benayahu, Meir. *Rabbi Elijah Capsali of Crete.* (Heb.) Tel Aviv: Tel Aviv University Press, 1983.

Benayahu, Meir. "Rabbi Joseph Hamon—The Physician and Counselor of Sultan Selim and His Companion in His Advance on Egypt, the Land of Israel, and Syria." (Heb.) In *Nation and History: Studies in the History of the Jewish People*, edited by Menachem Stern, 281–87. Jerusalem: The Zalman Shazar Center for the Furtherance of Study of Jewish History, 1983.

Bercovy, David. "La famille Delmedigo." *Révue d'histoire de la médecine hébraïque* 22 (1969): 13–20.

Berzon, Todd S. *Classifying Christians.* Berkeley: University of California Press, 2016.

Binbaş, İ. Evrim. *Intellectual Networks in Timurid Iran: Sharaf al-Dīn ʿAlī Yazdī and the Islamicate Republic of Letters.* Cambridge: Cambridge University Press, 2016.

Bisaha, Nancy. "European Cross-Cultural Contexts before Copernicus." In *Before Copernicus: The Cultures and Contexts of Scientific Learning in the Fifteenth Century*, edited by Rivka Feldhay and F. Jamil Ragep, 29–41. Montreal, QC: McGill-Queen's University Press, 2017.

Biṭrūjī, Nūr al-Dīn al-. *On the Principles of Astronomy.* Edition, translation, and commentary by Bernard R. Goldstein. 2 volumes. New Haven, CT: Yale University Press, 1971.

Bland, Kalman. "Elijah Del Medigo, Unicity of Intellect, and Immortality of Soul." *Proceedings of the American Academy for Jewish Research* 61 (1995): 1–22.

Bland, Kalman. "Elijah del Medigo's Averroist Response to the Kabbalahs of Fifteenth-Century Jewry and Pico della Mirandola." *Journal of Jewish Thought and Philosophy* 1 (1991): 23–53.

Blåsjö, Viktor. "A Critique of the Arguments for Marāgha Influence upon Copernicus." *Journal for the History of Astronomy* 45, no. 2 (2014):183–95.

Blåsjö, Viktor. "A Rebuttal of Recent Arguments for Maragha Influence on Copernicus." *Historia Scientiarum* 17 (2018): 479–97.

Bloom, Harold. *The Anxiety of Influence: A Theory of Poetry.* Oxford: Oxford University Press, 1973; reprinted 1997.

Bobzin, Hartmut. "Guillaume Postel (1510–1581) und die Terminologie der arabischen Nationalgrammatik." In *Studies in the History of Arabic Grammar II: Proceedings of the Second Symposium on the History of Arabic Grammar*, Nijmegen, April 27–May 1, 1987, edited by Kees Versteegh and Michael G. Carter, 57–71. John Benjamins Publishing Company, 1990.

Bos, Gerrit. *Novel Medical and General Hebrew Terminology from the 13th Century.* Leiden: Brill, 2019.

Bowman, Steven. *The Jews of Byzantium.* Tuscaloosa: University of Alabama Press, 1985.

Brentjes, Sonja. "Cross-Cultural Exchanges in the Mediterranean." *Journal for the History of Astronomy* 42, no. 3 (2011): 411–14.

Brentjes, Sonja. "Courtly Patronage of the Ancient Sciences in Post-Classical Islamic Societies." *Al-Qanṭara* 29, no. 2 (2008): 403–36.

Brentjes, Sonja. "Early Modern Western European Travellers in the Middle East and Their Reports about the Sciences." In *Sciences, techniques et instruments dans le monde iranien*, edited by Nasrollah Pourjavady and Živa Vesel, 379–420. Tehran: Presses universitaires d'Iran, 2004.

Brentjes, Sonja. "Patronage of the Mathematical Sciences in Islamic Societies." In *The Oxford Handbook of the History of Mathematics*, edited by Eleanor Robson, 301–27. Oxford: Oxford University Press, 2009.

Brentjes, Sonja. *Travellers from Europe in the Ottoman and Safavid Empires, 16th-17th Centuries: Seeking, Transforming, Discarding Knowledge*. Farnham: Ashgate, 2010.

Brethren of Purity. *On "Astronomia": An Arabic Critical Edition and English Translation of EPISTLE 3*. Edition and translation by F. Jamil Ragep and Taro Mimura. Oxford: Oxford University Press in association with The Institute of Ismaili Studies, 2015.

Bullard, Melissa Meriam. "The Inward Zodiac: A Development in Ficino's Thought on Astrology." *Renaissance Quarterly* 3 (1990): 687–708.

Burman, Thomas. *Reading the Qurʾān in Latin Christendom*. Philadelphia: University of Pennsylvania Press, 2009.

Cantera Burgos, Francisco. *Abraham Zacut Siglo XV*. Madrid: M. Aguilar, 1935.

Capsali, Elijah. *Seider Eliyahu Zuṭa*. Edition by Aryeh Shmuelevitz. 2 volumes. Jerusalem: Ben Tzvi Institute, 1975.

Carpi, Daniel. *L'Individuo e la collettività*. Florence: Leo Olschki, 2002.

Cassuto, Umberto. *Gli ebrei a Firenze nell'età del Rinascimento*. Florence: Tipografia Galletti e Cocci, 1918; reprinted Florence: Leo Olschki, 1965.

Cassuto, Umberto. *I manoscritti palatini ebraici della Biblioteca Apostolica Vaticana e la loro storia*. Vatican City: Biblioteca Apostolica Vaticana, 1935.

Casulleras, Josep. "Mathematical Astrology in the Medieval Islamic West." *Zeitschrift für Geschichte der arabisch-islamischen Wissenschaften* 18 (2009): 241–68.

Casulleras, Josep, and Jan Hogendijk. "Progressions, Rays and Houses in Medieval Islamic Astrology: A Mathematical Classification." *Suhayl* 11 (2012): 33–102.

Caudano, Anne-Laurence. "Astronomy and Astrology." In *A Companion to Byzantine Science*, edited by Stavros Lazaris, 202–30. Leiden: Brill, 2020.

Chabás, José. "The Astronomical Tables of Jacob Ben David Bonjorn." *Archive for History of Exact Sciences* 42 (1991): 279–314.

Chabás, José. "Astronomy for the Court in the Early Sixteenth Century Alfonso de Córdoba and His Tabule Astronomice Elisabeth Regine." *Archive for History of Exact Sciences* 58 (2004): 183–217.

Chabás, José. "Interactions between Jewish and Christian Astronomers in the Iberian Peninsula." In *Science in Medieval Jewish Cultures*, edited by Gad Freudenthal, 147–54. Cambridge: Cambridge University Press, 2011.

Chabás, José, and Bernard R. Goldstein. *The Astronomical Tables of Giovanni Bianchini*. Leiden: Brill, 2009.

Chabás, José, and Bernard R. Goldstein. *Essays on Medieval Computational Astronomy*. Leiden: Brill, 2015.

Chabás, José, and Bernard R. Goldstein. "Transmission of Computational Methods within the Alfonsine Corpus: The Case of the Tables of Nicholaus de Heybech." *Journal for the History of Astronomy* 39 (2008): 345–55.

Chabás, José, and Antoni Roca. "Early Printing of Astronomy: The *Lunari* of Bernat de Granollachs." *Centaurus* 40, no. 2 (1998): 124–34.

Chajes, J. H. *The Kabbalistic Tree*. University Park: Pennsylvania State University Press, 2022.

Chajes, J. H. "Spheres, Sefirot, and the Imaginal Astronomical Discourse of Classical Kabbalah." *Harvard Theological Review* 113, no. 2 (2020): 230–62.

Charette, François. *Mathematical Instrumentation in Fourteenth-Century Egypt and Syria: The Illustrated Treatise of Najm al-Dīn al-Miṣrī*. Boston: Brill, 2003.

Ciscato, Antonio. *Gli Ebrei in Padova (1300–1800)*. Padua: Società Cooperativa Tipografica, 1901; reprinted Arnaldo Forni, 2004.

Cohen Ashkenazi, Saul. *Sh*ᵉʾ*eilot* (Questions). Venice: Juan DiGara, 1574.

Comes, M. "The Possible Scientific Exchange between the Courts of Hulaghu of Maragha and Alphonse 10th of Castille." In *Sciences, techniques et instruments dans le monde iranien*, edited by N. Pourjavady and Ž. Vesel, 29–50. Tehran: Presses universitaires d'Iran, 2004.

Conforte, David. *Qoreiʾ ha-dorot*. Edited by D. Cassel. Berlin: Abraham b. Asher & Co., 1846.

Cook, Harold J. *Matters of Exchange: Commerce, Medicine and Science in the Dutch Golden Age*. New Haven, CT: Yale University Press, 2007.

Cooper, Glen. "Astrology: The Science of Signs in the Heavens." In *The Oxford Handbook to Science and Medicine in the Classical World*, edited by P. T. Keyser and J. Scarborough, 381–407. Oxford: Oxford University Press, 2018.

Cooper, Glen. "Galen and Astrology: A *Mésalliance*?" *Early Science and Medicine* 16 (2011): 120–46.

Copernicus, Nicholas. *Nicolaus Copernicus Gesamtausgabe*. Edited by Heribert Nobis. 9 volumes. Hildesheim: H. A. Gerstenberg, 1974–.

Corazzol, Giacomo. "Gli ebrei a Candia nei secoli XIV-XVI: l'impatto dell'immigrazione sulla cultura della comunità locale." PhD Diss., Università di Bologna and École Pratique des Hautes Études, 2015.

Costil, Pierre. "Le Mécénat humaniste des Fugger I." *Humanisme et Renaissance* 6 (1939): 20–40.

Coudert, Allison. "Christian Kabbalah." In *Jewish Mysticism and Kabbalah: New Insights and Scholarship*, edited by Frederick E. Greenspahn, 159–74. New York: NYU Press, 2011.

Dan, Joseph. "Christian Kabbalah from Mysticism to Esotericism." In *Jewish Mysticism*, edited by Joseph Dan, 191–207. Northvale, NJ: Jason Aronson, 1998–99.

David, Abraham. "Moses Esrim ve-Arba." In *Encyclopaedia Judaica*, 2nd ed., edited by Michael Berenbaum and Fred Skolnik, 556. Vol. 14. Detroit, MI: Macmillan Reference USA, 2007.

David, Abraham. "New Information about Caleb Afondopolo." (Heb.) *Qiryat Seiper* 48 (1973): 180.

David, Abraham. "Taʿalumot ḥokmah by R. Moses ben Judah Galeano." (Heb.) *Qiryat Seiper* 63 (1990): 1338–40.

Davidson, Herbert. *Alfarabi, Avicenna, and Averroes on Intellect: Their Cosmologies, Theories of Active Intellect and Theories of Human Intellect*. Oxford: Oxford University Press, 1992.

Davidson, Herbert. "John Philoponus as a Source of Medieval Islamic and Jewish Proofs of Creation." *Journal of the American Oriental Society* 89, no. 2 (1969): 357–91.

de Abano, Petrus. *Expositio problematum Aristotelis cum textu*. 2 volumes. Venice: Johannes Herbort, 1482.

De Balmes, Abraham Ben Meir. *The Letter of Ten*. Edition and introduction by Raphael Kohen. Jerusalem: n.p., 1998.

de Lange, Nicholas. "Abraham Ibn Ezra and Byzantium." In *Abraham Ibn Ezra y Su Tiempo: Actas Del Simposio Internacional*, edited by Fernando Díaz Esteban, 181–92. Madrid: Asociación Española de Orientalistas, 1990.

de Lange, Nicholas. "Hebrew Scholarship in Byzantium." In *Hebrew Scholarship and the Medieval World: Festschrift for Raphael Loewe*, edited by Nicholas de Lange, 23–37. Cambridge: Cambridge University Press, 2001.

De Young, Gregg "Sibṭ al-Māridīnī: Muḥammad b. Muḥammad b. Aḥmad Abū ʿAbd Allāh Badr [Shams] al-Dīn al-Miṣrī." In *The Biographical Encyclopedia of Astronomers, Springer Reference*, edited by Thomas Hockey et al., 1058. New York: Springer, 2007.

Delmedigo, Elijah. *Examen de la religion*. Translation by Maurice. R. Hayoun. Paris: *Éditions de Cerf*, 1992.

Demaitre, Luke E. *Doctor Bernard de Gordon: Professor and Practitioner*. Toronto: Pontifical Institute of Mediaeval Studies, 1980.

Di Bono, Mario. "Copernicus, Amico, Fracastoro, and Ṭūsī's Device: Observations on the Use and Transmission of a Model." *Journal for the History of Astronomy* 26 (1995): 133–54.

Di Bono, Mario. *Le sfere omocentriche di Giovan Battista Amico nell'astronomia del cinquecento con il testo de "De motibus corporum coelestium."* Genoa: CNR - Centro di studio sulla storia della tecnica, presso l'Università degli studi, 1990.

Di Segni, Diana. "Early Quotations from Maimonides's *Guide of the Perplexed* in the Latin Middle Ages." In *Interpreting Maimonides: Critical Essays*, edited by Charles Manekin and Daniel Davies, 190–207. Cambridge: Cambridge University Press, 2018.

Dietrich, A., and E. Wiedemann. "Maghnāṭīs." In *Encyclopaedia of Islam New Edition Online (EI-2 English)*, edited by P. Bearman et al. Leiden: Brill, 2012.

Dukas, Jules. *Recherches sur l'histoire littéraire du quinzième siècle*. Paris: Léon Techener, 1876.

Dursteler, Eric. *Venetians in Constantinople: Nation, Identity, and Coexistence in the Early Modern Mediterranean*. Baltimore, MD: Johns Hopkins University Press, 2006.

Eamon, William. "Court, Academy, and Printing House: Patronage and Scientific Careers in Late Renaissance Italy." In *Patronage and Institutions: Science, Technology, and Medicine at the European Court, 1500–1750*, edited by Bruce T. Moran, 25–50. Suffolk, UK: Boydell Press, 1991.

Eisenmann, Esti. "Scientific Aspects in Mordechai Kumaṭiano's *Commentary on 'The Guide of the Perplexed.'*" (Heb.) *Peʿamim* 148 (2016–7): 95–116.

Elior, Ofer. "Rabbi Yedidyah Rakh on Ezekiel's 'I Heard': A Case Study in Byzantine Jews' Reception of Spanish-Provençal Jewish Philosophical-Scientific Culture." In *Texts in*

Transit in the Medieval Mediterranean, edited by Y. Tzvi Langermann and Robert Morrison, 29–46. University Park: Pennsylvania State University Press, 2016.

Endreß, Gerhard. "Averroes' *De Caelo*: Ibn Rushd's Cosmology in His Commentaries on Aristotle's *On the Heavens*." *Arabic Sciences and Philosophy* 5 (1995): 9–49.

Engel, Michael. *Elijah Delmedigo and Paduan Aristotelianism*. London: Bloomsbury Academic, 2017.

Erismann, Christophe. "Logic in Byzantium." In *The Cambridge Intellectual History of Byzantium*, edited by Anthony Kaldellis and Niketas Siniossoglou, 362–80. Cambridge University Press, 2017.

Euclid. *The Thirteen Books of The Elements*. Translation, introduction, and commentary by Thomas L. Heath. 3 volumes. New York: Dover Publications, 1956.

Evans, James, and J. Lennart Berggren. *Geminos's Introduction to the Phenomena: A Translation and Study of a Hellenistic Survey of Astronomy*. Princeton, NJ: Princeton University Press, 2007.

Fahd, T. "Ḥurūf." In *Encyclopaedia of Islam New Edition Online (EI-2 English)*, edited by P. Bearman. Leiden: Brill, 2012.

Fahd, Taoufic. *La divination arabe*. Leiden: E. J. Brill, 1966.

Fārābī, al-. *Al-Fārābī's Commentary and Short Treatise on Aristotle's "De Interpretatione."* Translation, introduction, and annotations by F. W. Zimmermann. London: Oxford University Press for the British Academy, 1981.

Fazlıoğlu, İhsan. "Hendese: Osmanlı Dönemi." In *Türkiye Diyanet Vakfı İslâm Ansiklopedisi*, 17:199–208. İstanbul: Türkiye Diyanet Vakfı, İslâm Ansiklopedisi Genel Müdürlüğü, 1988.

Fazlıoğlu, İhsan. "Osmanlı felsefe-biliminin arkaplanı: Semerkand matematik-astronomi okulu." *Dîvân İlmî Araştırmalar* 14, no. 1 (2003): 1–66.

Feldhay, Rivka, with F. Jamil Ragep. "Introduction." In *Before Copernicus: The Cultures and Contexts of Scientific Learning in the Fifteenth Century*, edited by Rivka Feldhay and F. Jamil Ragep, 3–13. Montreal, QC: McGill-Queens University Press, 2017.

Fenton, Paul. "Joseph Ibn Waqār et sa tentative de concilier mystique et philosophie." In *Mystique et philosophie dans les trois monothéismes*, edited by Danielle Cohen-Levinas, Géraldine Roux, and Meryem Sebti, 319–39. Paris: Hermann Éditeurs, 2015.

Ferrari, L. C. "Augustine and Astrology." *Laval théologique et philosophique* 33, no. 3 (1977): 241–51.

Ficino, Marsilio. *The Book of Life (Liber de vita)*. Translation by Charles Boer. University of Dallas, TX: Spring Publications, 1980.

Ficino, Marsilio. *The Letters of Marsilio Ficino*, vol. 7. Translation by members of the Language Department of the School of Economic Science, London. London: Shepheard-Walwyn, 1975.

Findlen, Paula. "The Economy of Scientific Exchange in Early Modern Italy." In *Patronage and Institutions: Science, Technology and Medicine at the European Court 1500–1700*, edited by Bruce T. Moran, 5–24. Suffolk, UK: The Boydell Press, 1991.

Fleischer, Cornell H. "Seer to the Sultan: Haydar-i Remmal and Sultan Süleyman." *Cultural Studies* 1 (2001): 290–99.

Fleming, Barbara, F. Babinger, and Christine Woodhead. "Ṭāshköprüzāde." In *Encyclopaedia of Islam New Edition Online*, edited by P. Bearman et al. Leiden: Brill, 2012.

Forin, E. Martellozzo. "A Padova col Caiado: Marco Fugger e Pandolfo Rem (1504–8)." *Quaderni per la Storia dell'Università di Padova* 4 (1971): 76–97.

Fornaciari, Paolo Edoardo. "Elementi di Astrologia nelle *Conclusiones* di Giovanni Pico Della Mirandola." In *Nella luce degli astri: L'astrologia nella cultura del Rinascimento*, edited by Ornella Pompeo Faracovi, 25–37. La Spezia: Agorà Edizioni, 2004.

Fracastoro, Girolamo. *The Sinister Shepherd: A Translation of Girolamo Fracastoro's "Syphilidis sive De Morbo Gallico libri tres."* Translation by William van Wyck. Los Angeles: The Primavera Press, 1934.

Fracastoro, Girolamo. *Syphilis sive morbus gallicus*. Edition by Christine Dussin. Paris: Éditions classiques Garnier, 2009.

Fracastoro, Hieronymus. *Homocentrica: Eiusdem de causis criticorum dierum per eaquae in nobis sunt*. Venice, 1538.

Fraenkel, Carlos. "Considering the Case of Elijah Delmedigo's Averroism and Its Impact on Spinoza." In *Renaissance Averroism and Its Aftermath: Arabic Philosophy in Early Modern Europe*, edited by Anna Akasoy and Guido Giglioni, 213–36. Dordrecht: Springer, 2013.

Frank, Daniel. "Karaite Exegetical and Halakhic Literature in Byzantium and Turkey." In *Karaite Judaism: A Guide to Its History and Literary Sources*, edited by Meira Polliack, 529–58. Leiden: Brill, 2003.

Friedlander, Gerald. *Pirḳê De Rabbi Eliezer*. London: Kegan Paul; and New York: The Bloch Publishing Company, 1916.

Galante, Avram. *Histoire des juifs de Turquie*. 9 volumes. Istanbul: Éditions Isis/Isis Yayıncılık, 1985.

Galante, Avram. *Türkler ve Yahudiler*. Istanbul: Gözlem, 1995.

García-Arenal, Mercedes. *A Man of Three Worlds: Samuel Pallache, A Moroccan Jew in Catholic and Protestant Europe*. Baltimore, MD: Johns Hopkins University Press, 2007.

García-Ballester, Luis, Lola Ferre, and Eduard Feliu. "Jewish Appreciation of Fourteenth-Century Scholastic Medicine." *Osiris*, second series, 6 (1990): 85–117.

Gardette, Philippe. "Judaeo-Provençal Astronomy in Byzantium and Russia." *Byzantinoslavica* 63 (2005): 195–209.

Garin, Eugenio. *Astrology in the Renaissance: The Zodiac of Life*. Translation by Carolyn Jackson and June Allen. Revised by Clare Robertson. London: Routledge and Kegan Paul, 1983.

Geffen, David. "Insights into the Life and Thought of Elijah Medigo Based on His Published and Unpublished Works." *Proceedings of the American Academy for Jewish Research* 41–42 (1973–4): 69–86.

Ghazālī, Abū Ḥāmid al-. *Maqāṣid al-falāsifa* (The Intentions of the Philosophers). Edition by Muḥyī al-Dīn al-Kurdī. Paris: Dār Bībliyyūn, 2005.

Giahi Yazdi, Hamid-Reza. "Naṣīr al-Dīn al-Ṭūsī on Lunar Crescent Visibility and an Analysis with Modern Altitude-Azimuth Criteria." *Suhayl* 3 (2002–3): 231–43.

Gigante, Federica. "A Medieval Islamic Astrolabe with Hebrew Inscriptions in Verona." *Nuncius* 39 (2024): 163–92.

Glasner, Ruth. *Averroës' Physics: A Turning Point in Medieval Natural Philosophy*. Oxford: Oxford University Press, 2009.

Glasner, Ruth. *Gersonides: A Portrait of a Fourteenth-Century Philosopher-Scientist*. Oxford: Oxford University Press, 2015.

Glasner, Ruth. "The Question of Celestial Matter in the Hebrew Encyclopedias." In *Medieval Hebrew Encyclopedias of Science and Philosophy*, edited by Steven Harvey, 313–34. Dordrecht: Kluwer Academic Publishers, 2000.

Goddu, André. "Birkenmajer's Copernicus: Historical Context, Original Insights, and Contributions to Current Debates." *Science in Context* 31, no. 2 (2018): 189–222.

Goddu, André. *Copernicus and the Aristotelian Tradition*. Leiden: Brill, 2010.

Goddu, André. "Ludwik Antoni Birkenmajer and Curtis Wilson on the Origin of Nicholas Copernicus's Heliocentrism." *Isis* 107 (2016): 225–53.

Goetschel, Roland. "Elie Ḥayyim de Genazzano et la Kabbale." *Revue des études juives* 92 (1983): 91–108.

Goitein, S. D. *A Mediterranean Society: The Jewish Communities of the Arab World as Portrayed in the Documents of the Cairo Geniza*. 6 volumes. (Berkeley: University of California Press, 1967).

Goldberg, Jessica. *Trade and Institutions in the Medieval Mediterranean*. Cambridge: Cambridge University Press, 2012.

Goldstein, Bernard R. "Abraham Zacut's Signature: A Mystery Solved." *Aleph* 11, no. 1 (2011): 159–67.

Goldstein, Bernard R. *The Astronomical Tables of Levi ben Gerson*. New Haven, CT: Transactions of the Connecticut Academy of Arts and Sciences, 1974.

Goldstein, Bernard R. "Astronomy among Jews in the Middle Ages." In *Science in Medieval Jewish Cultures*, edited by Gad Freudenthal, 136–46. Cambridge: Cambridge University Press, 2011.

Goldstein, Bernard R. "Astronomy and Astrology in the Works of Abraham Ibn Ezra." *Arabic Sciences and Philosophy* 6 (1996): 9–21.

Goldstein, Bernard R. "Astronomy and the Jewish Community in Early Islam." *Aleph* 1 (2001): 17–57.

Goldstein, Bernard R. "Astronomy in Hebrew in Istanbul: Abraham ben Yom Ṭov Yerushalmi (fl. 1510)." *Aleph* 20 (2020): 303–7.

Goldstein, Bernard R. *The Astronomy of Levi ben Gerson*. New York: Springer-Verlag, 1985.

Goldstein, Bernard R. "Levi ben Gerson and the Cross Staff Revisited." *Aleph* 11, no. 2 (2011): 365–83.

Goldstein, Bernard R. "Levi Ben Gerson's Astrology in Historical Perspective." In *Gersonide en son temps*, edited by Gilbert Dahan, 287–300. Louvain-Paris: E. Peeters, 1990.

Goldstein, Bernard R. "Levi ben Gerson's Contributions to Astronomy." In *Studies on Gersonides: A Fourteenth-Century Jewish Philosopher-Scientist*, edited by Gad Freudenthal, 3–19. Leiden: E. J. Brill, 1992.

Goldstein, Bernard R. "The Origins of Copernicus' Heliocentric System." *Journal for the History of Astronomy* 33 (2002): 219–35.

Goldstein, Bernard R. "The Survival of Arabic Astronomy in Hebrew." *Journal for the History of Arabic Science* 3 (1979): 31–39.

Goldstein, Bernard R., and José Chabás. *Astronomy in the Iberian Peninsula: Abraham Zacut and the Transition from Manuscript to Print*. Philadelphia: American Philosophical Society, 2000.

Goldstein, Bernard R., and José Chabás. "Isaac Ibn al-Ḥadib and Flavius Mithridates: The Diffusion of an Iberian Astronomical Tradition in the Late Middle Ages." *Journal for the History of Astronomy* 37 (2006): 147–72.

Goldstein, Bernard R., and José Chabás. "New Evidence on Abraham Zacut's Astronomical Tables." *Archive for History of Exact Sciences* 72 (2018): 21–62.

Goldstein, Bernard R., and José Chabás. "Ptolemy, Bianchini, and Copernicus: Tables for Planetary Latitudes." *Archive for History of Exact Sciences* 58 (2004): 453–73.

Goldstein, Bernard R., and David Pingree. "Levi ben Gerson's Prognostication for the Conjunction of 1345." *Transactions of the American Philosophical Society* 80, no. 6 (1990): 1–60.

Goldstein, Bernard R., and George Saliba. "A Hispano-Arabic Astrolabe with Hebrew Star Names." *Annali dell'Istituto e Museo di Storia della Scienza di Firenze* 8 (1983): 19–29.

Gómez-Aranda, Mariano. "Science and Jewish Identity in the Works of Abraham Zacut (1452–1515)." In *Late Medieval Jewish Identities: Iberia and Beyond*, edited by Carmen Caballero-Bavas and Esperanza Alfonso, 157–70. New York: Palgrave-MacMillan, 2010.

Gottlieb, Ephraim. "The Metempsychosis Controversy in Candia." (Heb.) *Seḇunot* 9 (1971–77): 43–66.

Grafton, Anthony. *Cardano's Cosmos: The Worlds and Works of a Renaissance Astrologer.* Cambridge: Harvard University Press, 1999.

Granada, Miguel, and Dario Tessicini. "Copernicus and Fracastoro: The Dedicatory Letters to Pope Paul III, the History of Astronomy, and the Quest for Patronage." *Studies in the History and Philosophy of Science* 36 (2005): 431–76.

Grant, Edward. *Planets, Stars, and Orbs.* Cambridge: Cambridge University Press, 1996.

Grätz, Heinrich. *The History of the Jewish People.* (Heb.) 8 volumes. Warsaw: n.p., 1897.

Greenberg, Moshe. "Urim and Thummim." In *Encyclopaedia Judaica*, 2nd ed., edited by Michael Berenbaum and Fred Skolnik, 422–23. Vol. 20. Detroit, MI: Macmillan Reference USA, 2007.

Grévin, Benoit. "Flavius Mithridate au travail sur le Coran." In *Flavio Mitridate mediatore*, edited by G. Corazzol and M. Perani, 26–46. Palermo: Officina di Studi Medievali, 2012.

Griffel, Frank. *Al-Ghazālī's Philosophical Theology.* Oxford: Oxford University Press, 2009.

Grupe, Dirk. "Stephen of Pisa's Theory of the Oscillating Deferents of the Inner Planets." *Archive for History of Exact Sciences* 71, no. 4 (2017): 379–407.

Guidi, Angela. "L'Astrologia nei Dialoghi d'Amore." In *Nella luce degli astri: L'astrologia nella cultura del Rinascimento*, edited by Ornella Pompeo Faracovi, 39–62. Sarzana: Agorà Edizioni, 2004.

Gurland, Hayyim Jonah. "The History of Rabbi Mordechai Kumaṭiano." (Heb.) Translation from Russian by Judah Levy Lewick. In *Talpiyyot*. Berditchev, 1895, 1–34.

Gutas, Dimitri. *Greek Thought, Arabic Culture.* London: Routledge, 1998.

Hacker, Joseph. "The Immigration of the Jews of Spain to the Land of Israel and Their Connection to It between 1391 and 1492." (Heb.) *Shaleim* 1 (1974): 105–56.

Hacker, Joseph. "The Intellectual Activity of the Jews of the Ottoman Empire during the Sixteenth and Seventeenth Centuries." In *Jewish Thought in the Seventeenth Century*, edited by Isadore Twersky and Bernard Septimus, 95–115. Cambridge, MA: Harvard University Press, 1987.

Hacker, Joseph. "Local Patriotism of Spanish Exiles in the Sixteenth-Century Ottoman Empire." (Heb.) In *Meiʾah Sheʿarim: Studies in Medieval Jewish Spiritual Life in Memory of*

Isadore Twersky, edited by E. Fleischer, G. Blidstein, C. Horowitz, and B. Septimus, 349–69. Jerusalem: Magnes Press, 2001.

Hacker, Joseph. "Mizraḥi, Elijah." In *Encyclopaedia Judaica*, 2nd ed., edited by Michael Berenbaum and Fred Skolnik, 393–95. Vol. 14. Detroit, MI: Macmillan Reference USA, 2007.

Hacker, Joseph. "The Rise of Ottoman Jewry." In *The Cambridge History of Judaism*. Vol. 7, *The Early Modern World*, edited by Jonathan Karp and Adam Sutcliffe, 77–112. Cambridge: Cambridge University Press, 2018.

Ḥalamish, Moshe. *An Introduction to the Kabbalah*. Albany: SUNY Press, 1999.

Halbertal, Moshe. *Maimonides: Life and Thought*. Princeton, NJ: Princeton University Press, 2014.

Halevi, Judah. *The Kuzari: An Argument for the Faith of Israel*. Translation by H. Hartwig. Introduction by H. Slominsky. New York: Schocken Books, 1964.

Ḥamawī, Yāqūt al-. *Muʿjam al-buldān*. 5 volumes. Beirut: Dār Ṣādir, n.d.

Hames, Harvey. "A Seal within a Seal: The Imprint of Sufism in Abraham Abulafia's Teachings." *Medieval Encounters* 12, no. 2 (2006): 159–72.

Harari, D. "Some Lost Writings of Judah Abravanel (1465?–1535?) Found in the Works of Giordano Bruno (1548–1600)." (Heb.) *Shoḇar* 10 (1992): 62–89.

Hartig, Otto. *Die Gründung der Münchner Hofbibliothek durch Albrecht V, und Johann Jakob Fugger*. Munich: Bavarian Academy of Sciences, 1917.

Hasan, Moiz. "Foundations of Science in the Post-Classical Islamic Era: The Philosophical Historical and Historiographical Significance of al-Sayyid al-Sharīf al-Jurjānī's Project." PhD Diss., University of Notre Dame, 2017.

Hasse, Dag N. "Arabic Philosophy and Averroism." In *Cambridge Companion to Renaissance Philosophy*, edited by James Hankins, 113–36. New York: Cambridge University Press, 2007.

Hasse, Dag N. "*Averroica Secta*: Notes on the Formation of Averroist Movements in Fourteenth-Century Bologna and Renaissance Italy." In *Averroes et les Averroïsmes juif et latin*, edited by J. B. Brenet, 307–31. Turnhout: Brepols, 2007.

Hasse, Dag N. *Success and Suppression: Arabic Sciences and Philosophy in the Renaissance*. Cambridge: Harvard University Press, 2016.

Hayton, Darin. *The Crown and the Cosmos: Astrology and the Politics of Maximilian I*. Pittsburgh, PA: University of Pittsburgh Press, 2015.

Hernández Pérez, Azucena. *Catálogo razonado de los astrolabios de la España medieval*. Madrid: Ergástula, 2018.

Hippolytus. *Refutatio omnium haeresium*. Edition by Miroslav Marcovich. Berlin: Walter de Gruyter, 1986.

Hogendijk, J. P. *Ibn al-Haytham's "Completion of the Conics."* New York: Springer-Verlag, 1985.

Holton, David. "The Cretan Renaissance." In *Literature and Society in Renaissance Crete*, edited by David Holton, 1–16. Cambridge: Cambridge University Press, 1991.

Horowitz, Yehoshua. "Balbo, Michael ben Shabbetai Cohen." In *Encyclopaedia Judaica*, 2nd ed., edited by Michael Berenbaum and Fred Skolnik, 84–85. Vol. 3. Detroit, MI: Macmillan Reference USA, 2007.

Howard, Deborah. "Cultural Transfer between Venice and the Ottomans." In *Cultural Ex-*

change in Early Modern Europe. Vol. 4, *Forging European Identities, 1400–1700*, edited by Robert Muchembled and William Monter, 138–77. Cambridge: Cambridge University Press, 2007.

Hughes, Aaron. "Transforming the Maimonidean Imagination: Aesthetics in the Renaissance Thought of Judah Abravanel." *Harvard Theological Review* 97, no. 4 (2004): 461–84.

Hyman, Arthur. *Averroes' "De substantia orbis": Critical Edition of the Hebrew Text with English Translation and Commentary*. Cambridge, MA: The Medieval Academy of America and the Israel Academy of Arts and Sciences, 1986.

Ibn al-Haytham. *Al-Shukūk ʿalā Baṭlamyūs* (Doubts about Ptolemy). Edition by A. I. Sabra and Nabil Shehaby. Cairo: Dār al-kutub, 1996.

Ibn Ezra, Abraham. *Abraham Ibn Ezra on Elections, Interrogations, and Medical Astrology: A Parallel Hebrew-English Critical Edition of the "Book of Elections" (3 Versions), the "Book of Interrogations" (3 Versions), and the "Book of the Luminaries."* Edition, translation, and annotations by Shlomo Sela. Leiden: Brill, 2011.

Ibn Ezra, Abraham. *Abraham Ibn Ezra on Nativities*. Edition, translation, and annotations by Shlomo Sela. Leiden: Brill, 2013.

Ibn Ezra, Abraham. *Abraham Ibn Ezra's Introductions to Astrology*. Edition, translation, and annotations by Shlomo Sela. Leiden: Brill, 2017.

Ibn Ezra, Abraham. *The Book of the World*. Edition, translation, and annotations by Shlomo Sela. Leiden: Brill, 2010.

Ibn Naḥmias, Joseph. *The Light of the World: Astronomy in Al-Andalus*. Edition, translation, and commentary by Robert Morrison. Berkeley: University of California Press, 2016.

Idel, Moshe. *Abraham Abulafia's Esotericism*. Berlin: De Gruyter, 2020.

Idel, Moshe. *Kabbalah in Italy*. New Haven, CT: Yale University Press, 2011.

Idel, Moshe. "Maimonides' *Guide of the Perplexed* and the Kabbalah." *Jewish History* 18 (2004): 197–226.

Idel, Moshe. *Mystical Experience in Abraham Abulafia*. Translation by Jonathan Chipman. Albany: State University of New York Press, 1988.

Idel, Moshe. "Mysticism: Kabbalah." In *New Dictionary of the History of Ideas*, edited by Maryanne Cline Horowitz, 4:1552–59. 6 volumes. New York: Charles Scribner's Sons, 2005.

İhsanoğlu, Ekmeleddin, ed. *Osmanlı Tıbbi Bilimler Literatürü Tarihi*. 4 volumes. Istanbul: Ircica, 2008.

İhsanoğlu, Ekmeleddin. "Some Remarks on Ottoman Science." In *Science, Technology, and Learning in the Ottoman Empire*, edited by Ekmeleddin İhsanoğlu, 45–73. Aldershot, UK: Ashgate, 2004.

Ījī, ʿAḍud al-Dīn al-. *Al-Mawāqif bi-Sharḥ al-Jurjānī* (The Stations in the Science of Theology with the Commentary of al-Jurjānī). Edition by ʿAbd al-Raḥmān ʿUmayra. 3 volumes. Beirut: Dār al-Jīl, 1997.

İlgürel, Mücteba. "Subaşı." In *Türkiye Diyanet Vakfı İslâm Ansiklopedisi*, 37:447–48. İstanbul: Türkiye Diyanet Vakfı, İslâm Ansiklopedisi Genel Müdürlüğü, 2009.

Ivry, Alfred. "Jewish Philosophy." In *The Cambridge History of Judaism*. Vol. 5, *Jews in the Medieval Islamic World*, edited by Philip Lieberman-Ackerman, 796–824. Cambridge: Cambridge University Press, 2021.

Ivry, Alfred. "Remnants of Jewish Averroism in the Renaissance." In *Jewish Thought in the*

Sixteenth Century, edited by Bernard Dov Cooperman, 243–65. Cambridge: Harvard University Center for Jewish Studies, 1983.

İzgi, Cevat. *Osmanlı Medreselerinde İlim*. 2 volumes. Istanbul: İz Yayıncılık, 1997.

İzgöer, Ahmet Zeki., ed. *16. Yüzyıl Osmanlı Tabibi Musa Bin Hamon ve Diş Tababetine Katkısı*. Istanbul: Merkezefendi Geleneksel Tıp Derneği, 2012.

Jacoby, David. "David Mavrogonato from Candia: Merchant, Lobbyist, and Jewish Spy." (Heb.) *Tarbiṣ* 33 (1964): 388–402.

Jacoby, David. "Jewish Physicians and Surgeons on Crete under Venetian Rule." (Heb.) *Tarbut we-ḥeb̲rah be-toldot Yisra'eil be-yemei ha-beinayyim* (1988–9): 431–44.

Jacoby, David. "Production et commerce de l'alun oriental en Méditerranée, XIe-XVe siècles." In *L'alun de Méditerranée*, edited by Philippe Borgard, Jean-Pierre Brun, and Maurice Picon, 219–67. Naples, Italy: Assessorato ai Beni Culturali, 2005.

Jacoby, David. "Un agent juif au service de Venise: David Mavrogonato de Candie." *Thesaurismata* 9 (1972): 68–96.

Janos, Damien. "Moving the Orbs: Astronomy, Physics, and Metaphysics, and the Problem of Celestial Motion According to Ibn Sīnā." *Arabic Sciences and Philosophy* 21 (2011): 165–214.

Jean de Jandun. *Quaestiones Joannis de Janduno de physico auditu: Noviter emendate: Helie hebrei Cretensis questiones: De primo motore, De efficientia mundi, De esse essentia [et] uno, Annotationes in plurima dicta co[m]mentatoris*. Venice, 1506.

Katz, Steven. "The Qabbalah." In *The Essential Agus: The Writings of Jacob B. Agus*, edited by Steven T. Katz, 192–206. New York: NYU Press, 2020.

Keim, Katharina. "*Pirqei de Rabbi Eliezer*: Structure, Coherence, Intertextuality, and Historical Context." PhD Diss., University of Manchester, 2014.

Kennedy, E. S., and Victor Roberts. "The Planetary Theory of Ibn al-Shāṭir." *Isis* 50 (1959): 227–35.

Kennedy, E. S., and George Saliba. "The Spherical Case of the Ṭūsī Couple." *Arabic Sciences and Philosophy* 1 (1991): 285–91.

Kieszkowski, Bohdan. "Les rapports entre Elie del Medigo et Pic de la Mirandole." *Rinascimento* 4 (1964): 41–91.

King, David. *In Synchrony with the Heavens*. Vol. 2, *Instruments of Mass Calculation*. Leiden: Brill, 2005.

King, David. "Spherical Astrolabes in Circulation: From Baghdad to Toledo and to Tunis and Istanbul." https://www.academia.edu/37957299/King_2018_Spherical_astrolabes_in_circulation_From_Baghdad_to_Toledo_and_to_Tunis_and_Istanbul.

King, Margaret L. "Bibliotheca graeca manuscripta cardinalis Dominici Grimani (1461–1523) (review)." *Renaissance Quarterly* 58 (2005): 1305–7.

Kohen, Elli. *History of the Turkish Jews and Sephardim: Memories of a Past Golden Age*. Lanham, MD: University Press of America, 2007.

Kotansky, Roy. "The Magic 'Crucifixion Gem' in the British Museum." *Greek, Roman, and Byzantine Studies* 57 (2017): 631–59.

Kozodoy, Maud. *The Secret Faith of Maestre Honoratus: Profayt Duran and Jewish Identity in Late Medieval Iberia*. Philadelphia: University of Pennsylvania Press, 2015.

Kremer, Richard L., Matthieu Husson, and José Chabás. "Introduction." In *Alfonsine As-*

tronomy: The Written Record, edited by Richard L. Kremer, Matthieu Husson and José Chabás, 7–17. Turnhout: Brepols, 2022.

Küçükdağ, Yusuf. "Muhyiddin Mehmed Şah." In *Türkiye Diyanet Vakfı İslâm Ansiklopedisi*, 31:84–85. İstanbul: Türkiye Diyanet Vakfı, İslâm Ansiklopedisi Genel Müdürlüğü, 2020.

Kumaṭiano, Mordechai. *An Ancient Commentary on "The Guide of the Perplexed."* (Heb.) Edition and introduction by Dov Schwartz and Esti Eisenmann. Ramat Gan: Bar Ilan University Press, 2016.

Kumaṭiano, Mordechai. *An Ancient Commentary on "Yesod Moraʾ."* (Heb.) Edition and introduction by Dov Schwartz. Ramat Gan: Bar Ilan University Press, 2010.

Kuntz, Marion. *Guillaume Postel: Prophet of the Restitution of All Things, His Life*. The Hague: Martinus Nijhoff, 1981.

Kupfer, Ephraim. "Comtino, Mordecai ben Eliezer." In *Encyclopaedia Judaica*, 2nd ed., edited by Michael Berenbaum and Fred Skolnik, 132–33. Vol. 5. Detroit, MI: Macmillan Reference USA, 2007.

Lacerenza, Giancarlo. "A Rediscovered Autograph Manuscript by Mordekay Finzi." *Aleph* 3 (2003): 301–25.

Langermann, Y. Tzvi. "Abraham Ibn Ezra." In *The Stanford Encyclopedia of Philosophy* (Winter 2018 Edition), edited by Edward N. Zalta. https://plato.stanford.edu/archives/win2018/entries/ibn-ezra/.

Langermann, Y. Tzvi. "Another Andalusian Revolt? Ibn Rushd's Critique of al-Kindī's *Pharmacological Computus*." In *The Enterprise of Science in Islam: New Perspectives*, edited by Jan P. Hogendijk and A. I. Sabra, 351–72. Cambridge, MA: MIT Press, 2003.

Langermann, Y. Tzvi. "The Astral Connections of Critical Days: Some Late Antique Sources Preserved in Hebrew and Arabic." In *Astromedicine, Astrology, and Medicine, East and West*, edited by Anna Akasoy, Charles Burnett, and Ronit Yoeli-Tlalim, 99–118. Florence: SISMEL edizioni del Galluzzo, 2008.

Langermann, Y. Tzvi. "A Compendium of Renaissance Science: *Taʿalumot ḥokmah* by Moses Galeano." *Aleph* 7 (2007): 285–318.

Langermann, Y. Tzvi. "Criticisms of Authority in the Writings of Moses Maimonides and Fakhr al-Dīn al-Rāzī." *Early Science and Medicine* 7, no. 3 (2002): 255–75.

Langermann, Y. Tzvi. "From My Notebooks: Medicine, Mechanics and Magic from Moses ben Judah Galeano's *Taʿalumot Ḥokmah*." *Aleph* 9, no. 1 (2009): 353–77.

Langermann, Y. Tzvi. "Gersonides on Astrology." In Levi Ben Gershom, *The Wars of the Lord*, translation and annotations by Seymour Feldman, 3:505–19. 3 volumes. Philadelphia: The Jewish Publication Society and New York and Jerusalem: The Jewish Theological Seminary of America, 1999.

Langermann, Y. Tzvi. "Gradations of Light and Pairs of Opposites: Two Theories and their Role in Abraham Bar Hiyya's Scroll of the Revealer." In *Texts in Transit in the Medieval Mediterranean*, edited by Y. Tzvi Langermann and Robert Morrison, 47–66. University Park: Pennsylvania State University Press, 2016.

Langermann, Y. Tzvi. *Ibn al-Haytham's "On the Configuration of the World."* New York: Garland Publishing, 1990; reprinted London: Routledge, 2017.

Langermann, Y. Tzvi. "Medieval Hebrew Texts on the Quadrature of the Lune." *Historia Mathematica* 23, no. 1 (1996): 31–53.

Langermann, Y. Tzvi. "Science in the Jewish Communities of the Byzantine Cultural Orbit: New Perspectives." In *Science in Medieval Jewish Cultures*, edited by Gad Freudenthal, 438–53. Cambridge: Cambridge University Press, 2011.

Langermann, Y. Tzvi. "The Scientific Writings of Mordekhai Finzi." *Italia* 7 (1988): 7–44.

Langermann, Y. Tzvi. "Some Astrological Themes in the Thought of Abraham ibn Ezra." In *Rabbi Abraham Ibn Ezra: Studies in the Writings of a Twelfth-Century Jewish Polymath*, edited by I. Twersky and J. M. Harris, 28–85. Cambridge, MA: Harvard University Press, 1993.

Langermann, Y. Tzvi, and Almog Kasher. "The Critical Notes of Muḥammad b. Ḥasan al-Nihmī on *The Guide of the Perplexed*." (Heb.) *Daʿat* 74–75 (2013): 237–66.

Lasker, Daniel. *From Judah Hadassi to Elijah Bashyatchi*. Leiden: Brill, 2008.

Lasker, Daniel. "Medieval Karaism and Science." In *Science in Medieval Jewish Cultures*, edited by Gad Freudenthal, 427–37. Cambridge: Cambridge University Press, 2011.

Lasker, Daniel J., Eli Citonne, Haggai Ben-Shammai, Joseph Elijah Heller, Leon Nemoy, and Shlomo Hofman. "Karaites." In *Encyclopaedia Judaica*, 2nd ed., edited by Michael Berenbaum and Fred Skolnik, 785–802. Vol. 11. Detroit, MI: Macmillan Reference USA, 2007.

Latour, Bruno. *We Have Never Been Modern*. Translation by Catherine Porter. Cambridge, MA: Harvard University Press, 1993.

Lauer, Rena. *Colonial Justice and the Jews of Venetian Crete*. Philadelphia: University of Pennsylvania Press, 2019.

Lauer, Rena. "Cretan Jews and the First Sephardic Encounter in the Fifteenth Century." *Mediterranean Historical Review* 27, no. 2 (2012): 129–40.

Lauer, Rena. "Venice's Colonial Jews: Community, Identity, and Justice in Late Medieval Venetian Crete." PhD Diss., Harvard University, 2014.

Lawee, Eric. *Isaac Abarbanel's Stance toward Tradition: Defense, Dissent, and Dialogue*. Albany: State University of New York Press, 2001.

Lay, Juliane. "*L'Abrégé de l'Almageste*: Un inédit d'Avveroès en version hébraïque." *Arabic Sciences and Philosophy* 6 (1996): 23–61.

Lehmann, Paul. *Eine Geschichte der alten Fuggerbibliotheken*. 2 volumes. Tübingen: J. C. B. Mohr, 1956.

Leicht, Reimund. "The Planets, the Jews, and the Beginnings of 'Jewish Astrology.'" In *Continuity and Innovation in the Magical Tradition*, edited by Gideon Bohak, Yuval Harari, and Shaul Shaked, 271–88. Leiden: Brill, 2011.

Leichter, Joseph G. "The Zīj al-Sanjarī of Gregory Chioniades." PhD Diss., Brown University, 2004.

Lelli, Fabrizio. "Cabbalà e aristotelismo in Italia tra XV e XVI secolo: Le 'radici' nell *Iggeret ha-ʿasiriya* (*Lettera della decade*) di Avraham ben Meʾir de Balmes." In *Aristotle and the Aristotelian Tradition: Innovative Contexts for Cultural Tourism; Proceedings of the International Conference, Lecce, June 12, 13*, edited by Ennio De Bellis, 209–22. Soveria Mannelli: Rubbettino Editore, 2008.

Lelli, Fabrizio. "'Prisca Philosophia' and 'Docta Religio': The Boundaries of Rational Knowledge in Jewish and Christian Humanist Thought." *The Jewish Quarterly Review* 91, no. 1/2 (2000): 53–99.

Lemay, Richard. *Abu Maʿshar and Latin Aristotelianism in the Twelfth Century*. Beirut: The Catholic Press, 1962.

Leone Ebreo, *Dialogues of Love*. Translation by Cosmos Damian Bacich and Rossella Pescatori. Introduction and annotations by Rossella Pescatori. Toronto: University of Toronto Press, 2009.

Lesley, Arthur. "The Place of the Dialoghi d'Amore in Contemporaneous Jewish Thought." In *Ficino and Renaissance Platonism*, edited by Konrad Eisenbichler and Olga Zorzi Pugliese, 69–86. Ottawa: Dovehouse Editions Canada, 1986.

Lévy, Tony. "The Establishment of the Mathematical Bookshelf of the Medieval Hebrew Scholar: Translations and Translators." *Science in Context* 10, no. 3 (1997): 431–51.

Lévy, Tony. "Gersonide, Commentateur d'Euclide." In *Studies on Gersonides, A Fourteenth-Century Jewish Philosopher-Scientist*, edited by Gad Freudenthal, 83–148. Leiden: E. J. Brill, 1992.

Lévy, Tony. "The Hebrew Mathematics Culture (Twelfth-Sixteenth Centuries)." In *Science in Medieval Jewish Cultures*, edited by Gad Freudenthal, 155–71. Cambridge, MA: Cambridge University Press, 2011.

Licata, Giovanni. *Secundum Avenroem: Pico della Mirandola, Elia del Medigo e la "seconda rivelazione" di Averroè*. Palermo: Officina di studi medievali, 2022.

Lindberg, David C. *The Beginnings of Western Science*. Chicago: University of Chicago Press, 2007.

Littmann, Meir. "Relations between Egypt and Candia in the 16th and 17th Centuries." (Heb.) *Sinai* 88 (1980): 48–59.

Long, Pamela. "Trading Zones in Early Modern Europe." *Isis* 106 (2015): 840–47.

Lonie, Iain. *The Hippocratic Treatises "On Generation," "On the Nature of the Child," "Diseases IV": A Commentary*. Berlin: De Gruyter, 1981.

Louth, Andrew. "Knowing the Unknowable: Hesychasm and the Kabbalah." *Sobornost* 16 (1994): 9–23.

Luigi Olivieri, *Pietro d'Abano e il pensiero neolatino*. Padua: Antenore, 1988.

Maclean, Ian. *Logic, Signs and Nature in the Renaissance: The Case of Learned Medicine*. Cambridge: Cambridge University Press, 2002.

Magdalino, Paul. "Astrology." In *The Cambridge Intellectual History of Byzantium*, edited by Anthony Kaldellis and Niketas Siniossoglou, 198–214. Cambridge: Cambridge University Press, 2017.

Mahoney, Edward. "Giovanni Pico della Mirandola and Elia del Medigo, Nicoletto Vernia and Agostino Nifo." In *Giovanni Pico della Mirandola: Convegno internazionale di studi nel cinquicentesimo anniversario della morte (1494–1994)*, edited by Gian Carlo Garfagnimi, 127–56. Florence: Leo S. Olschki, 1997.

Maimonides, Moses. *Guide of the Perplexed*. Translation by Shlomo Pines. Introduction by Leo Strauss. 2 volumes. Chicago: University of Chicago Press, 1963.

Maimonides, Moses. "Letter on Astrology." In *Medieval Political Philosophy*, edition and translation by Ralph Lerner and Muhsin Mahdi, 227–36. Ithaca, NY: Cornell University Press, 1972, reprinted 1991.

Maimonides, Moses. *Mishneh Torah: The Book of Knowledge*. Edition and translation by Moses Hyamson. Jerusalem: Feldheim Publishers, 1981.

Maimonides, Moses. "On Poisons and the Protection against Lethal Drugs." In *The Medical Works of Moses Maimonides: New English Translations Based on the Critical Editions of the*

Arabic Manuscripts, translation and introduction by Gerrit Bos, 77–104. Leiden: Brill, 2022.

Maimonides, Moses. *Pirkei Avot with the Rambam's Commentary including Shemoneh Perakim*. Translation and annotation by Rabbi Eliyahu Touger. Brooklyn, NY: Moznaim Publishing Company, 1994.

Maimonides, Moses. *Sanctification of the New Moon*. Translation by Solomon Gandz. Introduction by Julius Obermann. Commentary by Otto Neugebauer. New Haven, CT: Yale University Press, 1956.

Maimonides, Moses. *Sh^{e}moneh p^{e}raqim l^{e}ha-Rambam, zal: W^{e}hu haqdamah l^{e}-Maseḵet Aḇot ʿim beiʾur b^{e}-lashon qalah*. Edition by Mordechai b. Ephraim Moses Argaman. Jerusalem: Mordechai Argaman, 2000.

Malachi, Zvi. "The Balbo Family—Scholars of Hebrew Literature in Candia (15th Century)." (Heb.) *Michael: On the History of the Jews in the Diaspora* 7 (1981): 255–70.

Malkiel, David J. "The Ghetto Republic." In *The Jews of Early Modern Venice*, edited by Robert Davis and Benjamin Ravid, 117–42. Baltimore, MD: Johns Hopkins University Press, 2001.

Malpangotto, Michela. "The Original Motivation for Copernicus's Research: Albert of Brudzewo's *Commentariolum super Theoricas novas Georgii Purbachii*." *Archive for History of Exact Sciences* 70 (2016): 361–411.

Mancha, J. L. "Al-Biṭrūjī's Theory of the Motions of the Fixed Stars." *Archive for History of Exact Sciences* 58 (2004): 143–82.

Mancha, Jose Luis. "Ibn al-Haytham's Homocentric Epicycles in Latin Astronomical Texts of the XIVth and XVth Centuries." *Centaurus* 33 (1990): 70–89.

Mancha, Jose Luis, and Gad Freudenthal. "Levi ben Gershom's Criticism of Ptolemy's Astronomy." *Aleph* 5 (2005): 35–167.

Manekin, Charles. "Logic in Medieval Jewish Culture." In *Science in Medieval Jewish Cultures*, edited by Gad Freudenthal, 114–35. Cambridge: Cambridge University Press, 2011.

Manoussacas, M. "Le recueil de privileges de la famille juive Mavrogonato de Crète (1464–1642)." *Byzantinische Forschungen* 12 (1987): 345–66.

Marcus, Simon. "The Composition of the Jewish Settlement on the Island of Crete in the Period of Venetian Rule." (Heb.) *Sinai* 60 (1967): 63–76.

Marcus, Simon. "Crete." In *Encyclopaedia Judaica*, 2nd ed., edited by Michael Berenbaum and Fred Skolnik, 289–291. Vol. 5. Detroit, MI: Macmillan Reference USA, 2007.

Marcus, Simon, and Leah Bornstein-Makovetsky. "Hamon." In *Encyclopaedia Judaica*, 2nd ed., edited by Michael Berenbaum and Fred Skolnik, 311–12. Vol. 8. Detroit, MI: Macmillan Reference USA, 2007.

Margoliouth, George S. *Catalogue of the Hebrew and Samaritan Manuscripts in the British Museum*. London: The British Museum, 1905.

Marino, John. "Economic Encounters and the First Stages of a World Economy." In *A Companion to the Worlds of the Renaissance*, edited by Guido Ruggiero, 279–95. Oxford, UK: Blackwell Publishing, 2002.

Markham, Clements R. *The Story of Majorca and Minorca*. London: Smith, Elder & Co., 1908.

Markovits, Claude. *The Global World of Indian Merchants, 1750–1947: Traders of Sind from Bukhara to Panama*. Cambridge: Cambridge University Press, 2000.

Marquet, Yves. "La Détermination astrale de l'évolution selon les frères de la pureté." *Bulletin d'études orientales* 44 (1992): 127–46.

Marx, Alexander. "Additions and Corrections to 'The Correspondence between Rabbis of Southern France and Maimonides about Astrology.'" *Hebrew Union College Annual* 4 (1927): 493–94.

Marx, Moses. "Gershom (Hieronymus) Soncino's Wanderyears in Italy, 1498–1527: Exemplar Judaicae Vitae." *Hebrew Union College Annual* 11 (1936): 427–501.

Matsen, Herbert. *Alessandro Achillini (1463–1512) and His Doctrine of "Universals" and "Transcendentals": A Study in Renaissance Ockhamism.* Lewisburg, PA: Bucknell University Press, 1974.

Matsen, Herbert. "Alessandro Achillini (1463–1512) as Professor of Philosophy in the 'Studio' of Padua (1506–1508)." *Quaderni per la storia dell'Università di Padova* 1 (1968): 91–109.

McLean, Matthew. *The "Cosmographia" of Sebastian Münster: Describing the World in the Reformation.* Oxon, England: Routledge, 2016.

McVaugh, Michael R. "Quantified Medical Theory and Practice at Fourteenth-Century Montpellier." *Bulletin of the History of Medicine* 43 (1969): 397–413.

Meadow, Mark. "Merchants and Marvels: Hans Jacob Fugger and the Origin of the Wunderkammer." In *Merchants and Marvels: Commerce, Science and Art in Early Modern Europe*, edited by Pamela Smith and Paula Findlen, 182–200. New York: Routledge, 2002.

Medan, Meir. "Levita, Elijah." In *Encyclopaedia Judaica*, 2nd ed., edited by Michael Berenbaum and Fred Skolnik, 730–732. Vol. 12. Detroit, MI: Macmillan Reference USA, 2007.

Mehren, A. F. *Vues d'Avicenne sur l'astrologie et sur rapport de la responsabilité humaine avec le destin.* Louvain: C. Peeters, 1885.

Melvin-Koushki, Matthew. "Geomancy in the Islamic World." In *Prognostication in the Medieval World*, edited by Matthias Helduk, Klaus Herbers, and Hans-Christian Lehner, 2:788–93. 2 volumes. Berlin: Walter de Gruyter, 2021.

Melvin-Koushki, Matthew. "In Defense of Geomancy: Šaraf al-Dīn Yazdī Rebuts Ibn Ḫaldūn's Critique of the Occult Sciences." *Arabica* 64 (2017): 346–403.

Melvin-Koushki, Matthew. "Powers of One: The Mathematicalization of the Occult Sciences in the High Persianate Tradition." *Intellectual History of the Islamicate World* 5 (2017): 127–99.

Melvin-Koushki, Matthew. "Toward a Neopythagorean Historiography: Kemālpaşazāde's (d. 1534) Lettrist Call for the Conquest of Cairo and the Development of Ottoman Occult-Scientific Imperialism." In *Islamicate Occult Sciences in Theory and Practice*, edited by Liana Saif, Francesca Leoni, Matthew Melvin-Koushki, and Farouk Yahya, 380–419. Leiden: Brill, 2021.

Ménage, V. L. "Kemāl Pa<u>sh</u>a-Zāde." In *Encyclopaedia of Islam New Edition Online (EI-2 English)*, edited by P. Bearman et al. Leiden: Brill, 2012.

Mercier, Raymond. *An Almanac for Trebizond for the Year 1336.* Louvain: Academia, 1994.

Mercier, Raymond. "The Astronomical Tables of George Gemistus Plethon." *Journal for the History of Astronomy* 29 (1998): 117–27.

Mercier, Raymond. "The Greek 'Persian Syntaxis' and the Zīj-i Īlkhānī." *Archives Internationales d'histoire des Sciences* 34 (1984): 35–60.

Mercier, Raymond. "Shams al-Dīn al-Bukhārī." In *The Biographical Encyclopedia of Astronomers*, edited by Thomas Hockey et al., 1047–48. New York, Springer, 2007.

Merlan, Philip. *Monopsychism, Mysticism, and Metaconsciousness: Problems of the Soul in the Neoaristotelian and Neoplatonic Tradition*. The Hague: Martinus Nijhoff, 1963.

Michael Toch. "Economic Activities." In *The Cambridge History of Judaism*. Vol. 6, *The Middle Ages: The Christian World*, edited by Robert Chazan, 357–79. Cambridge: Cambridge University Press, 2018.

Michot, Yahya, *Avicenne: Réfutation de l'astrologie*. Beirut: Dar al-Bouraq, 2006.

Millás Vallicrosa, J. M. *Tractat de l'assafea d'Azarquiel*. Barcelona, 1933.

Mizraḥi, Elijah. *T^eshubot, Sh^eʾeilot* (Responses, Questions). Istanbul: Menaḥem Qabuli, 1560.

Molà, Luca. *The Silk Industry of Renaissance Venice*. Baltimore, MD: Johns Hopkins University Press, 2000.

Monfasani, John, ed. *Collectanea Trapezuntiana: Texts, Documents, and Bibliographies of George of Trebizond*. Binghamton, NY: The Renaissance Society of America, 1984.

Montada, Josep Puig. "Eliahu del Medigo, the Last Averroist." In *Exchange and Transmission Across Cultural Boundaries: Philosophy, Mysticism and Science in the Mediterranean World*, edited by Haggai Ben-Shammai, Shaul Shaked, and Sarah Stroumsa, 155–86. Jerusalem: Keter Press for the Israel Academy of Arts and Sciences, 2013.

Morel, Teymour, and Maroun Aouad. "Un fragment retrouvé (sur la composition des éléments) du troisième Livre perdu du Grand commentaire au *De Caelo* par Averroès." *Mélanges de l'Université Saint Joseph* 65 (2013–14): 195–205.

Morrison, Robert. "An Astronomical Treatise by Mūsā Jālīnūs alias Moses Galeano." *Aleph* 10, no. 2 (2011): 385–413.

Morrison, Robert. *Islam and Science: The Intellectual Career of Niẓām al-Dīn al-Nīsābūrī*. London: Routledge, 2007.

Morrison, Robert. "Mūsā Cālīnūs' *Treatise on the Natures of Medicines and Their Use*." *Nazariyat* 3, no. 1 (2017): 77–136.

Morrison, Robert. "Quṭb al-Dīn al-Shīrāzī's Hypotheses for Celestial Motions." *Journal for the History of Arabic Science* 13 (2005): 21–140.

Morrison, Robert. "The Role of Oral Transmission for Astronomy among Romaniot Jews." In *Texts in Transit*, edited by Y. Tzvi Langermann and Robert Morrison, 10–28. University Park: Pennsylvania State University Press, 2016.

Morrison, Robert. "A Scholarly Intermediary between the Ottoman Empire and Renaissance Europe." *Isis* 105, no. 1 (2014): 32–57.

Morrison, Robert. "Selective Appropriation: Jewish Scholars and Lunar Crescent Visibility Prediction in the Ottoman Empire." *Suhayl* 19 (2022): 151–74.

Morrison, Robert. "Tables for Computing Lunar Crescent Visibility in *Adderet Eliyahu*." *SCIAMVS* 20 (2019): 157–201.

Mozaffari, S. Mohammad. "Muḥyī al-Dīn al-Maghribī's Lunar Measurements at the Maragha Observatory." *Archive for History of Exact Sciences* 68 (2014): 67–120.

Mozaffari, S. Mohammad. "Muḥyī al-Dīn al-Maghribī's Measurements of Mars at the Maragha Observatory." *Suhayl* 16–17 (2018–19): 149–249.

Mufti, Aamir. *Forget English! Orientalisms and World Literatures*. Cambridge, MA: Harvard University Press, 2016.

Munk, Salomon. *Catalogues des manuscrits hébreux et samaritains de la Bibliothèque Impériale*. Paris: Imprimerie Impériale, 1866.

Munk, Salomon. *Manuscrits hebreux de l'Oratoire à la Bibliotheque nationale de Paris*. Frankfurt am Main: Kauffmann, 1911.

Munns, David P. D. "The Challenge of Variations: The Observational Traditions of Ptolemy, Aristotle, and Copernicus' Heliocentric Solution." *Nuncius* 22, no. 2 (2007): 223–59.

Münster, Sebastian. *Compendium aritmetices, decerptum ex libro arithmeticarum institutionum magistri Eliae Orientalis*. Basel, 1546; reprinted Jerusalem: Maqor, 1970.

Murano, Giovanna. *La biblioteca arabo-ebraica di Giovanni Pico della Mirandola*. Vatican City: Biblioteca Apostolica Vaticana, 2022.

Muthannā, Aḥmad b. al-. *Ibn al-Muthannā's Commentary on the Astronomical Tables of al-Khwārizmī: Two Hebrew Versions*. Edition, translation, and commentary by Bernard R. Goldstein. New Haven: Yale University Press, 1967.

Necipoğlu, Gülru. "Visual Cosmopolitanism and Creative Translation." *Muqarnas* 29 (2012): 1–81.

Necipoğlu, Gülru, Cemal Kafadar, and Cornell H. Fleischer, eds. *Treasures of Knowledge: An Inventory of the Ottoman Palace Library (1502/3–1503/4)*. 2 volumes. Leiden: Brill, 2019.

Neugebauer, Otto. *History of Ancient Mathematical Astronomy*. 3 volumes. Berlin: Springer-Verlag, 1975.

Neusner, Jacob. *A History of the Jews in Babylonia*. 5 volumes. Leiden: Brill, 1970.

Nicol, Donald *The Last Centuries of Byzantium, 1261–1453*. Cambridge: Cambridge University Press, 1993.

Nifo, Agostino. See Niphus, Augustinus.

Nikfahm-Khubravan, Sajjad. "The Reception of Ptolemy's Latitude Theories in Islamic Astronomy." PhD Diss., McGill University, 2022.

Nikfahm-Khubravan, Sajjad, and F. Jamil Ragep. "The Mercury Models of Ibn al-Šāṭir and Copernicus." *Arabic Sciences and Philosophy* 29 (2019): 1–59.

Niphus, Augustinus. *Commentationes in librum Averrois "De substantia orbis."* Venice: Hieronymus Scotum, 1559.

Niphus, Augustinus. *Expositiones in Aristotelis Libros "Metaphysices."* Venice: Hieronymus Scotum, 1559; reprinted Frankfurt am Main: Minerva, 1967.

Nīsābūrī, Niẓām al-Dīn al-. *Gharāʾib al-Qurʾān wa-raghāʾib al-furqān*. 30 volumes in 12. Printed in the margins of al-Ṭabarī, *Jāmiʿ al-bayān*. Beirut: Dār al-Maʿrifa, 1992; reprint of the 1905–11 Būlāq edition.

Noiret, Hippolyte. *Documents inédits pour servir a l'histoire de la domination Vénitienne en Crète de 1380 à 1485*. Paris: Thorin & Fils, Éditeurs, 1892.

North, John. *Cosmos: An Illustrated History of Astronomy and Cosmology*. Chicago: University of Chicago Press, 2008.

Norton, Claire. "Blurring the Boundaries: Intellectual and Cultural Interactions between the Eastern and Western; Christian and Muslim Worlds." In *The Renaissance and the Ottoman World*, edited by Anna Contadini and Claire Norton, 3–21. Farnham, England: Ashgate, 2013.

Novak, B. C. "Giovanni Pico della Mirandola and Jochanan Alemanno." *Journal of the Warburg and Courtauld Institutes* 45 (1982): 125–47.

Öçal, Şamil. *Kemal Paşazade'nin Felsefi ve Kelami Görüşleri*. Istanbul: T. C. Kültür Bakanlığı, 2000.

Ogden, Stephen R. *Averroes on Intellect: From Aristotelian Origins to Aquinas' Critique*. Oxford: Oxford University Press, 2022.

Ogren, Brian. *The Beginning of the World in Renaissance Jewish Thought: Ma'aseh Bereshit in Italian Jewish Philosophy and Kabbalah, 1492–1535*. Boston: Brill, 2016.

Ogren, Brian. *Renaissance and Rebirth: Reincarnation in Early Modern Italian Kabbalah*. Leiden: Brill, 2009.

Ogren, Brian. "Sefirotic Depiction, Divine Noesis, and Aristotelian Kabbalah: Abraham ben Meir de Balmes and Italian Renaissance Thought." *Jewish Quarterly Review* 104, no. 4 (2014): 573–99.

Olszowy-Schlanger, Judith. "The Science of Language among Medieval Jews." In *Science in Medieval Jewish Cultures*, edited by Gad Freudenthal, 359–424. Cambridge: Cambridge University Press, 2011.

Özervarlı, M. Sait. "Arbitrating between al-Ghazālī and the Philosophers." In *Islam and Rationality: The Impact of al-Ghazālī. Papers Collected on His Nine-Hundredth Anniversary*, edited by Georges Tamer and Frank Griffel, 1:375–97. 2 volumes. Leiden: Brill, 2015–16.

Özervarlı, M. Sait. "Theology in the Ottoman Lands." In *The Oxford Handbook of Islamic Theology*, edited by Sabine Schmidtke, 567–86. Oxford: Oxford University Press, 2016.

Panagiotakes, Nikolaios. *El Greco: The Cretan Years*. Translation by John C. Davis. Abingdon, Oxon: Routledge, 2016.

Papadia-Lala, Anastasia. "The Jews in Early Modern Venetian Crete: Community and Identities." In *Mediterranean Historical Review* 27 (2012): 141–50.

Park, Katharine. *Doctors and Medicine in Early Renaissance Florence*. Princeton, NJ: Princeton University Press, 1985.

Parra Pérez, Maria José. "Estudio y edición de las traducciones al árabe del Almanach perpetuum de Abraham Zacuto." PhD Diss., University of Barcelona, 2013.

Partington, J. R. *A History of Greek Fire and Gunpowder*. Cambridge, England: W. Heffer, 1960.

Paschos, Emmanuel, and Panagiotis Sotiroudis. *Schemata of the Stars*. World Scientific Publishing Company, 1998.

Paudice, Aleida. *Between Several Worlds: The Life and Writings of Elia Capsali*. Munich: M Press, 2010.

Pfeifer, Helen. *Empire of Salons: Conquest and Community in Early Modern Ottoman Lands*. Princeton, NJ: Princeton University Press, 2022.

Pfeiffer, Judith. "Emerging from the Copernican Eclipse: The Mathematical and Astronomical Sciences in Müʾeyyedzade ʿAbdurrahman Efendi's Private Library (fl. ca. 1480–1516)." In *The 1st International Prof. Dr. Fuat Sezgin Symposium on History of Science in Islam Proceedings Book*, edited by M. Kaçar, C. Kaya and A. Z. Furat, 151–91. Istanbul: Istanbul University Press, 2020.

Pfeiffer, Judith. "Teaching the Learned: Jalāl al-Dīn al-Dawānī's Ijāza to Muʾayyadzāda." In *The Heritage of Arabo-Islamic Learning*, edited by Maurice Pomerantz and Aram Shahin, 284–332. Leiden: Brill, 2016.

Pico della Mirandola, Giovanni. *Disputationes adversus astrologiam divinatricem*. Edition and Italian translation by Eugenio Garin. 2 volumes. Florence: Vallecchi Editore, 1943.

Pico della Mirandola, Giovanni. *Oration on the Dignity of Man*. Edition, translation, and commentary by Francesco Borghesi, Michael Papio, and Massimo Riva. Cambridge: Cambridge University Press, 2012.

Pico della Mirandola, Giovanni. *Syncretism in the West: Pico's 900 Theses*. Translation and commentary by S. A. Farmer. Tempe, AZ: Medieval and Renaissance Texts and Studies, 1998.

Pingree, David. "The Astrological School of Abramios." *Dumbarton Oaks Papers* 25 (1971): 189–215.

Pingree, David. *The Astronomical Works of Gregory Chioniades*. Vol. 1, part 1, *The Zīj al-ʿAlāʾī*. Amsterdam: J. C. Gieben 1985.

Pingree, David. *The Astronomical Works of Gregory Chioniades*. Vol. 1, part 2, *Tables*. Amsterdam: J. C. Gieben, 1986.

Pingree, David. "In Defence of Gregory Chioniades." *Archives Internationales d'Histoire des Sciences*. 35 (1985): 436–38.

Poliak, Abraham N., and Yehuda Slutsky. "Crimea." In *Encyclopaedia Judaica*, 2nd ed., edited by Michael Berenbaum and Fred Skolnik, 298–301. Vol. 5. Detroit, MI: Macmillan Reference USA, 2007.

Pormann, Peter E. "Case Notes and Clinicians: Galen's *Commentary* on the Hippocratic *Epidemics* in the Arabic Tradition." *Arabic Sciences and Philosophy* 18 (2008): 247–84.

Poulle, Emmanuel. "The Alfonsine Tables and Alfonso X of Castille." *Journal for the History of Astronomy* 19 (1988): 97–113.

Puig, Roser. *Los Tratados de Construcción y Uso de la Azafea de Azarquiel*. Madrid: Instituto Hispano-Árabe de Cultura, 1987.

Rabinowicz, H. "Joseph Colon and Moses Capsali." *Jewish Quarterly Review* 46, no. 4 (1957): 336–44.

Raby, Julian. "Mehmed the Conqueror's Greek Scriptorium." *Dumbarton Oaks Papers* 37 (1983): 15–34.

Ragep, F. Jamil. "ʿAlī Qushjī and Regiomontanus: Eccentric Transformations and Copernican Revolutions." *Journal for the History of Astronomy* 36 (2005): 359–71.

Ragep, F. Jamil. "From Tūn to Toruń: The Twists and Turns of the Ṭūsī-Couple." *Before Copernicus: The Cultures and Contexts of Scientific Learning in the Fifteenth Century*, edited by Rivka Feldhay and F. Jamil Ragep, 161–97. Montreal, QC: McGill-Queen's University Press, 2017.

Ragep, F. Jamil. "Ibn al-Shāṭir and Copernicus: The Uppsala Notes Revisited." *Journal for the History of Astronomy* 47 (2016): 395–415.

Ragep, F. Jamil. "New Light on Shams: The Islamic Side of Σὰμψ Πουχάρης." In *Politics, Patronage, and the Transmission of Knowledge in 13th–15th Century Tabriz*, edited by Judith Pfeiffer, 231–47. Leiden: E. J. Brill, 2014.

Ragep, F. Jamil, and Sally P. Ragep, eds., with Steven Livesey. *Tradition, Transmission, Transformation: Proceedings of Two Conferences on Pre-Modern Science Held at the University Of Oklahoma*. Leiden: Brill, 1996.

Ragep, Sally. *Jaghmīnī's "Mulakhkhaṣ": An Islamic Introduction to Ptolemaic Astronomy*. Cham: Springer International Publishing, 2016.

Ravid, Benjamin. "The Legal Status of the Jews in Venice to 1509." *Proceedings of the American Academy for Jewish Research* 54 (1987): 169–202.

Ravitzky, Aviezer. "The God of the Philosophers and the God of the Kabbalists: A Controversy in Fifteenth Century Crete." In Aviezer Ravitzky, *History and Faith: Studies in Jewish Philosophy*, 115–53. Amsterdam: J. C. Gieben, 1996.

Rāzī, Fakhr al-Dīn al-. *Manṭiq al-Mulakhkhaṣ* (The Logic of "The Mulakhkhaṣ"). Edited by Aḥad Farāmarz Qarāmalikī and Ādīna Aṣgharīnizhād. Tehran: Intishārāt-i Dānishgāh-i Imām Ṣādiq, 2002.

Raz-Krakotzkin, Amnon. "Censorship, Editing, and Jewish Identity." In *Hebraica Veritas*, edited by Allison Coudert and Jeffrey Shoulson, 125–55. Philadelphia: University of Pennsylvania Press, 2004.

Recanati, Menaḥem. *Commentary on the Daily Prayers: Flavius Mithridates' Latin Translation, the Hebrew Text, and an English Version*. Edition, introduction, and annotations by Giacomo Corazzol. 2 volumes. Turin: Nino Aragno Editore, 2008.

Reichmuth, Stefan. *The World of Murtaḍā al-Zabīdī (1732-91): Life, Networks and Writings*. Edinburgh: Edinburgh University Press, 2009.

Reifenberg, A., and A. Schwabe. "A Judaeo-Greek Amulet from Syria." *Bulletin of the Jewish Palestine Exploration Society* 12 (1945–6): 68–72.

Renan, Ernest. *Averroès et l'Averroïsme*. Paris: Auguste Duran: 1852.

Ricci, Ronit. *Islam Translated: Literature, Conversion, and the Arabic Cosmopolis of South and Southeast Asia*. Chicago: University of Chicago Press, 2011.

Richler, Benjamin, with Malachi Beit-Arié and Nurit Pasternak, eds. *Hebrew Manuscripts in the Vatican Library*, Studi e Testi 458. Vatican City: Biblioteca Apostolica Vaticano, 2008.

Roberts, Steven V. "Not a Slave to the Zodiac, Reagan Says." *New York Times*, May 18, 1988, A22.

Roberts, Victor. "The Solar and Lunar Theory of Ibn ash-Shāṭir: A Pre-Copernican Copernican Model." *Isis* 48 (1957): 428–32.

Robinson, James T. "The First References in Hebrew to al-Biṭrūjī's *On the Principles of Astronomy*." *Aleph* 3 (2003): 145–63.

Rochberg, Francesca. *The Heavenly Writing: Divination, Horoscopy and Astronomy in Mesopotamian Culture*. Cambridge: Cambridge University Press, 2004.

Rodríguez-Arribas, Josefina. "Divination According to Goralot." In *Geomancy and Other Forms of Divination*, edited by Alessandro Palazzo and Irene Zavattero, 243–70. Florence: Galluzzo, 2017.

Rodríguez-Arribas, Josefina. "The Terminology of Historical Astrology according to Abraham Bar Ḥiyya and Abraham Ibn Ezra." *Aleph* 11 (2011): 11–54.

Rosenthal, Franz. "From Arabic Books and Manuscripts V: A One-Volume Library of Arabic Philosophical and Scientific Texts in Istanbul." *Journal of the American Oriental Society* 75 (1955): 14–23.

Roth, Cecil. "Capsali, Elijah." In *Encyclopaedia Judaica*, 2nd ed., edited by Michael Berenbaum and Fred Skolnik, 455–456. Vol. 4. Detroit, MI: Macmillan Reference USA, 2007.

Rothman, E. Natalie. *Brokering Empire: Trans-Imperial Subjects between Venice and Istanbul*. Ithaca: Cornell University Press, 2012.

Rouayheb, Khaled El-. *Islamic Intellectual History in the Seventeenth Century: Scholarly Currents in the Ottoman Empire and the Maghreb*. Cambridge: Cambridge University Press, 2015.

Rozen, Minna. *A History of the Jewish Community in Istanbul: The Formative Years, 1453–1566.* Leiden: Brill, 2002.

Ruderman, David. "The Italian Renaissance and Jewish Thought." In *Renaissance Humanism: Foundations, Forms, and Legacy*, edited by Albert Rabil, 1:382–431. 3 volumes. Philadelphia: University of Pennsylvania Press, 1988.

Ruderman, David. *Jewish Thought and Scientific Discovery in Early Modern Europe.* New Haven, CT: Yale University Press, 1995.

Ruderman, David B. *The World of a Renaissance Jew: The Life and Thought of Abraham ben Mordecai Farissol.* New York: Hebrew Union College - Jewish Institute of Religion, 1981.

Russell, Norman. "The Hesychast Controversy." In *The Cambridge Intellectual History of Byzantium*, edited by Anthony Kaldellis and Niketas Siniossoglou, 494–508. Cambridge: Cambridge University Press, 2017.

Ruths, Fredi. "Das homozentrische Sphärensystem des Girolamo Fracastoro." PhD Diss., Johann Wolfgang Goethe Universität (Frankfurt am Main), 1977.

Rutkin, H. Darrel. "Astrology, Natural Philosophy and the History of Science, c. 1250–1700." PhD Diss., Indiana University, 2002.

Saliba, George. "Arabic Science in Sixteenth-Century Europe: Guillaume Postel (1510–1581) and Arabic Astronomy." *Suhayl* 7 (2007): 115–64.

Saliba, George. "The Astronomical Tradition of Maragha: A Historical Survey and Prospects for Future Research." *Arabic Sciences and Philosophy* 1 (1991): 67–99.

Saliba, George. "Cometary Theory and Prognostications in the Islamic World and Their Relationship to Renaissance Europe." In *The Occult Sciences in Pre-Modern Islamic Cultures*, edited by Nader El-Bizri and Eva Orthmann, 105–35. Beirut: Orient-Instut Beirut; Würzburg: Ergon Verlag, 2018.

Saliba, George. "The Development of Astronomy in Medieval Islamic Society." *Arab Studies Quarterly* (1982): 211–25.

Saliba, George. "Early Arabic Critique of Ptolemaic Cosmology: A Ninth-Century Text on the Motion of the Celestial Spheres." *Journal for the History of Astronomy* 25 (1994): 115–141.

Saliba, George. *Islamic Science and the Making of the European Renaissance.* Cambridge: MIT Press, 2007.

Saliba, George. "An Observational Notebook of a Thirteenth-Century Astronomer." *Isis* 74 (1983): 388–401.

Saliba, George. "Theory and Observation in Islamic Astronomy: The Work of Ibn al-Shāṭir of Damascus." *Journal for the History of Astronomy* 18, no. 1 (1987): 35–43.

Samsó, Julio. "Abraham Zacut and José Vizinho's *Almanach perpetuum* in Arabic (16th–19th centuries)." *Centaurus* 46 (2004): 82–97.

Samsó, Julio. "Alfonso X and Arabic Astronomy." In *De Astronomia Alphonsi Regis*, edited by Mercè Comes, Roser Puig, and Julio Samsó, 23–33. Barcelona: University of Barcelona, 1987.

Samsó, Julio. *On Both Sides of the Straits of Gibraltar.* Leiden: Brill, 2020.

Sandal, Ennio. "I libri scolastici." In *L'attività editoriale di Gershom Soncino 1502–1527*, edited by Giuliano Tamani, 99–109. Soncino: Edizioni dei Soncino, 1997.

Saraç, M. A. Yekta. *Şeyhülislam Kemal Paşazade: Hayatı, Şahsiyeti, Eserleri ve Bazı Şiirleri.* Istanbul: Risale Basın-Yayın Ltd, 1995.

Savadi, Fateme. "The Historical and Cosmographical Context of *Hay'at al-arḍ* with a Focus on Quṭb al-Dīn Shīrāzī's *Nihāyat al-Idrāk*." PhD Diss., McGill University, 2019.

Savage-Smith, Emilie. "Geomancy in the Islamic World." In *Encyclopaedia of the History of Science, Technology, and Medicine in Non-Western Cultures*, edited by Helaine Selin, 998–99. Dordrecht, Netherlands: Springer, 2008.

Savage-Smith, Emilie, and Marion B. Smith. "Islamic Geomancy and a Thirteenth-Century Divinatory Device: Another Look." In *Magic and Divination in Early Islam*, edited by Emilie Savage-Smith, 211–76. Aldershot, UK: Ashgate, 2004.

Schmitt, Charles. "Science in the Italian Universities in the Sixteenth and Early Seventeenth Centuries." *The Emergence of Science in Western Europe*, edited by M. O. Crosland, 35–56. London: Macmillan Press, 1975. Reprinted in Schmitt, *The Aristotelian Tradition and Renaissance Universities*. London: Variorum Reprints, 1984.

Schmitt, Charles. "Thomas Linacre and Italy." In *Essays on the Life and Work of Thomas Linacre, c. 1460–1524*, edited by F. Maddison, M. Pelling, and C. Webster, 36–75. Oxford: Oxford University Press, 1977. Reprinted in Schmitt, *The Aristotelian Tradition and Renaissance Universities*. London: Variorum Reprints, 1984.

Scholem, Gershom. "The Beginnings of the Christian Kabbalah." In *The Christian Kabbalah*, edited by Joseph Dan, 17–51. Cambridge, MA: Harvard College Library, 1997.

Scholem, Gershom. "Gilgul." In *Encyclopaedia Judaica*, 2nd ed., edited by Michael Berenbaum and Fred Skolnik, 602–4. Vol. 7. Detroit, MI: Macmillan Reference USA, 2007.

Scholem, Gershom. *On the Kabbalah and Its Symbolism*. Translation by Ralph Mannheim. Introduction by Bernard McGinn. New York: Schocken Books, 1996.

Scholem, Gershom. *Origins of the Kabbalah*. Translation by Allan Arkush. Philadelphia: Jewish Publications Society, 1987; reprinted Princeton, NJ: Princeton University Press, 1991.

Scholem, Gershom, Jonathan Garb, and Moshe Idel. "Kabbalah." In *Encyclopaedia Judaica*, 2nd ed., edited by Michael Berenbaum and Fred Skolnik, 586–692. Vol. 11. Detroit, MI: Macmillan Reference USA, 2007.

Schub, Pincus. "A Mathematical Text by Mordecai Comtino." *Isis* 17, no. 1 (1932): 54–70.

Schwartz, Dov. "Conceptions of Astral Magic within Jewish Rationalism in the Byzantine Empire." *Aleph* 3 (2003): 165–211.

Schwartz, Dov. "From Theurgy to Magic: The Evolution of the Magical-Talismanic Justification of Sacrifice in the Circle of Nahmanides and His Interpreters." *Aleph* 1 (2001): 165–213.

Schwartz, Dov. "To Understand Something through Something." (Heb.) *Peʿamim* 133–34 (2013): 127–83.

Sebastien, Peter Mario Luciano. "Turkish Prosopography in the *Diarii* of Mario Sanuto." 2 volumes. PhD Diss., University of London, 1988.

Segre, Renata. "Juifs à Venise et Juifs en Crète." In *Les Juifs méditerranéens au Moyen Age: Culture et prosopographie*, edited by Danièle Iancu-Agou with Élie Nicola, 67–80. Paris: Éditions Cerf, 2010.

Sela, Shlomo. *Abraham Ibn Ezra and the Rise of Medieval Hebrew Science*. Leiden: Brill, 2003.

Sela, Shlomo. "The Abraham Ibn Ezra-Peter of Limoges Astrological-Exegetical Connection." *Aleph* 19, no. 1 (2019): 9–57.

Sela, Shlomo. *Asṭrologyah u-p̱arshanut ha-miqra bᵉ-haguto shel Aḇraham Ibn ʿEzra* (Astrology and Scriptural Interpretation in the Thought of Abraham Ibn Ezra). Ramat Gan, Israel: Bar Ilan University Press, 1999.

Sela, Shlomo, Carlos Steel, C. Philipp E. Nothaft, David Juste, and Charles Burnett. "A Newly Discovered Treatise by Abraham Ibn Ezra and Two Treatises Attributed to al-Kindī in a Latin Translation by Henry Bate." *Mediterranea* 5 (2020): 193–305.

Seller, Fabio. *Scientia Astrorum: La fondazione epistemologica dell'astrologia in Pietro d'Abano.* Naples: Giannini Editore, 2009.

Şen, A. Tunç. "Astrology in the Service of the Empire Knowledge, Prognostication, and Politics at the Ottoman Court, 1450s–1550s." PhD Diss., University of Chicago, 2016.

Şen, A. Tunç. "Reading the Stars at the Ottoman Court: Bāyezīd II (r. 886/1481–918/1512) and His Celestial Interests." *Arabica* 64 (2017): 557–608.

Serjeant, R. B. "Notices on the 'Frankish Chancre' (Syphilis) in Yemen, Egypt, and Persia." *Journal of Semitic Studies* 10 (1965): 241–52.

Setton, Kenneth M. *The Papacy and the Levant.* 4 volumes. Philadelphia: The American Philosophical Society, 1976–84.

Shank, Michael. "The Geometrical Diagrams in Regiomontanus's Edition of His Own *Disputationes* (c. 1475): Background, Production, and Diffusion." *Journal for the History of Astronomy* 43 (2012): 27–55.

Shank, Michael. "The 'Notes on al-Biṭrūjī' Attributed to Regiomontanus: Second Thoughts." *Journal for the History of Astronomy* 23 (1992): 15–30.

Shank, Michael. "Regiomontanus and Astronomical Controversy." *Before Copernicus: The Cultures and Contexts of Scientific Learning in the Fifteenth Century*, edited by Rivka Feldhay and F. Jamil Ragep, 79–109. Montreal, QC: McGill-Queen's University Press, 2017.

Shank, Michael. "Regiomontanus as a Physical Astronomer: Samplings from *The Defence of Theon against George of Trebizond.*" *Journal for the History of Astronomy* 38 (2007): 325–49.

Shank, Michael. "Regiomontanus versus George of Trebizond on Planetary Order, Distances, and Orbs (Almagest 9.1)." In *Ptolemy's Science of the Stars in the Middle Ages*, edited by David Juste, Benno van Dalen, Dag Nikolaus Hasse, and Charles Burnett, 345–426. Turnhout: Brepols, 2020.

Shank, Michael. "Setting up Copernicus? Astronomy and Natural Philosophy in Giambattista Capuano da Manfredonia's *Expositio* on the *Sphere.*" *Early Science and Medicine* 14 (2009): 290–315.

Shefer-Mossensohn, Miriam. *Science among the Ottomans: The Cultural Creation and Exchange of Knowledge.* Austin: University of Texas Press, 2015.

Shemesh, Yonatan. "Averroes as Intertext: Moses Narboni's Commentary on Maimonides' *Guide of the Perplexed.*" PhD Diss., University of Chicago, 2021.

Shmuelevitz, Aryeh. "Capsali as a Source for Ottoman History." *International Journal of Middle East Studies* 9, no. 3 (1978): 339–44.

Shmuelevitz, Aryeh. *The Jews of the Ottoman Empire in the Late Fifteenth and the Sixteenth Centuries: Administrative, Economic, Legal and Social Relations as Reflected in the Responsa.* Leiden: E. J. Brill, 1984.

Silvaticus, Matthaeus. *Pandectarum medicinae.* Venice, 1540.

Siraisi, Nancy. "Medicine and the Renaissance World of Learning." *Bulletin of the History of Medicine* 68 (2004): 1–36.

Siraisi, Nancy. *Medieval and Early Renaissance Medicine: An Introduction to Knowledge and Practice*. Chicago: University of Chicago Press, 1990.

Siraisi, Nancy. "Renaissance Readers and Avicenna's Organization of Medical Knowledge." In *Medicine and the Italian Universities 1250–1600*, edited by Nancy Siraisi, 203–25. Leiden: Brill, 2001.

Siraisi, Nancy. *Taddeo Alderotti and his Pupils*. Princeton, NJ: Princeton University Press, 1981.

Siraisi, Nancy. "Two Models of Medical Culture." In *Medicine and the Italian Universities 1250–1600*, edited by Nancy Siraisi, 79–99. Leiden: Brill, 2001.

Smithius, Renate. "Abraham Ibn Ezra's Astrological Works in Hebrew and Latin: New Discoveries and Exhaustive Listing." *Aleph* 6 (2006): 239–338.

Smoller, Laura Ackermann. *History, Prophecy, and the Stars: The Christian Astrology of Pierre d'Ailly, 1350–1420*. Princeton, NJ: Princeton University Press, 1994.

Socrates Scholasticus. *The Ecclesiastical History of Socrates, Surnamed Scholasticus or the Advocate*. Translation by Henri de Valois. London: G. Bell, 1904.

Solon, Peter. "The Six Wings of Immanuel Bonfils and Michael Chrysokkokes." *Centaurus* 15 (1970): 1–20.

Stahl, Alan M. *Zecca: The Mint of Venice in the Middle Ages*. Baltimore, MD: Johns Hopkins University Press, 2001.

Stam, David H., ed. *International Dictionary of Library Histories*. 2 volumes. Chicago: Fitzroy Dearborn, 2001.

Stearns, Justin. *Infectious Ideas: Contagion in Premodern Christian and Islamic Thought*. Baltimore, MD: Johns Hopkins University Press, 2011.

Stearns, Justin. "Review of *Cross-Cultural Scientific Exchanges in the Eastern Mediterranean, 1560–1660*." *International Journal of Middle East Studies* 43 (2011): 760–63.

Steimann, Ilona. "The Story of One Acquisition: Hebrew Manuscripts from Venetian Candia." *Mediterranean Historical Review* 38, no. 1 (2023): 25–70.

Steinschneider, Moritz. *Die hebräischen Übersetzungen des Mittelalters und die Juden als Dolmetscher*. Berlin: Kommissionsverlag des Bibliographischen Bureaus, 1893. https://archive.org/details/diehebraeischen00steigoog/page/n5/mode/2up.

Steinschneider, Moritz. "Die mathematischer Wissenschaften bei den Juden, 1441–1500." *Bibliotheca Mathematica* 3 (1901): 58–76.

Steinschneider, Moritz. *Mathematik bei den Juden*. Hildesheim: George Olms, 1964.

Strieder, Jacob. *Jacob Fugger the Rich: Merchant and Banker of Augsburg*. Translation by Mildred Hartsough and N. S. B. Gras. New York: The Adelphi Company, 1931.

Stroumsa, Sarah. *Maimonides in His World: Portrait of a Mediterranean Thinker*. Princeton, NJ: Princeton University Press, 2009.

Svachula, Amanda. "Like a Virgo: How The Times Covers Astrology." *New York Times*, August 31, 2019.

Swerdlow, Noel M. "Aristotelian Planetary Theory in the Renaissance: Giovanni Battista Amico's Homocentric Spheres." *Journal for the History of Astronomy* 3 (1972): 36–48.

Swerdlow, Noel M. "Copernicus and Astrology, with an Appendix of Translations of Primary Sources." *Perspectives on Science* 20 (2012): 353–78.

Swerdlow, Noel M. "Copernicus' Derivation of the Heliocentric Theory from Regiomontanus's Eccentric Models of the Second Inequality of the Superior and Inferior Planets." *Journal for the History of Astronomy* 48, no. 1 (2017): 33–61.

Swerdlow, Noel M. "The Derivation and First Draft of Copernicus's Planetary Theory: A Translation of the Commentariolus with Commentary." *Proceedings of the American Philosophical Society* 117, no. 6 (1973): 423–512.

Swerdlow, Noel M. "The Most Complex Homocentric Astronomy." *Journal for the History of Astronomy* 48, no. 3 (2017): 358–61.

Swerdlow, Noel M. "Nature, Experiment, and the Sciences." In *Nature, Experiment and the Sciences: Essays on Galileo and the History of Science*, edited by Trevor H. Levere and William R. Shea, 165–95. Dordrecht: Kluwer Academic Publishers, 1990.

Swerdlow, Noel M. "Regiomontanus's Concentric-Sphere Models for the Sun and the Moon." *Journal for the History of Astronomy* 30 (1999): 1–23.

Swerdlow, Noel M., and Otto Neugebauer. *Mathematical Astronomy in Copernicus' "De Revolutionibus."* New York: Springer-Verlag, 1984.

Sylla, Edith Dudley. "The Status of Astronomy." In *Before Copernicus: The Cultures and Contexts of Scientific Learning in the Fifteenth Century*, edited by Rivka Feldhay and F. Jamil Ragep, 45–78. Montreal, QC: McGill-Queen's University Press, 2017.

Taftāzānī, Saʿd al-Dīn. *Sharḥ al-Maqāṣid*. Edition by ʿAbd al-Raḥmān ʿUmayra. 5 volumes. Beirut: ʿĀlam al-kutub, 1989.

Tamani, G. "Le traduzioni ebraico latine di Abraham de Balmes." In *Biblische und judaistische Studien: Festschrift für Paolo Sacchi*, edited by Johann Maier and Angelo Maier, 613–35. Frankfurt am Main: Peter Lang, 1990.

Ṭāsh Kubrī Zādah, *Miftāḥ al-saʿāda*. 3 volumes. Beirut: Dār al-kutub al-ʿilmiyya, n.d.

Terzioğlu, Arslan. "Un Traité Turc inconnu de Moses Hamon sur l'art dentaire du début du XVIe siècle." In *Mélanges d'histoire de la médecine hébraïque*, edited by Gad Freudenthal and Samuel Kottek, 111–22. Leiden: Brill, 2003.

Thorndike, Lynn. *History of Magic and Experimental Science*. 8 volumes. New York, Macmillan, 1923–58.

Tihon, Anne. "Astronomy." In *The Cambridge Intellectual History of Byzantium*, edited by Anthony Kaldellis and Niketas Siniossoglou, 183–97. Cambridge: Cambridge University Press, 2017.

Tihon, Anne. "The Astronomy of George Gemistus Plethon." *Journal for the History of Astronomy* 29 (1998): 109–16.

Tihon, Anne. "L'Astronomie Byzantine à l'Aube de la Renaissance." *Byzantion* 66, no. 1 (1996): 244–80.

Tihon, Anne. "Science in the Byzantine Empire." In *The Cambridge History of Science*. Vol. 2, *Medieval Science*, edited by David C. Lindberg and Michael Shank, 190–206. Cambridge: Cambridge University Press, 2013.

Tihon, Anne, and Raymond Mercier. *Georges Gémiste Pléthon: Manuel d'Astronomie*. Louvain-la-Neuve, 1998.

Tirmidhī, Muḥammad b. ʿĪsā b. Sūra al-. *Al-Jāmiʿ al-ṣaḥīḥ*. 5 volumes. Edition and commentary by Aḥmad Muḥammad Shākir. (Beirut: Dār al-kutub al-ʿilmiyya, n.d.)

Tirosh-Rothschild, Havah. *Between Worlds: The Life and Thought of Rabbi David ben Messer Leon*. Albany, NY: State University of New York Press, 1991.

Toepel, Alexander. "Yonton Revisited: A Case Study in the Reception of Hellenistic Science within Early Judaism." *Harvard Theological Review* 9 (2006): 235–45.

Toomer, G. J. *Ptolemy's "Almagest."* London: Duckworth, 1984.

Töyrylä, Hannu. *Abraham Bar Hiyya on Time, History, Exile and Redemption: An Analysis of "Megillat ha-Megalleh."* Leiden: Brill, 2014.

Trigg, Scott. "From Samarqand to Istanbul: Astronomy and Scientific Education in the Commentaries of Fatḥallāh al-Shirwānī." PhD Diss., University of Wisconsin-Madison, 2016.

Trivellato, Francesca. *The Familiarity of Strangers: The Sephardic Diaspora, Livorno, and Cross-Cultural Trade in the Early Modern Period.* New Haven, CT: Yale University Press, 2009.

Trivellato, Francesca. "Jews and the Early Modern Economy." In *The Cambridge History of Judaism.* Vol. 7, *The Early Modern World*, edited by Jonathan Karp and Adam Sutcliffe, 139–67. Cambridge: Cambridge University Press, 2017.

Truitt, E. A. *Medieval Robots: Mechanism, Magic, Nature, and Art.* Philadelphia: University of Pennsylvania Press, 2013.

Ṭūsī, Naṣīr al-Dīn al-. *The Memoir on Astronomy (al-Tadhkira fī ʿilm al-hayʾa).* Edition, translation, and commentary by F. Jamil Ragep. 2 volumes. New York: Springer-Verlag, 1993.

Twersky, Isadore. *Introduction to the Code of Maimonides (Mishneh Torah).* New Haven, CT: Yale University Press, 1980.

Università degli Studi di Padova. *Archivio Antico dell' Università.* Vol. 669.

ʿUrḍī, Muʾayyad al-Dīn al-. *The Astronomical Work of Muʾayyad al-Dīn al-ʿUrḍī: A Thirteenth Century Reform of Ptolemaic Astronomy (Kitāb al-hayʾa).* Edition and introduction by George Saliba. Center for Arab World Unity Studies, 1990.

van Bladel, Kevin. *The Arabic Hermes.* Oxford: Oxford University Press, 2009.

van der Werf, Siebren. "History and Critical Analysis of Fifteenth and Sixteenth Century Nautical Tables." *Journal for the History of Astronomy* 48 (2017): 207–32.

van Ess, Josef. *Frühe muʿtazilitische Häresiographie.* Beirut: Beiruter Texte und Studien, 1971.

Vanden Broecke, Steven. *The Limits of Influence: Pico, Louvain, and the Crisis of Renaissance Astrology.* Leiden: Brill, 2003.

Varlık, Nükhet. *Plague and Empire in the Early Modern Mediterranean World.* Cambridge: Cambridge University Press, 2015.

Vescovini, Graziella. "Peter of Abano and Astrology." In *Astrology, Science, and Society: Historical Essays*, edited by Patrick Curry, 19–39. Woodbridge, England: Boydell and Brewer, 1987.

Vescovini, Graziella. "The Theological Debate." In *A Companion to Astrology in the Renaissance*, edited by Brendan Dooley, 99–140. Leiden: Brill, 2014.

Vesel, Matjaž. *Copernicus: Platonist Astronomer-Philosopher: Cosmic Order, the Movement of the Earth, and the Scientific Revolution.* Frankfurt am Main: Peter Lang, 2014.

Vesel, Živa. *Les encyclopédies Persanes: Essai de typologie et de classification des sciences.* Paris: Éditions Recherche sur les civilisations, 1986.

Vincent, Alfred L. "Money and Coinage in Venetian Crete, c. 1400–1669: An Introduction." *Thesaurismata* 37 (2007): 267–326.

von Stuckrad, Kocku. "Christian Qabbalah and Anti-Jewish Polemics: Pico in Context."

In *Polemical Encounters: Esoteric Discourse and Its Others*, edited by Olav Hammer and Kocku von Stuckrad, 3–24. Leiden: Brill, 2007.

Wakin, Daniel J. "Pluto, Paradoxically, Joins 'The Planets.' " *New York Times*, September 8, 2006, E1.

Wallis, Faith, and Robert Wisnovsky, eds. *Medieval Textual Cultures: Agents of Transmission, Translation and Transformation*. Boston: De Gruyter, 2016.

Westfall, Richard. "Science and Patronage: Galileo and the Telescope," *Isis* 76 (1985): 11–30.

Westman, Robert S. *The Copernican Question: Prognostication, Skepticism, and Celestial Order*. Berkeley: University of California Press, 2011.

Wilkinson, Robert. *Orientalism, Aramaic, and Kabbalah in the Catholic Reformation*. Leiden: Brill, 2007.

Wilson, Curtis. "Rheticus, Ravetz and the Copernican Revolution." In *The Copernican Achievement*, edited by Robert S. Westman, 17–39. Berkeley: University of California Press, 1975.

Wirszubski, Chaim. *Pico della Mirandola's Encounter with Jewish Mysticism*. Cambridge, MA: Harvard University Press, 1989.

Wolfson, Elliot. "The Doctrine of Sefirot in the Prophetic Kabbalah of Abraham Abulafia." *Jewish Studies Quarterly* 2 (1995): 336–71.

Wolfson, Elliot. "The Doctrine of Sefirot in the Prophetic Kabbalah of Abraham Abulafia (Part 2)." *Jewish Studies Quarterly* 3 (1996): 47–84.

Yisraeli, Oded. "The Mezuzah as an Amulet: Directions and Trends in the Zohar." *Jewish Studies Quarterly* 22, no. 2 (2015): 137–61.

Zambelli, Paola. "Astrologers' Theories of History." In *Astrologi hallucinati: Stars and the End of the World in Luther's Time*, edited by Paolo Zambelli, 1–28. Berlin: De Gruyter, 1984.

Zonta, Mauro. "Al-Fārābī's Commentaries on Aristotelian Logic: New Discoveries." In *Philosophy and Arts in the Islamic World*, edited by U. Vermeulen and D. De Smet, 219–32. Leuven: Peeters, 1998.

Zonta, Mauro. "Fonti antiche e medievali della logica ebraica nella Provenza del Trecento." *Medioevo* 23 (1997): 515–94.

Zonta, Mauro. *Hebrew Scholasticism in the Fifteenth Century: A History and Source Book*. Springer, 2006.

Zonta, Mauro. *La filosofia antica nel Medioevo ebraico*. Brescia: Paideia, 1996.

Zonta, Mauro. "Medieval Hebrew Translations of Philosophical and Scientific Texts: A Chronological Table." In *Science in Medieval Jewish Cultures*, edited by Gad Freudenthal, 17–73. Cambridge: Cambridge University Press, 2011.

Zonta, Mauro. "Un'ignota versione ebraica delle *Quaestiones in De anima* di Jean de Jandun e il suo traduttore." *Annali di Ca' Foscari* 32, no. 3 (1993): 5–34.

INDEX

Book titles with known authors are listed by author. Book titles are listed in the language in which they are initially mentioned in the main text; the translation follows.

Abano, Pietro d': critical days and medical astrology, 208–9; *Expositio problematum Aristotelis*, 194–95

ʿAbd al-Raḥmān (astrologer): *Kitāb ḥifẓ al-ṣiḥḥa* (Book on the Preservation of Health), 33

Abravanel, Isaac, 44–45, 58, 141–42, 148

Abravanel, Judah (Leone Ebreo): astrology, 45; *Dialoghi d'amore*, 45, 58; reputation, 142

Abū al-Khayr, Isaac b. Samuel, 173

Abū Maʿshar: *Great Introduction* (Ar. *al-Mudkhal al-kabīr*), 44

Abulafia, Abraham: *Seiper ha-meliṣ* (Book of the Interpreter), 126; *sᵉpirot*, nonhypostatic, 124

Abulafia, Samuel, 203

Accademia dei Lincei, 187

Achillini, Alessandro: *De orbibus*, 112, 156–57

Adramitteno, Manuel, 99

Adrianople. *See* Edirne

Afendopolo, Caleb: *Adderet Eliyahu* (Elijah's Mantle), 77; *Almagest*, study of, 166–67; Averroism, 90–91; and David b. Judah Messer Leon, 167, 192; *Gan ha-melek* (The Garden of the King), 166; *Seiper kᵉli robaʿ ha-shaʿot* (The Book of the Horary Quadrant), 64–65. *See also* Bashyatchi, Elijah and Qaraites

Ahī Çelebī: patron of Galeano, Moses b. Judah (Mūsā Jālīnūs), 23, 185, 200–201, 206; *Risāla fī al-ṭibb* (Epistle on Medicine), 201. *See also* Bayezit II: chief physician

Akkach, Samer: *ʿAbd al-Ghani al-Nabulusi: Islam and the Enlightenment*, 5

Albalag, Isaac, 131

Albert of Brudzewo, 164

Albo, Joseph: *Seiper ha-ʿiqqarim* (The Book of Roots), 16

alchemy, 54

Aldine Press, 16

Alemanno, Yoḥanan, 62–63, 67

Alexander of Aphrodisias, 97

Alfonso X (king of Castile): Alfonsine Tables, 67–68, 80, 85

Algazi, Abraham, 22, 142, 144–45, 194
Algazi family, 17, 22, 26
almanac (*taqwīm*), 36, 39–40, 55–56, 79
Almuli, Moses, 147
Alum, 3
Amico, Giovanni Battista: *De motibus corporum coelestium*, 162; and Ibn Naḥmias, 182; Ṭūsī couple, use of, 161–62
ʿAnabi, Shalom, 125
Anastasia, widow of Judah Balbo, 20
Andalus, al-, 84, 161. *See also* Iberian Peninsula
Andronicus II Palaeologos (Byzantine emperor), 72
Apollonios: *Conics*, 95, 177
Apostolis, Arsenios, 16
Apulia, 63
Aquinas, Thomas, 90, 99, 116
Argyros, Isaac, 75–76
Aristotle: anatomy, 53; astrology, 208; *De anima*, 97; heavens, composition and motions of, 110–11, 128, 142, 157; *Metaphysics*, 157. *See also* Averroism and Avicennism
Arnaldo da Villanova: *Aspects in Judgment*, 172; cited by Galeano, Moses b. Judah (Mūsā Jālīnūs), 201, 205–6, 209, 213; *De gradibus*, 201, 205
Ashkenazi, Israel, 20
Ashqar, Rabbi al-, 20
Aslanian, Sebouh, 27
astrology: ascendant, 31–32, 37, 39, 61; aspects, 32, 39, 45, 82–83, 207; battle deaths, 31, 54; Christian discussions of, 43–44; conjunction, planetary, 31–32, 39–40, 42–43, 68–70, 177; Enoch's rectification (*Moʾznei Ḥanok*), 37; forecasting, 12, 31–45, 51–56, 61, 67–68, 110, 113, 180 (*see also* astrology: conjunction, planetary); Islamic discussions of, 41; Jewish discussions of, 42–43, 51–52; medical, 208; natural, 31, 42, 54, 56; political role, 29, 32–33, 38–41, 57, 60, 150; and *Qabbalah*, 123, 128, 140–41, 149; rays, 32, 56–57; *topoi* (astrological places or houses), 31–32, 39, 45, 188, 191; twins paradox, 36–37, 43, 56; zodiacal signs, 31–32, 35–37, 43, 48, 74, 191–92. *See also* Averroism: astrology, denial of; Cohen Ashkenazi, Moses: forecasts for Vittoris and Cohen Ashkenazi, Moses: *Urim wᵉ-ṯummim*; della Mirandola, Pico: *Disputationes*; Delmedigo, Elijah: *Commentary on "De substantia orbis"*; Galeano, Moses b. Judah: astrology
astronomy, tables and instruments: astrolabe, 61, 72, 81; astrolabe, universal, 63–65, 67, 84–85, 117; eclipse, 39, 68–71, 73–74, 78, 109–11; ecliptic, obliquity of, 74; equation of time, 70; equatorium, 63–64; *Paradosis*, 74–76, 78–79, 84; quadrant, horary, 61, 65–66; syzygy calculation, 69–74, 76- 77, 79, 83; visibility, lunar crescent, 77, 81, 91; *zīj*, 39, 63, 66–67, 70–75, 77–79, 85. *See also* Galeano, Moses b. Judah: *Kitāb al-Zīj*; Kumaṭiano, Mordechai: *Peirush luḥot Paras*; Levi b. Gerson: Jacob staff; timekeeping: calendars, lunar
astronomy, theoretical: apogee of lunar epicycle, 164–65; apogee, motion of solar, 75; equant, 153, 164–65; *hayʾa basīṭa* (plain astronomy), 170–72; latitudes, 61–62, 64, 75–77, 80, 83, 161; longitudes, 70, 75, 83, 152–53, 156, 167; models, heliostatic, 179–82; models, homocentric, 151–65, 175–79, 182; models, lunar, 154–55, 162, 164–65, 167–69, 174–75, 179, 181; models, planetary, 163–66, 170, 173–76, 178, 181; models, precision of, 102–8, 129, 151–57, 162–67, 173–74, 178 (*see also* astronomy, theoretical: planets, sizes and distances of); models, solar, 151–55, 161–62; planets, sizes and distances of, 74, 152, 169, 176–77; prosneusis point,

165; stars, fixed, 102–5, 107–8, 110, 129, 158, 209; *theorica planetarum*, 152–54, 164, 170. *See also* Averroism: denial of epicycle and eccentric; Copernicus, Nicholas; and Ṭūsī: Ṭūsī couple
Astruc family, 17, 25, 27–28
Augsburg, 15
Augustine: *City of God*, 43; *Confessions*, 44
Averroës: *Compendium of the "Almagest,"* 80, 163, 165; *De anima* (middle commentary), 92; *De substantia orbis*, 94, 106–8, 110–11; *Epistle on the Possibility of Conjunction*, 94; *Faṣl al-maqāl* (The Decisive Treatise), 134; *al-Kulliyyāt fī al-ṭibb* (The Colliget), 205; *Long Commentary on Aristotle's "Metaphysics,"* 130, 141; *Long Commentary on Aristotle's "Physics,"* 101; *Long Commentary on "De anima,"* 97–98, 100; *Meteorology*, 110, 135, 141. *See also* Averroism
Averroism: agent intellect, conjunction with, 92, 98, 100, 109, 138; astrology, denial of, 110–13; della Mirandola, Pico, 99–100, 106–7, 109, 111–12, 135, 141; emanation, denial of, 128–33; epicycle and eccentric, denial of, 106–7, 129–31, 151–53, 157, 178; God as prime mover, 95, 101–5, 108–9, 130, 142; Jean de Jandun, 99–101; particulars, God's knowledge of, 50–53, 96–98, 108–14; unicity of the intellect, 95–100, 102, 109, 112, 127–28, 132–33, 138. *See also* Delmedigo, Elijah and Galeano, Moses b. Judah: particulars, God's knowledge of
Avicenna: and Aristotelianism, 87; Averroës, comparison/contrast with, 87, 95, 109, 128–29; *Canon on Medicine* (al-Qānūn fī al-ṭibb), 87, 92; influence in Ottoman Empire, 116–17
Avicennism: conjunction of opposites, 94; emanation, 52, 128–33, 137; *nafs al-amr* (the fact of the matter), 94
Azalino, 8
Balaza, Sabathi, 10
Balbo family, 17, 20–21, 25–26
Balbo, Michael: and Cohen Ashkenazi, Moses, 21–22; *Commentary on "Yᵉsod Mora',"* ownership, 46, 59, 93; forecast of redemption, copying, 63; *Wikkuaḥ* (Debate), 123–5, 127–33, 148, 150
Balmes, Abraham de: *Iggeret ha'asiriyya* (Epistle of the ten), 141; scholasticism, Hebrew, 101
Bar Ḥiyya, Abraham: *Scroll of the Revealer* (Mᵉgillat ha-mᵉgalleh), 44–45, 52; *zīj*, 70
Barbarossa, Khayr al-Dīn, 193
Bari, 63
Barlaam of Calabria, 126
Bartolocci, Giulio: *Bibliotheca magna rabbinica de scriptoribus*, 155
Bashyatchi, Elijah: *Adderet Eliyahu* (Elijah's Mantle), 77; new moon, calculations of, 90. *See also* Qaraites
Bashyatchi, Menaḥem, 64
Battānī, Muḥammad b. Jābir b. Sinān al-: *Zīj*, 70, 77
Battista, 8
Bayḍāwī, Nāṣir al-Dīn, al- 55
Bayezit II (sultan): chief physician (Ahī Çelebī), 23, 185, 200–201; and Hamon, Joseph, 204; library inventory, 116; *ṣāḥib-i qırān*, 39; science, study of, 33, 80–81, 85, 213; welcoming Jews, 3
Ben Verga, Judah, 80
Ben-Zaken, Avner: *Cross-Cultural Scientific Exchanges in the Eastern Mediterranean, 1560–1660*, 6
Bernard de Gordon: cited by Galeano, Moses b. Judah (Mūsā Jālīnūs), 201; *De gradibus*, 205; *Seiper ha-gᵉbulim* (On Prognosis), 209
Bessarion (cardinal), 78
Bianchini, Giovanni: *Tabulae astronomiae*, 68
Binbaş, İ. Evrim: *Intellectual Networks in Timurid Iran*, 6

Birkenmaijer, Ludwik, 181
Bisaha, Nancy, xiii
Biṭrūjī, Nūr al-Dīn al-: *Kitāb fī al-hayʾa* (On the Principles of Astronomy), improvement on, 153–56; as *marʿīsh*, 173, 175; motion of zodiac, 158
Black Death, 4. *See also* medicine: plague
Bland, Kalman, 98
Blåsjö, Viktor, 181–82
Bloom, Harold: *The Anxiety of Influence*, 182
Bomberg, Daniel, 16
Bona, 25
Bonbari, Bernolai, 9
Bonfils, Emmanuel b. Jacob: *Six Wings* (Sheish k^{e}napayim), 69–70, 78–80
Bonjorn, Jacob b. David: *Luḥot ha-Poʿeil* (Tables), 69–70, 80, 83
Breastplate of High Priest, 35, 47–48
Brethren of Purity, 41
Bricot, Thomas, 101
Bruno, Giordano, 58
Bukhārī, Shams al-Dīn al-, 72
Burley, Walter, 116

calendar, lunar. *See* timekeeping: calendars, lunar
Callippos, 157, 161–62
camerarius (chamberlain), 20
Cana, Mustafa, 11
Canalis, Jacob, 9
Candia (Heraklion): ducal court, 8–11, 21, 24–27, 211; as entrepôt, 2, 11, 72; Galeano, Moses b. Judah (Mūsā Jālīnūs), 23–24, 193–95, 211; merchant-scholar families, x, 15–6, 18–19, 21–26, 28, 46; and Michael Balbo, 93, 124; migration, Jewish, 4; and Muslims, 11–12, 17; and Ottoman Empire, 11, 15, 93, 211; student of Afendopolo from, 166; *Urim w^{e}-ṯummim*, composition of, 49, 58; and Venetian Republic, 4–5, 9, 99
Canea, 25–26. *See also* Chania
Capsali, David, 17, 20
Capsali, Elijah: career, 19–22; and Elijah Mizraḥi, 167; sale of manuscripts, 8, 16–17, 20; *Seider Eliyahu zuṭa*, 3, 20;
Capsali, Elqanah, 16, 19, 21
Capsali, Moses, 19, 21, 90, 93, 125
Capsali, Samuel, 19–20
Capuano, Francesco, 156
Caro, Joseph, 19
Casani family, 17–18, 24
Cassuto, Umberto, 8
cause: astral, 32–38, 40–43, 47–48, 54–57, 109, 143, 208–10; customary, 56; God as first cause, 103–10, 113, 136, 141–44, 148, 166; intermediate, 54, 56, 187, 194, 196–99, 208–9. *See also* intellect, agent and Avicennism: emanation
Celichi, Mustapha, 11
Chabás, José, 81
chamberlain (*camerarius*), 20
Chania, 25–26. *See also* Canea
Chioniades, Gregory: *Persian Syntaxis*, 71; *Schemata of the Stars*, 71, 85; trip to Tabrīz, 71–72
Chios, 11, 19
chirograph, 2, 9, 27–28
Chrysococcès, George: *Persian Syntaxis*, 72–74, 76
Chrysococcès, Michael, 70
climes, 35–36, 43
Cohen, Judah b. Moses ha-, 67. *See also* Alfonso X: Alfonsine Tables)
Cohen Ashkenazi, Moses: forecasts for Vittoris, 21–22, 33; *Urim w^{e}-ṯummim*, 21–22, 35–36, 40, 44, 49–59, 92; *Wikkuaḥ* (Debate), 123–24, 127–28, 131–33, 148
Cohen Ashkenazi, Samuel, 22
Cohen Ashkenazi, Saul, 22, 44–45, 57–8, 135, 141, 148
condestabulo, 18–22, 24–25
Constantinople: Chioniades, 71; Christians, 3, 196–97; Ibn Ezra's works in, 89; merchant-scholar families, 18–9; Venetians, 9. *See also* Istanbul
Copernicus, Nicholas: *Commentariolus*, 154, 162, 179; *De revolutionibus*, 162; and

Islamic astronomy, ix-xi, 85, 174, 179–83; homocentric theories, 154, 162; tables, 68; University of Bologna, 156; University of Cracow, 164; University of Padua, 13
Corazzol, Giacomo, 23
Corersi, Nicholas, 18
Corfu, 11
court, Ottoman (*dīwān*): astrology at, 33–34, 36, 39–40, 56, 83; astronomy at, 23, 81–82, 155, 170–72, 174; machinations at, 14, 187, 194; medicine at, 3, 202–4, 212–13
Crete: Galeano, Moses b. Judah (Mūsā Jālīnūs) on, 23, 35, 194, 211; Greek to Latin translations on, 73, 76; immigration to, 3–4; Jewish-Christian relations, 10; merchant-scholar families, x, 18–21, 23, 26, 215; Moses Cohen Ashkenazi on, 35–36;
cross staff. *See* Levi b. Gerson: Jacob staff

Damascus, 50, 79, 174
David b. Shushan, 101
David Kalonymos b. Maestro Jacob, 63
Dawānī, Jalāl al-Dīn al-, 80, 170
de Salaya, Juan, 80
de Zuñiga, Juan, 79
della Mirandola, Giovanni Pico: *900 Theses*, 56, 100, 138, 140–41; astronomy, 78–79, 84; *Disputationes adversus astrologiam divinatricem*, 40, 44–45, 56, 112, 180; and Islam, 146–47; patron and student of Elijah Delmedigo in Averroism, 71, 99–100, 102, 106–7, 109–12; patron and student of Elijah Delmedigo in *Qabbalah*, 120, 134–41, 148–49; patron and student of Yoḥanan Alemanno, 62–63, 67, 139–40; "Philosophical Conclusions," 111
Delmedigo, Eliezer, 19
Delmedigo, Elijah: *Beḥinat ha-dat* (The Examination of Religion), 134–35; *Commentary on "De substantia orbis,"* 106–7, 110–12, 135, 163–64, 213; *"De Primo motore"* (Question on the Prime Mover), 101–5, 108, 158; debate with Flavius Mithridates, 57, 99, 139; *Investigations in Accordance with the Principles of the Philosophers* (Derushim ke-pi shoreshei ha-pilosopim), 99, 102, 109–10; *Letter* to Pico, 135–39; Scholasticism, Hebrew, 117, 179, 182; teacher of Domenico Grimani, 17, 101, 141, 178; teacher of Pico della Mirandola in Averroism, 98–100, 106, 109–12; teacher of Pico della Mirandola in *Qabbalah*, 120, 134–41, 148–49; teacher of Saul Cohen Ashkenazi, 22; translation of Averroës' epitome of Plato's *Republic*, 100
Delmedigo, Ephraim, 20
Delmedigo, Julio, 19
Delmedigo, Menaḥem (Manuel), 18–19
Delmedigo, Menasheh, 10, 19–21, 24
Delmedigo, Meyuḥas, 19
Delmedigo, Moses, 18, 25
Delmedigo, Moses b. Abba ha-Zakein, 18
Delmedigo, Mossaninus, 21
Delmedigo, Pothula, 19, 21
Delmedigo, Salamo, 19
Delmedigo, Solomon, 25
Diameters. *See* astronomy, theoretical: planets, sizes and distances of
Donato, Girolamo, 141
Donortae, 27
ducat, Venetian, 11, 18, 20, 24
duke. *See* Candia: ducal court
Duns Scotus, John, 116
Duran, Profayṭ, 24, 165

Easter computus. *See* timekeeping: Easter computus
Edirne (Adrianople), 76, 90–91
Egypt, 18, 28, 31, 88, 104
Ein sop̱, 121, 138
Eip̱od, 47–48
Eisenmann, Esti, 126

Emanation. *See* Avicennism: emanation and Averroism: emanation, denial of
Emmanuel b. Jacob. *See* Bonfils, Emmanuel b. Jacob
Empire, Byzantine, 3–4, 44, 48, 61, 71, 75, 178
Empire, Sassanian, 1, 31
Euclid: parallels postulate, 177
Eudoxos: accounts of in Arabic, 157; Eudoxan Couple, 159–161, 164
Eugenikos, Marcos 69
Ezekiel: chariot, vision of, 120, 142

Fahhād, Farīd al-Dīn al-: *al-Zīj al-ʿalāʾī*, 71–73, 75
Faji, Elijah al-: *Mikṯaḇ Eliyahu*, 173
Fārābī, Abū Naṣr al-: *Long Commentary on "De interpretatione,"* 114–15
Fazlıoğlu, İhsan, 174
Ferdinand (king), 58
Ficino, Marsilio: correspondent of Pico, 56; debate with Elijah Delmedigo, 57, 99, 139; *Liber de vita*, 57–59
Finzi, Mordechai, 68, 85
fixed stars. *See* astronomy, theoretical: stars, fixed and orb: outermost
Florence, 15, 99–100, 139
Forecasting. *See* astrology: forecasting and astrology: planetary conjunctions
Fracastoro, Girolamo: *De causis criticorum dierum libellus*, 209; *Homocentrica*, 152, 161–62; *Syphilis, sive morbus gallicus*, 209–10
Francis I (king), 147
Fugger, Hans Jakob (Johann Jakob), 15–16
Fugger, Jakob, 8, 15
Fugger, Ulrich, 8, 12, 15–16

Galeano, Elijah, 196
Galeano, Jonah b. Moses b. Judah, 23
Galeano, Moses b. Elijah: relative of Galeano, Moses b. Judah (Mūsā Jālīnūs), 23, 65; *Seiper Mᵉzuqqaq* (A Refined Book), 170–72
Galeano, Moses b. Judah (Mūsā Jālīnūs): astrology, 34–40, 45, 49, 55–56, 208–10; astronomy, homocentric, 155–56, 158, 175–78; critical days, 207–9; *Dhikr baʿḍ al-maḥallāt* (An Account of Some of the Impossibilities), 176; fallacies, logical and *Taʿalumot ḥoḵmah* (Puzzles of Wisdom), 38, 45, 176, 185–87, 195; fraud, medical, 186, 202–4; God's knowledge of particulars, 113–15, 17; and "Iggeret ha-maʿaseh ba-luaḥ ha-niqraʾ ṣapiḥah" (Treatise on How to Make a Universal Astrolabe), 65–67; Islamic theoretical astronomy, late medieval, 161, 170, 172–75, 179, 182–84; *Kitāb al-Zīj* (translation of *Almanach perpetuum*), 23, 79–85, 156; machinations, religious, 196–200; mechanics and technology, 185–95; *Melʾeḵet ha- higgayon* (The Art of Logic), 115–16; Oxford Calculators, 178; *Peirush ʿal ha-juyub* (Commentary on Sines), 66; *Qabbalah*, 142–46; residence on Crete, 23–4; *Seiper ha-goralot* (The Book of Lots), 33–34; "Treatise on the Natures of Medicines and Their Use," 185, 200–201, 204–6; *Treatise on the Quadrant* (translation from Arabic), 66, 174–75; trip to the Veneto, 23, 81, 84, 156, 174, 182–85
Galeano family, 17, 22–24
Galen, 53, 144, 204, 207–8
geomancy, 34, 38, 192
George of Trebizond, 81
Gersonides. *See* Levi b. Gerson
Ghana (sister of Pothula Delmedigo), 21
Ghazālī, Abū Ḥāmid al-: *Maqāṣid al-falāsifa* (The Intentions of the Philosophers), 90–91, 94, 128, 131; *Tahāfut al-falāsifa* (The Precipitance of the Philosophers), 41
Ghāzān (Ilkhanid sultan), 71
Gikatilla, Joseph, 121–22
Gilgul. *See* metempsychosis
Goddu, André, 180–81
Goitein, S. D.: *A Mediterranean Society*, 5
Goldberg, Jessica, 27

Goldstein, Bernard, 81
Gottlieb, Ephraim, 127
Granollachs, Bernat de: *Lunari*, 69
Gratian (Ḥen) family, 24
Grimani, Domenico, 17, 101, 141, 178
Grina, Stamtio, 11
Guglielmo, Raimondo Moncada. *See* Mithridates, Flavius

Hamon family, 204
Ḥarizi, R. Judah, 34
Hasse, Dag Nikolaus: *Success and Suppression: Arabic Sciences and Philosophy in the Renaissance*, 7
Ḥaydar-i Remmāl, 34
Ḥen (Gratian) family, 24
Henry of Langenstein, 153
Heraklion. *See* Candia
Hermann of Carinthia, 44
Heron of Alexandria, 95
Hestera, 10
Hesychasm, 126
Hippocrates, 53, 207
Hugo of Santillana, 67

Iberian Peninsula: expulsions from, 3–4, 58, 79, 101; *The Light of the World*, 155; merchant-scholar families from, 24–26; origins of *Qabbalah*, 119. *See also* Andalus, al-
Ibn al-Ḥadīb, Isaac: *Paved Way* (Oraḥ s^e^lulah), 70–71
Ibn al-Haytham: critic of Ptolemy, 163; *On the Configuration of the World* (Maqāla fī hay'at al-ʿālam), 154; *al-Shukūk ʿalā Baṭlamyūs* (Doubts about Ptolemy), 167
Ibn al-Muthannā 67
Ibn al-Shāṭir, ʿAlāʾ al-Dīn: and Copernicus, 179–82; lunar model, 174, 179, 181; *Nihāyat al-sūl fī taṣḥīḥ al-uṣūl* (The Ultimate Quest in the Rectification of the Hypotheses/Principles), 176; planetary model, 173, 175
Ibn al-Zarqāl: "Treatise on How to Make a Universal Astrolabe" (Iggeret ha-maʿaseh ba-luaḥ ha-niqraʾ ṣap̱iḥah), 63–65, 67; zodiac, motions of, 158
Ibn Ezra, Abraham: influence on Romaniot intellectual life, 35, 49, 89–90, 93, 117, 125; *Long Commentary* (on the Tanak̠), 42, 46–47, 54; Ps. Ibn Ezra, 44; *Seip̱er K^e^lī ha-n^e^ḥoshet* (The Brass Instrument), 64; *Seip̱er ha-Moladot* (On Nativities), 37, 42; *Seip̱er ha-ʿOlam* (The Book of the World), 35; *Y^e^sod moraʾ* (The Foundation of Awe), 46–47, 59, 91, 123 (*see also* Kumatiano, Mordechai: *Commentary on "Y^e^sod moraʾ"*); *Zīj al-sindhind*, translation of, 67
Ibn Gabirol, Solomon, 89
Ibn Naḥmias, Joseph b. Joseph: critic of Ptolemy, 163; and Galeano, Moses b. Judah (Mūsā Jālīnūs), 176, 182; Ibn al-Shāṭir, knowledge of, 174; *The Light of the World* (Nūr al-ʿālam), 155, 158–62, 164–65, 175
Ibn Rushd, Abū Walīd. *See* Averroës
Ibn Sīnā, Abū ʿAlī. *See* Avicenna
Ibn Tibbon, Jacob b. Mak̠ir, 64–65
Ibn Ṭufayl, Abū Bakr, 163
Ibn Waqar, Joseph: *al-Maqāla al-jāmiʿa bayn al-falsafa wa-l-sharīʿa* (The Treatise Combining Philosophy and Religious Law), 140
Ījī, ʿAḍud al-Dīn al-: *al-Mawāqif fī ʿilm al-kalām* (The Stations in the Science of Theology), 55
Ikhwān al-Ṣafāʾ. *See* Brethren of Purity
intellect, active, 98, 100, 109. *See also* intellect, agent
intellect, agent, 96–99, 124, 128 *See also* intellect, active
intellect, hylic (material), 92, 96–100
intellect, recipient, 98
intellects, celestial, 128, 132–34
iperpera, 19, 24
Iran, ix, 33, 68, 79, 85, 180. *See also* Marāgha and Tabrīz
Isaac b. Sid: Alfonsine Tables, 67–68, 80, 85. *See also* Alfonso X

Isaac of Acre, 121
Isabella (queen), 58
Israeli, Isaac, 88, 206
Istanbul: astronomy in, 13, 32, 77, 170–74; Candia, intellectual connections with, 93, 125; Christian scholars, 146–47; debate about metempsychosis reaching, 125; and Fuggers, 8; Galeano, Moses b. Judah (Mūsā Jālīnūs), 81–83, 184–85; immigration of Jews, 3; Kumaṭiano, Mordechai, 59, 76, 90, 93, 95; merchant-scholar families, 19–21, 23–24, 26–28. *See also* Constantinople
İzgi, Cevat, 174

Jābir b. Aflaḥ: critic of Ptolemy, 163; *Iṣlāḥ al-Majisṭī* (Heb. *Qiṣṣur al-Magisṭi*; Correction of the "Almagest"), 166–67
Jacob staff. *See* Levi b. Gerson: Jacob staff
Jaghmīnī, Muḥammad b. ʿUmar al-: *al-Mulakhkhaṣ fī al-hayʾa al-basīṭa* (The Summary of Plain Astronomy), 170–71. *See also* Galeano, Moses b. Elijah: *Seiper Mezuqqaq*
Jālīnūs, Mūsā. *See* Galeano, Moses b. Judah
Jean de Jandun: *Investigations*, target of, 99–100; *Quaestiones in libros physicorum Aristotelis*, 99, 101. *See also* Averroism
Jews, Ashkenazic, 4–5
Jews, Sephardic, 3, 23, 122
Jocuda b. Michael Turco, 11
John of Gmünden: *Marʾit ha-kokabim* (The Appearance of the Stars), 63
John of Lignères, 80
John of Sacrobosco: *De sphaera*, 107, 156
John Philoponus, 177
Joseph b. Yaʿish, 173, 175
Judah b. Asher, II 80
Jurjānī, al-Sayyid al-Sharīf al-, 55–56, 94

Kalām, 88, 94, 115. *See also* Ījī, ʿAḍud al-Dīn al-
Kamāl Pāsha, 37–38
Kamāl Pāshazādah, 38
Kamariotes, Matthew, 71
Katz, Steven, 120
Khāzinī, ʿAbd al-Raḥmān al-: *al-Zīj al-sanjarī*, 71–72, 74
Khwārizmī, Muḥammad b. Mūsā al-: *Zīj al-sindhind*, 67
Kindī, Yaʿqūb b. Isḥāq al-: astrology, 41; *De gradibus* (On Degrees), 204–5, 207; and della Mirandola, Pico, 147
Kumatiano, Mordechai: astrolabes, 63–64, 67, 84–85; astrology, 47–49; commentary on *Guide of the Perplexed*, 64, 93–95, 117, 126, 165–66; commentary on *Millot ha-higgayon* (The Terms of Logic), 116; *Peirush luḥot Paras* (Commentary on the Persian Tables), 73–76, 78; *Peirush Yesod moraʾ* (Commentary on "Yesod moraʾ"), 46–49, 58–59, 91–93, 95, 125–126, 133; *Qabbalah*, 125–26, 143; Qaraites, 76–77, 90–91; scriptural commentary, 46–48; syzygy tables and computations, 70–73, 76–77

Langermann, Y. Tzvi, 45, 116, 172
Latour, Bruno, 214
Lazari Turco, 11
Leichter, Joseph, 72
Leone Ebreo. *See* Abravanel, Judah
letter of exchange, 18
Lettrism (*ʿilm al-ḥurūf*), 38, 192
Levi, Elijah ha-, 20
Levi b. Gerson: astrology, 43; commentary on *Song of Songs*, 139; critique of Ptolemy, 165, 167, 173; Jacob staff (or cross staff), 62; *Wars of the Lord* (Heb. Milḥamot ha-Shem), 132, 172
Levita, Elijah, 16
Lima, George, 9
Logic: and astrology, 45, 208–9; and astronomy, 180; study by Jews in Islamic lands, 115–16. *See also* Galeano, Moses b. Judah: fallacies, logical and *Taʿalumot ḥokmah*
Lorqi, Joseph, 16

Lorqi, Joshua: *Gerem ha-maʿalot* (The Causes of the Degrees), 201
Lucae e Camarino, Johannes, 70

Machinations. *See taḥbulot*
Madanī, Zakariyyāʾ al-, 34
Maḥmūd Çelebī, 171
Mahón, 193
Maimonides, Moses: astronomy, 103–4, 129, 155, 163, 165–66, 177; commentary on *Pirqei Aḇot*, 114; *Guide of the Perplexed* (Ar. Dalālat al-ḥāʾirīn; Heb. Moreh la-neḇoḵim), 43, 52–53, 93–95, 125–26, 197; *Iggeret ha-sodot* (The Epistle of Secrets), 53; "Letter on Astrology," 42–43, 51–52; *Mishneh Torah: Book of Knowledge*, 127; *Mishneh Torah: Hilḵot ʿaḇodah zarah*, 51; *Mishneh Torah: Sanctification of the New Moon*, 91; "Treatise on Poison and Antidotes," 201
Majūsī, ʿAlī b. al-ʿAbbās al-: *Kāmil al-ṣināʿa* (The Complete Book of the Art), 186
Malbiegonato, Musetto b. Judah, 4
Mamlūks, 6, 174–75
Manoli, 11
manuscripts: acquisition by Balbo, 93; acquisition by Muslims, 93, 117; multi-directional exchange, 58–59; purchase by Afendopolo, 167; purchase by Fuggers, 8, 15–17, 19–26, 46, 49, 57, 90–1, 142; purchase by Postel, 147
Manutius, Aldus, 16
Marāgha: Andalusia, connection with, 161; observatory, 71–72, 160, 174, 180; Islamic theoretical astronomy, 174, 180–82; tables from, 84–5. *See also* Iran and Tabrīz
Māridīnī, Sibṭ al-, 66, 175
Markovits, Claude, 26
Marsilio of Inghen, 116
Mashriq (Islamic East), 87
Mavrogonato family, 9–10
Meʾati, Nathan ha-, 16
mechanical devices. *See taḥbulot*
Medici family, 78
medicine: critical days, 207–9; fraud, medical, 186, 202–4; materia medica,184, 200–202, 204, 206, 210–11; pharmacology, 14, 23, 204–7, 213; plague, 4, 22–23, 39, 90, 144, 202–3, 210; syphilis, 209–10. *See also* Padua: medicine at university
Mehmed (sailor), 11
Mehmed II (Ottoman sultan; a.k.a. Mehmed the Conqueror): interest in Europe, 3, 81, 93; and Moses Capsali, 19; and science, 33; *sürgün*, 3
Mercier, Raymond, 72
Messer Leon, David b. Judah, 5, 167, 192
Messiah, return of, 44–45
metempsychosis (*gilgul*; transmigration of souls), 20–21, 63, 123–24, 127–29, 131–33, 135
Mevlānā Aḥmeṭ, 171
Minz, Judah, 20
Mīrim Çelebī, 39, 81
Mithridates, Flavius: astronomy tables, 71; patronage from della Mirandola, Pico, 148–49; translations of qabbalistic texts, 139–40
Mizraḥi, Elijah: lunar model, 167–70, 172; *Peirush seiper al-Magisṭi* (Commentary on the "Almagest"), 167, 171–72; and Qaraites, 90; *Qiṣṣur melʾeḵet ha-mispar* (The Abridgment of the Operation of Number), 46, 148; teacher of Galeano, Moses b. Judah (Mūsā Jālīnūs), 23, 148, 167, 170, 204
money lending, 5, 20
Moses b. Maymon. *See* Maimonides, Moses
Moses de Léon: *The Zohar*, 120
Muʾayyadzādah, ʿAbd al-Raḥmān: *Ḥāshiya ʿalā al-Mawāqif fī ʿilm al-kalām* (Gloss on "The Stations"), 55–56, 59; patron of translation of *Almanach perpetuum*, 23, 80–85, 156, 170
Muḥammad (prophet), 41, 82, 207

Muhtadī, ʿAbd al-Salām al-: *Kitāb majannat al-ṭāʿūn wa-l-wabāʾ* (The Book of the Shield from Plague and Pestilence), 202
Müller von Königsberg, Johannes. *See* Regiomontanus
Münster, Sebastian: publisher of Mizraḥi, *Qiṣṣur melʾeḵet ha-mispar* (The Abridgment of the Operation of Number), 46, 148; translator of Ibn Ezra, *Long Commentary*, 46
mutakallimūn (Muslim theologians), 41, 88, 94. *See also* Ījī, ʿAḍud al-Dīn al-
mysticism, Christian, 119

Nābulusī, ʿAbd al-Ghānī, 5
Naples, 16
Narboni, Moses, 94, 126, 132
Naẓẓām, Ibrāhīm al-, 41
Negroponte, 18
Neoplatonism, 89, 138, 141–42
Nesitorisi, 36
network, definition of, 26–28
Neugebauer, Otto, 180
Nifo, Agostino: acknowledgment of Arabic sources, 182; commentary on *De substantia orbis*, 107–8; *Expositions of the "Metaphysics,"* 158
Niphus, Augustus. *See* Nifo, Agostino
Nīsābūrī, Niẓām al-Dīn al-, 41
Nomico family, 10, 17, 25
notaries, 4, 8–9, 18, 24–25

Ogren, Brian, 127
orb(s): complementary bodies, 163; concentric epicycles, 164–65; Eudoxan couple, 159–61, 164; outermost, 95, 102–5, 107–8, 128, 130, 142; reciprocating mechanism, 154–55, 158, 160. *See also* astronomy: models and Averroism: denial of epicycle and eccentric
Ottoman Empire: astrology, 32–33, 35–36, 39, 82; astronomy, 170–73, 180; Christian scholars, 147, 156; Galeano, Moses b. Judah (Mūsā Jālīnūs), 23, 81, 156, 200, 213; Jewish immigration, 3, 5, 79; merchant-scholar families, x, 19, 215–16; philosophy, 87, 95, 113, 117; Venetian Republic, relations, xii-xiii, 2, 9, 12, 19, 23, 192. *See also* Istanbul
ox skins, 18
Oxford Calculators, 178

Padua: astronomy, 107–8, 112, 152, 156, 158, 160–61, 173, 182; Delmedigo, Elijah, 99, 141; and Fugger family, 8, 15; Jewish settlement, 4–5; medicine at university, 4–5, 13, 18, 209–10; merchant-scholar families in, 18–20, 26–27; philosophy at university, 98–99, 102, 107–8, 112, 152, 178 (*see also* Averroism); yeshivah, 4–5, 20
Palaiologan period, 44, 48
Palamas, Gregory, 44, 126
Palatine Library, 8
Pallache, Samuel, 5
Paradosis, 74–76, 78–79, 84
Parra Perez, Maria José, 81
Paul of Venice, 98
Perpignan, 83
Peter of Limoges, 46
Peuerbach, Georg: *Epitome of the "Almagest,"* 78; *Theoricae novae planetarum*, 152, 164
Pfeifer, Helen: *Empire of Salons: Conquest and Continuity in Early Modern Ottoman Lands*, 6
Pharmacology. *See* medicine: pharmacology
Philosophy. *See* Averroism, Avicennism, intellect, logic, Padua: philosophy at university, and Platonism
Phocaea (modern Foça), 3
Phylletus, Francis, 25
Pingree, David, 72
Pirqei de-Rabbi Eliezer, 52, 132–33
Pizzamano, Antonio, 141

Plaidemo, Michaele, 11
Plato, 100, 128, 137, 139, 142
Platonism, 62, 87–88, 142, 180–81
Plecti, 10–11
Plethon, Gemistius: attacks on Aristotelianism, 116; Jewish sources, use of, 78; *Manual of Astronomy*, 78
Pomponazzi, Pietro, 178
Postel, Guillaume, 147
Providence, divine, 48–49, 52–54, 57, 114
Ptolemy: critiques of, 129, 152–53, 163–68, 176, 207; *Handy Tables*, 73; models, astronomical, 129, 173; syzygy calculation, 73–74; *Tetrabiblos*, 31. *See also* astronomy, theoretical

Qabbalah: *prisca theologia*, 122, 147; *Seip̱er ha-Bahir*, 123, 126, 147; *Seip̱er Y*ᵉ*ṣirah*, 120–21, 125, 147. *See also* della Mirandola, Pico: patron and student of Elijah Delmedigo in *Qabbalah*; metempsychosis; and *s*ᵉ*p̱irah*
Qāḍī Zādah, Mūsā b. Muḥammad b. Maḥmūd, 170
Qaraites, 76–77, 89–91. *See also* Afendopolo, Caleb and Bashyatchi, Elijah
Qurʾān, 41

Rabbanites, 76–77, 89–92, 166–67
Ragep, Jamil, 72, 179, 181
Raina, 28
Ravitzky, Aviezer, 127
Rāzī, Fakhr al-Dīn al-: *Jāmiʿ al-ʿulūm* (The Compendium of the Sciences), 41; *Manṭiq al- Mulakhkhaṣ* (The Logic of "The Mulakhkhaṣ"), 116, 213; *al-Mulakhkhaṣ fī al-ḥikma* (A Concise Exposition of Philosophy), 116–17
Recanati, Menaḥem, 140, 142, 144–45
Regiomontanus: and Alfonsine Tables, 68; *Epitome of the "Almagest,"* 78; homocentric astronomy, 152–54, 158, 160–61, 176; tables for Padua, 156
Reichmuth, Stefan: *The World of Murtaḍā al-Zabīdī (1732–91): Life, Networks and Writings*, 5
Renan, Ernest, 87–88
Rethymno, 11, 25
Rhodes, 193
Ricci, Ronit: *Islam Translated: Literature, Conversion, and the Arabic Cosmopolis of South and Southeast Asia*, 6–7
Rizo, Georgio, 19
Romano, Samuel b. Jonathan, 9
Rome, 15
Rothman, E. Natalie: *Brokering Empire: Trans-Imperial Subjects between Venice and Istanbul*, 6
Rouayheb, Khaled el-: *Islamic Intellectual History in the Seventeenth Century: Scholarly Currents in the Ottoman Empire and the Maghreb*, 5

Saadia Gaon: *Emunot w*ᵉ*-deiʿot*, 145; and *kalām*, 88
Sacellani, Thoma, 8
Sacerdote, Mechir, 9
Şāh Çelebī, Muhyiddin Mehmed, 201
ṣāḥib-i qırān, 39. *See also* Ottoman Empire: astrology
Salamanca, 79–80, 83
Salamon of Candia, 5
Salamone of Canea, 90
Saliba, George: *Islamic Science and the Making of the European Renaissance*, 7
Salonichico, Lazarus, 10–11
Samarqand, 85, 172, 174, 178
Samuel b. Nissim Abū al-Farāj. *See* Mithridates, Flavius
Sap̱orta, Ḥanok̲, 90
Sarulla, 10
Schirazzo, 11
Scholarios, Gennadios (George), 78, 116
Scholasticism, Hebrew, 100–2, 116, 169, 195, 201, 215
Schrenza, Jemalio, 11
Schwartz, Dov, 126, 143

Selim (Ottoman sultan), 38, 185, 200–202, 204
Şen, A. Tunç, 36
sᵉp̱irah (pl. *sᵉp̱irot*): Christian interpretations, 134, 139–40; hypostatic and nonhypostatic, 121–22, 124, 135–36, 138, 140–41, 150; Neoplatonic interpretations, 138; in qabbalistic scriptural commentary, 145–46; *shᵉḵinah* (divine presence), 126
Shabbetai b. Moses, 93
Sharḇiṭ ha-Zahaḇ, Solomon b. Elijah: *Luḥot Paras* (Persian Tables), 72–74, 78
Shefer-Mossensohn, Miri, 178, 212
Shem Ṭoḇ b. Falaquera ,142
Shīrāzī, Quṭb al-Dīn al-: *Nihāyat al-idrāk fī dirāyat al-aflāk*, (The Highest Attainment in Comprehending the Orbs) 171
Signolo, Marinus, 11
Silvaticus, Matthaeus: *Pandectarum medicinae*, 202
Simon bar Yoḥay, 120
Socrates of Constantinople, 4
Solomon b. Solomon, 90. *See also* Salamone of Canea
Soncino, Gershom, 16, 81, 156, 179, 182, 187
Soncinus Hieronymus. *See* Soncino, Gershom
Spagnola, wife of Galeano, Moses b. Judah (Mūsā Jālīnūs), 24
Spano, Moses, 28
Spanopoulo, Zachary, 19
Steinschneider, Moritz, 77
Sufism, 6, 119
Sunbāṭī, Aḥmad b. Aḥmad ʿAbd al-Ḥaqq al-, 66
Sürgün, 3, 23
Swerdlow, Noel, 174, 180–81

Tabrīz ,71–72. *See also* Iran
Tabrīzī, Muḥammad b. Kamāl al-. *See* Ahī Çelebī
Taftāzānī, Saʿd al-Dīn al-, 94
taḥbulot (machinations): amulet, 143; crucifix, 196; deception in, 186, 195, 203–4; flambeau, 192–93; siege of Rhodes, 193; transcription device(s), 14, 187–92, 195; wick soaked in turpentine, 186, 193; writing on palm, 195
talismans, 49, 54, 143, 147, 200
taqwīm. *See* almanac
Ṭarābulsī, Abū Saʿīd al-, 34
Taranto, 63
Ṭāsh Kubrī Zādah, 170–71
Tatars, 4
taxation: censuses, 19–20, 22, 24; inequality of, 10; *messitaria*, 25; rates of, 25; *sansarius*, 25
technology. *See taḥbulot*
Themistius, 97
theology, Islamic. *See kalām*
Theophrastus, 97
Theotonico, Samuel, 25
timekeeping: calendars, lunar, 60–61, 76–77, 90–91; Easter computus, 61; prayer time determination, 66. *See also* astronomy, tables and instruments: syzygy calculation; astronomy, tables and instruments: visibility, lunar crescent; Qaraites; and Rabbanites
Tire, 23, 82
Tirosh-Rothschild, 5, 192
Todesco, Isaac, 20
Topkapı Library, 64, 93, 155, 176
transcription device. *See taḥbulot*
translation: astrology, 44, 172; astronomy, tables and instruments, 23, 63–66, 70, 72–74, 77, 79–83; astronomy, theoretical, 154–55, 166, 171–72, 182; commercial/political, 2, 6, 9, 12, 18, 23, 27–29; logic, 115–16; mathematics, 148; medicine, 16, 201, 205; philosophy, 90, 101, 111, 141, 157 (*see also* translation: logic); *Qabbalah*, 122, 139–40; religion, 46, 139
Transmigration of souls. *See* metempsychosis
Trebizond: Chioniades' trip, 71–72; Chrysococcès, George, 72; George of, 81. *See also* Bessarion

Tunis, 79
Ṭūsī, Naṣīr al-Dīn al-: Marāgha Observatory, 71–72; recension of the *Almagest*, 167; *Risālah-yi muʿīniyya* (The Muʿīnian Epistle), 71; *al-Tadhkira fī ʿilm al-hayʾa* (The Memoir on Astronomy), 147, 171; Ṭūsī couple, 160–62, 171, 178–79, 181–82; *Zīj-i īlkhānī*, 71–72, 77, 79

Ulugh Beg: *Zīj-i Ulugh Beg*, 39, 77–79
unicity of the intellect. *See* Averroism
ʿUrḍī, Muʾayyad al-Dīn al-: *Kitāb al-Hayʾa* (The Astronomical Work), 171

Varlık, Nükhet, 203, 210
Venerio, D. Viti, 24
Venetian Republic, 4, 9–12, 15, 32, 215. *See also* Ottoman Empire, Veneto and Venice
Veneto: Delmedigo, Elijah in, 164; Galeano, Moses b. Judah's (Mūsā Jālīnūs) trip, 23–24, 84, 174, 182–85, 200, 209; and merchant-scholar families, x-xii, 19, 28. *See also* Venetian Republic and Venice
Venice: Delmedigo, Elijah in, 99; Fuggers in, 8, 15–16; Galeano, Moses b. Judah's (Mūsā Jālīnūs) trip, 13–14, 158, 174, 192, 201, 210; manuscript production and sale, 63, 77; printing, 16, 81, 101, 156; Soncino, Gershom, 81, 156, 179, 182, 187; St. Mark's Square, 32. *See also* Venetian Republic and Veneto
Vesel, Matjaž: *Copernicus: Platonist Astronomer-Philosopher*, 180–81
Vittori family, 21–22, 33
Vizinho (Vizinius), José, 80
Vrachuli, Michael, 11

Wābkanawī, Shams al-Dīn b. ʿAlī Khwāja al-. *See* Bukhārī, Shams al-Dīn al-.
Warsanīn, 174
Westman, Robert: astrology 32; *The Copernican Question*, 180–81
Wilson, Curtis, 181

Yacup Pasha, 3
Yahūdī, Ilyās b. Ibrāhīm al-. *See* Muhtadī, ʿAbd al-Salām al-
Yerushalmi, Abraham b. Yom Ṭob̲, 77
yibbum (levirate marriage), 124. *See also* Balbo, Michael: *Wikkuaḥ* and Cohen Ashkenazi, Moses: *Wikkuaḥ*

Zabīdī, Murtaḍā al-, 5
Zacut, Abraham: *ha-Ḥibbur ha-gadol* (The Grand Composition), 79–80, 83; travels, 79–80
Zafar, Abraham, 28
Zafer, 11
Zammis, 11
Zanātī, Muḥammad b. ʿUthman al-, 34
zīj. *See* astronomy, tables and instruments: *zīj*
Zilaiti, Demetrius, 24
Zodiac. *See* astrology: zodiacal signs
Zonta, Mauro, 195

STANFORD **OTTOMAN WORLD** SERIES

Critical Studies in Empire, Nature, and Knowledge

Nükhet Varlık and Ali Yaycioğlu, editors

EDITORIAL BOARD

Julia Phillips Cohen, Nahyan Fancy, John-Paul Ghobrial,
Mayte Green-Mercado, Tijana Krstić, Harun Küçük,
Dana Sajdi, Fatih Yeşil

The Stanford Ottoman World Series showcases cutting-edge interdisciplinary scholarship in Ottoman history from the thirteenth to the twentieth centuries. Books in the series are concerned with three major themes—empire, nature, and knowledge—and the connections among them. The books in this series foster ambitious and innovative scholarship and open new paths in Ottoman studies and beyond.

NIR SHAFIR, *The Order and Disorder of Communication: Pamphlets and Polemics in the Seventeenth-Century Ottoman Empire* **2024**

UĞUR ZEKERIYA PEÇE, *Island and Empire: How Civil War in Crete Mobilized the Ottoman World* **2024**

ELIZABETH R. WILLIAMS, *States of Cultivation: Imperial Transition and Scientific Agriculture in the Eastern Mediterranean* **2023**

The authorized representative in the EU for product safety and compliance is:
Mare Nostrum Group B.V.
Mauritskade 21D
1091 GC Amsterdam
The Netherlands
Email address: gpsr@mare-nostrum.co.uk

KVK chamber of commerce number: 96249943

The authorized representative in the EU for product safety and compliance is:
Mare Nostrum Group
B.V Doelen 72
4831 GR Breda
The Netherlands

www.ingramcontent.com/pod-product-compliance
Lightning Source LLC
Chambersburg PA
CBHW030606230426
43661CB00053B/1859
* 9 7 8 1 5 0 3 6 4 2 6 8 3 *